Human Fungal Diseases

This reference book reviews the latest advancements in fungal infections and their impact on human health. It presents epidemiology, diagnosis, pathogenesis, risk factors, virulence mechanisms, treatment, and strategies for the disease management and prevention of fungal infections. The book further reviews host-pathogen interactions, biofilm formation, and quorum sensing. It also covers the clinical manifestations, diagnostic approaches, and management strategies of opportunistic fungal infections, emerging fungal infections, and allergic fungal infections. It presents the latest advancements in diagnostic methods and therapeutic strategies, covering both conventional techniques and state-of-the-art approaches.

- Reviews pathophysiology, diagnosis, and treatment strategies against different fungal infections
- Examines the molecular mechanisms underlying fungal infections and their interactions with the host immune system
- Provides insights into conventional and modern techniques, including the latest diagnostic and therapeutic approaches
- Discusses the role of nanotechnology, omics technologies, and novel strategies for prevention and control of fungal diseases
- Presents antifungal drugs, their mechanisms of action, pharmacokinetics, and clinical efficacy
- Highlights the challenges posed by emerging fungal species and antifungal drug resistance

Further, the book elucidates antifungal stewardship, nanotechnology, and omics technologies, providing insights into cutting-edge strategies for prevention, control, and management of multidrug-resistant fungi. This book is useful for researchers, students, and health professionals working in the fields of mycology, infectious diseases, immunology, dermatology, and pulmonology.

Human Fungal Diseases
Diagnostics, Pathogenesis, Drug Resistance and Therapeutics

Edited by
Saif Hameed

CRC Press
Taylor & Francis Group
Boca Raton London New York

CRC Press is an imprint of the
Taylor & Francis Group, an **informa** business

Designed cover image: Shutterstock

First edition published 2025
by CRC Press
2385 NW Executive Center Drive, Suite 320, Boca Raton, FL 33431

and by CRC Press
4 Park Square, Milton Park, Abingdon, Oxon, OX14 4RN

ISBN: 9781032633022 (hbk)
ISBN: 9781032642857 (pbk)
ISBN: 9781032642864 (ebk)

DOI: 10.1201/9781032642864

Typeset in Times
by KnowledgeWorks Global Ltd.

Contents

Foreword

I am pleased to support the publication of the book titled *Human Fungal Diseases: Diagnostics, Pathogenesis, Drug Resistance and Therapeutics* from CRC Press/Taylor & Francis. The prevalence of invasive fungal infections is on the rise, garnering increased attention compared to the past, when they were underestimated in comparison to bacterial infections. Although fungi have long been utilized in food and food processing, the impact of disease-causing fungi outweighs that of beneficial fungi. Fungal infections affect over 150 million individuals annually, with more than 2 million cases leading to severe and life-threatening outcomes.

The incidence of fungal infections poses a significant public health challenge due to fungal tropism, enabling a single fungus to infect various cells and tissues in a host. These fungal infections range from mild skin and cutaneous infections to systemic infections, posing a serious threat to the population and escalating morbidity and mortality rates. In fact, mortality rates can reach up to 67% in patients hospitalized in the ICU with invasive fungal infections.

I am delighted to see that this book addresses these previously neglected yet crucial topics, presenting the latest research and developments in human fungal diseases. It covers key aspects such as epidemiology, diagnostics, drug resistance, and therapeutics of human fungal pathogens. I commend Dr. Saif Hameed, a young faculty member at Amity University, Haryana, for his two decades (approx.) of experience in this field. The editor has successfully curated chapters from distinguished scientists working in diverse areas of fungal diseases, addressing most of the essential aspects related to the book's title. I firmly believe that this work is a timely and significant addition to mycology research resources, contributing to the understanding and management of human fungal diseases both nationally and internationally. I extend my best wishes for the book to have a wide readership and great success in the field.

Rajendra Prasad, PhD, FNASc, FASc, FNA
Dean, Faculty of Science, Engineering, and Technology
Director, Amity Institute of Integrative Sciences and Health
Director, Amity Institute of Biotechnology
Amity University Haryana
Gurugram, India

Preface

This book *Human Fungal Diseases: Diagnostics, Pathogenesis, Drug Resistance and Therapeutics* aims to dwell broadly on various topics pertaining to the advancements and current research in the field of mycology particularly about human fungal diseases. The book presents an updated view of the recent developments in this field into a state-of-the-art framework to understand human fungal infection biology of most common fungal pathogens on a common platform. Research in fungal diseases is largely an underestimated area of healthcare settings around the globe, but more so in Asian countries and, particularly, on the Indian subcontinent. Invasive fungal infections have a serious effect on human health with approximately 1.5 million deaths annually, which is a mortality rate similar to renowned bacterial diseases. Despite the availability of antifungal drugs, the situation is complicated further by the emergence of the drug-resistance phenomenon, which is a major obstacle against efficient therapeutics. Considering the urgency to understand the epidemiology, diagnostics, drug resistance, and therapeutics of human fungal pathogens and gain better insights for rapid diagnosis, development of effective antifungal therapy and combating drug resistance, a book is needed to provide an update of developments in this field on a common platform.

This reference book aims to broadly cover recent advances in research on major human fungi. The book summarizes themes such as epidemiology, diagnostics, pathogenesis, virulence, host-pathogen interaction, drug resistance, and therapeutic aspects of fungal diseases. Selected experts and scientists of repute working in the related area with their cumulative efforts discuss their recent research, assisting in this compilation. It is a research-based reference book beneficial for biomedical scientists, researchers, and those in the healthcare industries involved in various aspects of human fungal diseases. The book will also be useful for students in medical lab technology, biomedicine, pathology, and pharmacy.

I am grateful to our esteemed contributors for their worthy and timely contributions, without which this compilation would not have become a ready reference for the researchers in this field. We are grateful for the blessings and constant motivation from Prof. Rajendra Prasad, Dean Research, Amity University Haryana for his overall support to enhance my academic rigor. I take pride to thank Dr. Prasad, who himself is a stalwart in this field for endorsing this book by giving his valuable foreword. Patience and support from Dr. Bhavik Sawhney, CRC Press/Taylor & Francis during the book's preparation is deeply acknowledged.

Dr. Saif Hameed
(Editor)

Editor

Dr. Saif Hameed is currently Associate Professor at Amity Institute of Biotechnology (AIB), Amity University Haryana (AUH). He received his bachelor's from the University of Delhi and master's from Jamia Hamdard with distinction in 2003 and 2005, respectively. He did his doctoral studies in life sciences from Jawaharlal Nehru University in 2010, where he also received CSIR Research Associateship for a one-year postdoctoral work. He also worked as Visiting Scholar in the Institut für Mikrobiologie, Heinrich-Heine-Universität, Düsseldorf, Germany, in 2008. He has been teaching biochemistry and microbiology courses to both undergraduates and postgraduates for 12 years. Dr. Hameed received the Young Scientist Award under the Fast Track Scheme from SERB, New Delhi, in 2012. He is life member of the International Society of Infectious Diseases (ISID), Association for Microbiologist (AMI), India, and Society of Biological Chemists (SBC), India. Dr. Hameed has been actively engaged in research for almost two decades in the field of infectious diseases, particularly multidrug resistance (MDR) in pathogenic fungi. He has 70 peer-reviewed papers to his credit in international journals of repute, 25 book chapters, and authored 7 books. He has supervised 6 PhD students and guided 25 undergraduate and post graduate-level students on their research projects.

Dr. Saif Hameed
Associate Professor
Amity Institute of Biotechnology
Amity University Haryana
Gurugram (Manesar), India

Contributors

Heba S. Abbas
Department of Microbiology and Immunology
Faculty of Pharmaceutical Science and Drug
  Manufacturing
Misr University for Science and Technology
Cairo, Egypt

and

Department of Microbiology
Egyptian Drug Authority
Previously National Organization for Drug
  Control and Research
Giza, Egypt

Saiema Ahmedi
Department of Biosciences
Jamia Millia Islamia
New Delhi, India

Khushboo Arya
Department of Biochemistry
University of Lucknow
Lucknow, India

Dibyendu Banerjee
Division of Biochemistry & Structural Biology
  and Division of Cancer Biology
CSIR–Central Drug Research Institute
Lucknow, India

Nitin Bhardwaj
Department of Zoology and Environmental
  Science
Gurukula Kangri (Deemed to be University)
Haridwar, India

Shikha Chandra
Department of Biochemistry
University of Lucknow
Lucknow, India

Saumya Chaturvedi
Department of Biochemistry
University of Lucknow
Lucknow, India

Gemini Gajera
Institute of Science
Nirma University
Ahmedabad, India

Shabana Khatoon
Department of Biosciences
Jamia Millia Islamia
New Delhi, India

Vijay Kothari
Institute of Science
Nirma University
Ahmedabad, India

Antresh Kumar
Department of Biochemistry
Central University of Haryana
Mahendergarh, India

Sushil Kumar
Division of Biochemistry & Structural Biology
  and Division of Cancer Biology
CSIR–Central Drug Research Institute
Lucknow, India

Deepika Kumari
Department of Biochemistry
Maharshi Dayanand University
Rohtak, India

Hemlata Kumari
Department of Biochemistry
Central University of Haryana
Mahendergarh, India

Pammi Kumari
Department of Biochemistry
Maharshi Dayanand University
Rohtak, India

Nikhat Manzoor
Department of Biosciences
Jamia Millia Islamia
New Delhi, India

Sudhir Mehrotra
Department of Biochemistry
University of Lucknow
Lucknow, India

Minakshi
Department of Biochemistry
Central University of Haryana
Mahendergarh, India

Manzoor Ahmad Mir
Department of Bioresources
School of Biological Sciences
University of Kashmir
Srinagar, India

Narender Mishra
Department of Botany
Guru Ghasidas Vishwavidyalaya
Bilaspur, India

Parveen
Department of Biosciences
Jamia Millia Islamia
New Delhi, India

Ritu Pasrija
Department of Biochemistry
Maharshi Dayanand University
Rohtak, India

Pranava Prakash
Department of Medical and Allied Health
    Sciences
GD Goenka University
Sohna, India

Shaurya Prakash
Department of Biochemistry
Central University of Haryana
Mahendergarh, India

Hafsa Qadri
Department of Bioresources
School of Biological Sciences
University of Kashmir
Srinagar, India

Nafis Raj
Department of Biosciences
Jamia Millia Islamia
New Delhi, India

Abdul Haseeb Shah
Department of Bioresources
School of Biological Sciences
University of Kashmir
Srinagar, India

Preeti Sharma
Department of Biochemistry
Maharshi Dayanand University
Rohtak, India

Ashutosh Singh
Department of Biochemistry
University of Lucknow
Lucknow, India

Nidhi Thakkar
Institute of Science
Nirma University
Ahmedabad, India

Sana Akhtar Usmani
Department of Biochemistry
University of Lucknow
Lucknow, India

Pratima Vishwakarma
St. Joseph's College for Women
Gorakhpur, India

Ashok Kumar Yadav
Department of Medical and Allied Health
    Sciences
GD Goenka University
Sohna, India

1 Unraveling the World of Human Fungal Infections

An Introductory Exploration

Pratima Vishwakarma* and Narendra Kumar

1.1 INTRODUCTION

Fungi are eukaryotic microorganisms that can be yeasts or molds, and some types can cause diseases. Yeasts reproduce through budding, while molds grow in long filaments called hyphae. Fungi are all heterotrophic and digest their food externally by releasing enzymes. They synthesize lysine and have a chitinous cell wall, plasma membranes containing sterol ergosterol, 80S rRNA, and microtubules composed of tubulin. Fungi use various carbon sources to make carbohydrates, lipids, nucleic acids, and proteins. They get energy from sugars, alcohols, proteins, lipids, and polysaccharides. Yeasts are categorized by their ability to use different carbon sources and their form. Fungi require nitrogen for amino acids, purines, pyrimidines, glucosamine, and vitamins. Nitrate is the most common nitrogen source for fungi, which is reduced to nitrite and ammonia (McGinnis and Tyring 1996).

1.2 STRUCTURE OF FUNGI

The cell walls of fungi are made up of chitinous microfibrils, along with a matrix of various compounds such as polysaccharides, proteins, lipids, salts, and pigments. It is worth noting that the composition of cell walls varies among different fungi. For instance, some fungi contain soluble peptidomannans, which play an essential role in the immunologic response. Meanwhile, *Cryptococcus neoformans*, a type of fungi, produces a capsular polysaccharide consisting of chitin, glucan, mannan, lipid, protein, chitosan, and inorganic ions. It is also important to note that the composition of cell walls is closely related to the taxonomy of fungi (Table 1.1). Fungi's plasma membrane has ergosterol, which regulates the passage of lipids, proteins, and some carbohydrates. Antifungal agents target ergosterol, inhibiting growth and causing abnormal chitin production. Fungi use microtubules made of tubulin to move organelles and chromosomes. Griseofulvin medication interferes with this process. Fungal nuclei have chromatin and a nucleolus mRNA attaches to ribosomes for protein synthesis (McGinnis and Tyring 1996).

Fungi, ubiquitous microorganisms, are found worldwide and can cause adverse health effects in individuals with predisposed conditions, respiratory conditions, or immunocompromised patients. Common outdoor fungi include *Alternaria*, *Cladosporium*, *Epicoccum*, ascospores, and basidiospores. Indoor fungi, such as *Penicillium*, *Aspergillus*, and *Stachybotrys*, are associated with water damage or decay (Baxi et al. 2016).

Moisture, in the form of dampness and mold, is a breeding ground for microbial growth. Its health effects are difficult to distinguish, but visible water damage, mold, unpleasant smells, and the proliferation of dust mites and actinomycetes are all telltale signs (Baxi et al. 2016).

---

* Corresponding Author: pratima.vishwakarma12@gmail.com

DOI: 10.1201/9781032642864-1

1

TABLE 1.1

Different Phyla of Fungi with Its Primary Cell Wall Polymers and Examples

S. No.	Phyla of Fungi	Principal Cell Wall Polymers	Examples
1.	Ascomycota	Chitin-Chitosan	*Aspergillus fumigatus, Pneumocystis jirovecii, Histoplasma capsulatum, Blastomyces dermatitidis, Paracoccidioides brasiliensis, Paracoccidioides lutzii, Coccidioides immitis, Coccidioides posadasii, Talaromyces marneffei*
2.	Basidiomycota	Chitin	*Cryptococcus neoformans, Cryptococcus gattii, Hormographiella aspergillata*
3.	Fungi Imperfecti	Chitin	*Candida albicans, Candida tropicalis, Blastomyces dermatitidis, Cryptococcus neoformans*
4.	Zygomycota	Chitin	*Rhizopus, Mucor, Saksenaea, Apophysomyces, Rhizomucor, Absidia, Cunninghamella, Cokeromyces, Syncephalastrum* spp.

1.3 CLASSIFICATION OF FUNGI ON THE BASIS OF FORM

1.3.1 YEAST

Yeasts grow as single cells that reproduce through budding. Size, shape, presence or absence of capsules, and spore formation distinguish between different yeast taxa. Morphology differentiates at the genus level, while carbon source assimilation and nitrate utilization identify species. *C. albicans* and *C. neoformans* produce blastoconidia through bud emergence, growth, and conidium separation. Mitosis occurs as the bud grows, and a chitin ring forms between the developing blastoconidium and its parent yeast cell, leaving a bud scar on the parent cell wall. *C. albicans* may also form true hyphae.

1.3.2 MOLDS

Identification of molds is reliant on the growth of hyphae, which elongate through apical elongation. This intricate process requires a delicate balance between cell wall lysis and new cell wall synthesis. To differentiate between molds, a critical analysis of structures such as conidiophores and conidiogenous cells is mandatory as they are often categorized based on conidiogenesis. It is important to note that some molds produce sporangia, sac-like cells containing sporangiospores formed from the entire protoplasm. Sporangia are typically formed on specialized hyphae called sporangiophores and require meticulous examination.

1.3.3 DIMORPHIC FUNGI

There exist specific fungi that hold medical significance and display two distinct growth forms, commonly known as dimorphic fungi. This term refers to fungi that exhibit mold growth in laboratory conditions, while *in vivo*, they form yeast cells or spherules. Notable examples of such fungi include *Blastomyces dermatitidis* and *Coccidioides immitis*.

1.4 CLASSIFICATION OF FUNGI ON THE BASIS OF SPORES FORMED

Fungi are a diverse group of organisms that are categorized based on their method of reproduction, which can be sexual, asexual, or both (Table 1.2). Two terms, anamorphs and teleomorphs, are used to classify these fungi. Anamorphs are used for practical purposes, such as communication, while teleomorphs are more useful for accurately reflecting the evolutionary relationships between different fungi. The term holomorph encompasses both forms. For example, *Ajellomyces dermatitidis* is the name given to the sexual form that develops when two compatible isolates mate. When referring to the entire fungus, the teleomorph's name is used.

TABLE 1.2

Classification of Fungi on the Basis of Spores

S. No.	Phyla of Fungi	Asexual Spore	Sexual Spore
1.	Ascomycota	Conidia	Ascospore
2.	Basidiomycota	Conidia	Basidiospore
3.	Chytridiomycota	Spore and conidia	Oospores
4.	Fungi Imperfecti	Conidia	Absent
5.	Zygomycota	Spore and conidia	Zygospore

1.5 FUNGAL INFECTIONS OF HUMAN

Mycosis, which is more commonly known as fungal infections, is caused by the proliferation of yeasts or molds. These infections can affect different parts of the body, such as the skin, nails, throat, lungs, urinary tract, and mouth. They may cause discomfort and inconvenience, and can even lead to serious health problems if left untreated. Having a fungal infection on skin or nails is usually not a major issue. Nevertheless, specific kinds of fungal infections can result in significant complications for individuals with compromised immune systems (Figure 1.1).

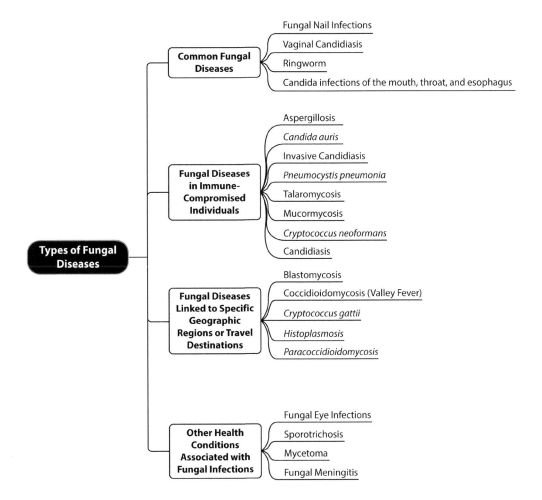

FIGURE 1.1 Types of fungal diseases in humans.

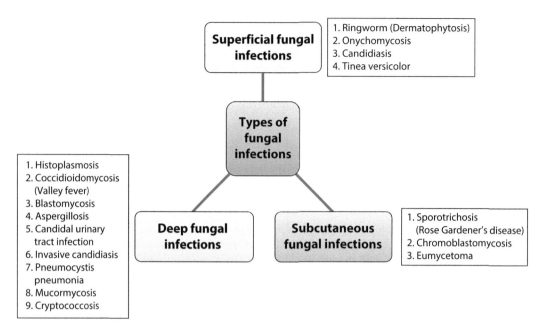

FIGURE 1.2   Types of fungal infections in humans.

Fungal infections in humans are categorized into superficial, deep mycosis, and subcutaneous, with distinct characteristics, clinical manifestations, and treatment approaches (Figure 1.2).

1.5.1 Superficial Fungal Infections

These infections are caused by dermatophytes, yeast, or mold, affect skin, hair, and nails, and are often not life-threatening and manageable with topical antifungal medications.

1.5.1.1 *Tinea* Infections (Ringworm)

It can affect various body parts, including the scalp (*Tinea capitis*), body (*Tinea corporis*), and feet (*Tinea pedis* or athlete's foot). They are brought on by dermatophytes like Trichophyton, Microsporum, and Epidermophyton (*Tinea cruris* or jock itch).

1.5.1.2 *Candida* Infections

This is also known as candidiasis, this condition is brought on by the yeast *Candida* and can cause vaginal yeast infections as well as thrush in the mouth and skin.

1.5.2 Deep Mycosis (Systemic Fungal Infections)

This is more serious disease and can affect internal organs and tissues. People with weakened immune systems or other underlying medical conditions are more prone to developing these infections. They may require systemic antifungal medications for treatment because they can spread throughout the body. Deep mycosis examples include:

1.5.2.1 Candidemia

A *Candida* species infection affects the entire body and spreads to different organs through the bloodstream.

1.5.2.2 Aspergillosis

This is caused by the fungus *Aspergillus* and commonly affects lungs, causing respiratory symptoms. In severe cases, it can spread to other organs, especially in immunocompromised individuals.

1.5.2.3 Cryptococcosis

This is caused by *Cryptococcus* species, this infection primarily affects lungs but can also involve the central nervous system in severe cases.

1.5.3 SUBCUTANEOUS FUNGAL INFECTIONS

The deeper layers of the skin, subcutaneous tissues, and occasionally even muscle and bone are all susceptible to subcutaneous fungal infections. These infections are frequently brought on by fungi that are present in the environment and they can enter the body through wounds or traumatic injuries. Surgery and long-term antifungal medications are frequently used in treatment. Examples of subcutaneous fungal infections include the following.

1.5.3.1 Sporotrichosis

Caused by the fungus Sporothrix, this infection is often associated with gardening or handling plants. It can lead to nodular skin lesions that may spread along the lymphatic vessels.

1.5.3.2 Chromoblastomycosis

Caused by various dematiaceous fungi, this chronic infection presents with wart-like nodules and plaques on the skin. It often occurs in individuals with frequent exposure to soil and plant material.

1.6 ROUTE OF TRANSMISSION OF FUNGAL PATHOGENS

The Dikarya sub-kingdom of fungi, which comprises the Ascomycota and Basidiomycota phyla, is responsible for causing the majority of fungal infections and diseases in humans. Ascomycota fungi are notorious for causing a range of infections affecting various parts of the body, including the mouth, throat, skin, eyes, nerves, genitals, heart, lungs, and entire body. Basidiomycota organisms, such as *Cryptococcus* and *Malassezia*, are commonly linked to invasive meningitis and mild skin infections, respectively.

Fungal pathogens have different ways of spreading, either by direct contact or inhalation. Each group of fungi has unique strategies to infect their host, causing various diseases with different levels of severity and symptoms.

Dermatophytes are a type of fungi that mainly spread through direct contact. These fungi belong to the genera of *Microsporum*, *Epidermophyton*, *Trichophyton*, and other species like *Sporothrix* and *Malassezia*. They target the skin, particularly when it is already damaged, to establish infections. By secreting different proteolytic enzymes, these pathogens break down the keratinized tissues of the skin, leading to superficial mycoses, known as fungal skin infections. Close contact with infected material or contaminated surfaces facilitates the transmission of these fungi, making personal hygiene and environmental cleanliness critical in preventing their spread.

There are various fungal species that infect their hosts through the inhalation of spores or conidia into the respiratory system. This mode of transmission is significant for several pathogenic fungi, which can cause mild respiratory issues or severe, life-threatening conditions. *Blastomyces dermatitidis* is one such fungus that causes Blastomycosis, and its spores can disseminate to other organs after infecting the lungs. Similarly, *Paracoccidioides brasiliensis* and *P. lutzii*, agents of Paracoccidioidomycosis, are also inhaled and infect the lungs, spreading to other parts of the body. *Histoplasma capsulatum*, another respiratory pathogen, causes Histoplasmosis, and its spores are found in environments with bird or bat droppings. Humans can inhale these spores and get infected when they become airborne.

It's crucial to know that certain fungi are extremely hazardous and can cause serious respiratory problems, particularly in immunocompromised individuals. *Pneumocystis jirovecii* is one such fungus that can lead to severe *Pneumocystis* pneumonia, which can be life threatening, and it only infects those with weakened immune systems, especially those with HIV/AIDS.

Aspergillus species, including *A. fumigatus* and *A. flavus*, are commonly found in the environment and can produce large amounts of spores, which can cause different types of Aspergillosis that affect various organs, including lungs and sinuses.

Coccidioides immitis and *C. posadasii* are endemic in specific regions, causing Coccidioidomycosis or Valley Fever. Inhaling their infectious form of fungi, arthroconidia, can lead to respiratory symptoms, and in some cases, the infection may spread to other parts of the body.

Cryptococcus species, such as *C. neoformans* and *C. gattii*, are encapsulated yeast fungi commonly found in soil and bird droppings. If their spores are inhaled, it can cause Cryptococcosis, primarily affecting the lungs and sometimes leading to severe central nervous system infections.

Talaromyces marneffei is a highly intriguing fungal pathogen due to its ability to spread through both direct contact and inhalation. This results in the development of talaromycosis, a condition that can infect the skin through contaminated materials or lead to respiratory infections when inhaled. Understanding the transmission methods of various fungal pathogens is essential for effectively preventing and controlling fungal infections. These fungi possess unique adaptations that enable them to infect specific niches and cause a variety of infections in humans and animals. Advancements in research will allow for improved comprehension of the complex interactions between fungi and their hosts, leading to better diagnostic methods, treatments, and preventative measures.

1.7 EPIDEMIOLOGY AND RISK FACTORS ASSOCIATED WITH FUNGAL INFECTIONS

Every year, more than 150 million cases of severe fungal infections occur worldwide, leading to around 1.7 million deaths. The numbers are only increasing due to social and medical developments in recent decades that have facilitated the spread of fungal infections. The long-term use of antifungal drugs in high-risk patients has also led to the emergence of drug-resistant fungi, including the lethal *Candida auris* strain. Fungal infections are a global threat that is becoming more severe, and it's crucial to conduct more effective research to counteract their consequences (Kainz et al. 2020).

Despite significant changes in the epidemiology of fungal diseases, the primary fungal pathogens responsible for serious fungal diseases remain *Aspergillus*, *Candida*, *Cryptococcus* species, *Pneumocystis jirovecii*, endemic dimorphic fungi like *Histoplasma capsulatum*, and *Mucormycetes*. *Candida albicans* is the main cause of mucosal disease, while *Aspergillus fumigatus* is responsible for most allergic fungal diseases. *Trichophyton* spp., particularly *T. rubrum*, are the primary cause of skin infections (Bongomin et al. 2017). For over four years, the LIFE portal has successfully estimated the prevalence of severe fungal infections for more than 5.7 billion individuals, which makes up over 80% of the world's population on a country-by-country basis. These studies have conclusively revealed the stark disparities in the global burden of fungal infections, not only between countries but also within the same regions of a country, and among populations at a higher risk (Denning 2017).

It is important to be aware that *Candida* and *Aspergillus* species are the primary causes of invasive fungal infections, but other fungi, including yeasts and filamentous fungi, are also becoming significant pathogens. In patient groups such as those with leukemia and bone marrow transplants, yeast-like fungi and molds such as Zygomycetes, *Fusarium* spp., and *Scedosporium* spp. are increasingly being recognized. It is critical to be aware of these changes in patient care and make informed decisions about antifungal treatment. Factors such as the type of patient (solid-organ or stem-cell transplant), the level of immunosuppression, the patient's history of prolonged exposure to antifungal drugs, and knowledge of the infecting pathogen's genera and species and its typical susceptibility pattern should be taken into consideration when selecting an appropriate antifungal agent (Richardson and Lass-Flörl 2008).

1.8 IMPORTANCE OF STUDYING AND UNDERSTANDING FUNGAL INFECTIONS

Fungi play a crucial role in agroindustry and animal diseases. However, fungal infections, especially those caused by *Candida*, *Cryptococcus*, and *Aspergillus* species, result in over a million deaths annually. These infections are difficult to cure, and mortality rates remain high even with antifungal treatments (Janbon et al. 2019). Fungal infections acquired in healthcare facilities have increased due to aging populations, intensive therapies for cancer patients, increased care for critically ill patients, and more organ and stem cell transplants. This trend is expected to continue and result in further increases in nosocomial fungal infections (Perlroth, Choi, and Spellberg 2007). Fungal infections invading the body are increasingly prevalent in hospitals and medical facilities. Patients with weakened immune systems are at risk of serious illness and death, even with antifungal medications. Given the growing risk factors in countries with advanced medical technology, it is inevitable that the number of cases will continue to rise. Prompt recognition of symptoms by doctors is vital to reduce mortality rates. In the case of severe disease, a combination of medications may be required. Diagnostic tests and new treatments require further research. Ultimately, our focus should be on preventing these infections from occurring in the first place.

1.9 TREATMENT OF HUMAN FUNGAL INFECTIONS

It is crucial to take action promptly in response to the increasing occurrences of fungal infections. These infections are often overlooked during their early stages, causing them to become more severe and difficult to treat. Fungal pathogens use various methods to avoid the immune system and worsen the infection. Doctors depend on the antifungal medications available to treat both surface level and systemic infections (Reddy et al. 2022). Antifungal therapies encompass both superficial and systemic treatments, which are supported by five well-known classes of drugs. These classes are azoles, polyenes, echinocandins, allylamines, and pyrimidine analogs (Hokken et al. 2019; Kainz et al. 2020).

The azole class of drugs, specifically imidazoles (such as miconazole and ketoconazole) and triazoles (like fluconazole and voriconazole), have demonstrated exceptional success in producing a vast array of antifungal compounds for clinical use. These drugs have proven to be highly effective against *Candida* spp. and other fungal pathogens, and are widely preferred due to their versatile administration routes (Nett and Andes 2016).

Azoles are a type of medication that inhibits the production of ergosterol, an important component of fungal cell membranes. They do this by blocking a key enzyme involved in sterol biosynthesis called sterol 14α-demethylase. In contrast, polyene antimycotics like amphotericin B and nystatin work by binding to membrane sterols, causing the formation of pores in the fungal cell membrane and ultimately leading to its death. Echinocandins, which are semisynthetic lipopeptides like caspofungin, micafungin, and anidulafungin, prevent the synthesis of key cell wall components, specifically 1,3-β-d-glucan, and are effective against *Candida* spp. and *Aspergillus* spp. They inhibit the enzyme 1,3-β-d-glucan synthase, which is essential for cell wall synthesis in various fungi, and display fungistatic properties. Finally, allylamine drugs are often used to control superficial dermatophytoses by inhibiting ergosterol biosynthesis, which slows down fungal growth (Abdel-Rahman and Newland 2009). Various drugs treat fungal infections. Examples include terbinafine and naftifine, which inhibit squalene epoxidase. Pyrimidine analogue drugs like 5-FC (5-fluorocytosine) work against *Candida* and *Cryptococcus* by affecting nucleic acid and protein synthesis. Amphotericin B and miconazole induce oxidative stress for enhanced antifungal activity. Other drugs target protein synthesis, microtubule assembly, and calcineurin signaling (Scorzoni et al. 2017).

1.10 CONCLUSION

Fungal infections caused by yeasts and molds are prevalent worldwide and can lead to various diseases in humans. Understanding the structure and physiology of fungi is crucial in developing effective antifungal treatments.

The cell wall of fungi consists of chitinous microfibrils and a matrix of compounds that vary among different fungi. The plasma membrane contains ergosterol, a target for antifungal agents. Fungal infections can be classified based on the form and spores produced by fungi. Superficial fungal infections affect the skin, hair, and nails, while deep mycoses and subcutaneous infections can be life threatening, particularly in immunocompromised individuals. Fungal pathogens can spread through direct contact or inhalation, with different fungi using distinct strategies to infect their hosts.

The epidemiology of fungal infections shows significant regional disparities and poses a growing global threat, especially with the emergence of drug-resistant fungi. Therefore, studying and understanding fungal infections are crucial to develop effective prevention and treatment strategies. Antifungal therapies include various drug classes, such as azoles, polyenes, echinocandins, allylamines, and pyrimidine analogs, which target different aspects of fungal physiology

In conclusion, fungi are diverse microorganisms capable of causing a wide range of diseases in humans. Further research is needed to improve the diagnosis, treatment, and prevention of fungal infections, given their increasing impact on global health. Prompt recognition of symptoms and early intervention are vital to reduce mortality rates and improve patient outcomes.

REFERENCES

Abdel-Rahman, S., and J. G. Newland. 2009. "Update on Terbinafine with a Focus on Dermatophytoses." *Clinical, Cosmetic and Investigational Dermatology* 49. doi: 10.2147/CCID.S3690.

Baxi, S. N., J. M. Portnoy, D. Larenas-Linnemann, W. Phipatanakul, C. Barnes, S. Baxi, C. Grimes, W. E. Horner, K. Kennedy, D. Larenas-Linnemann, E. Levetin, J. D. Miller, W. Phipatanakul, J. M. Portnoy, J. Scott, and P. B. Williams. 2016. "Exposure and Health Effects of Fungi on Humans." *Journal of Allergy and Clinical Immunology: In Practice* 4(3):396–404. doi: 10.1016/j.jaip.2016.01.008.

Bongomin, F., S. Gago, R. Oladele, and D. Denning. 2017. "Global and Multi-National Prevalence of Fungal Diseases—Estimate Precision." *Journal of Fungi* 3(4):57. doi: 10.3390/jof3040057.

Denning, D. W. 2017. "Calling upon All Public Health Mycologists." *European Journal of Clinical Microbiology & Infectious Diseases* 36(6):923–24. doi: 10.1007/s10096-017-2909-8.

Hokken, M. W. J., B. J. Zwaan, W. J. G. Melchers, and P. E. Verweij. 2019. "Facilitators of Adaptation and Antifungal Resistance Mechanisms in Clinically Relevant Fungi." *Fungal Genetics and Biology* 132:103254. doi: 10.1016/j.fgb.2019.103254.

Janbon, G., J. Quintin, F. Lanternier, and C. D'Enfert. 2019. "Studying Fungal Pathogens of Humans and Fungal Infections: Fungal Diversity and Diversity of Approaches." *Genes & Immunity* 20(5):403–14. doi: 10.1038/s41435-019-0071-2.

Kainz, K., M. A. Bauer, F. Madeo, and D. Carmona-Gutierrez. 2020. "Fungal Infections in Humans: The Silent Crisis." *Microbial Cell* 7(6):143–45. doi: 10.15698/mic2020.06.718.

McGinnis, M. R., and S. K. Tyring. 1996. "General Concepts of Mycology." in *Medical Microbiology*, edited by Baron S. Galveston (TX): University of Texas Medical Branch at Galveston.

Nett, J. E., and D. R. Andes. 2016. "Antifungal Agents." *Infectious Disease Clinics of North America* 30(1):51–83. doi: 10.1016/j.idc.2015.10.012.

Perlroth, J., B. Choi, and B. Spellberg. 2007. "Nosocomial Fungal Infections: Epidemiology, Diagnosis, and Treatment." *Medical Mycology* 45(4):321–46. doi: 10.1080/13693780701218689.

Reddy, G. K. K., A. R. Padmavathi, and Y. V. Nancharaiah. 2022. "Fungal Infections: Pathogenesis, Antifungals and Alternate Treatment Approaches." *Current Research in Microbial Sciences* 3:100137. doi: 10.1016/j.crmicr.2022.100137.

Richardson, M., and C. Lass-Flörl. 2008. "Changing Epidemiology of Systemic Fungal Infections." *Clinical Microbiology and Infection* 14:5–24. doi: 10.1111/j.1469-0691.2008.01978.x.

Scorzoni, L., A. C. A. de Paula e Silva, C. M. Marcos, P. A. Assato, W. C. M. A. de Melo, H. C. de Oliveira, C. B. Costa-Orlandi, M. J. S. Mendes-Giannini, and A. M. Fusco-Almeida. 2017. "Antifungal Therapy: New Advances in the Understanding and Treatment of Mycosis." *Frontiers in Microbiology* 08. doi: 10.3389/fmicb.2017.00036.

2 Emerging Prevalence of Fungal Infections and Challenges in Development of New Antifungals

*Gemini Gajera, Nidhi Thakkar, and Vijay Kothari**

2.1 BACKGROUND

Since antiquity, microbial infections have remained an important factor in human life. In the early years of the history of microbiology, most infection research remained focused on bacterial pathogens, and in recent years, the focus has been largely on bacterial and viral infections. Eukaryotic agents of infection (protozoa and fungi) have historically received less attention. In part, this is because the majority of infections in humans are caused by bacteria and viruses, and a lesser number are attributable to fungi or protozoa. Though some can act as opportunistic pathogens, fungi primarily are saprophytic organisms and have not evolved to be professional pathogens (Jampilek, 2022). Fungi do not pose much risk to a healthy individual with a functional immune system. Instead, they are more of a risk to immunocompromised people, hospitalized patients, and the elderly (Morton et al., 2022).

Compared to previous years, now we have better hospital facilities and an overall better healthcare infrastructure. This has conferred an extension of life on those immunocompromised or aged people, who otherwise would have died. But this also means that opportunistic pathogens have a large population with weaker immunity to attack. Organ transplantation and Human Immunodeficiency Virus (HIV) infection are examples of conditions which were almost absent from human life before the 1970s (Barin, 2022); however, now we have a large number of HIV-infected immunocompromised people, as well as people who have undergone organ transplantation and are put on immunosuppressive drugs to avoid organ rejection (Ngan et al., 2022). These are examples of populations more susceptible to opportunistic pathogens like fungi. In other words, people who would have died in earlier decades owing to limited treatment options are now surviving in an immunocompromised state for years, providing for fungal pathogens sufficient potential victims. Fungi may be lethal while causing a secondary infection in a person already infected with some other professional pathogen. This was witnessed recently when mucormycosis proved lethal to people with the coronavirus infection (Gunathilaka et al., 2023). During 2020–2021, patients hospitalized with COVID-19-associated fungal infections had higher (48.5%) in-hospital mortality rates than those with non-COVID-19-associated fungal infections (12.3%) (Gold et al., 2023).

The emerging prevalence of fungal infections makes it necessary for us to be ready with a weaponry containing a sufficient number of effective antifungal antibiotics (Fisher et al., 2022; Xin et al., 2023). Unfortunately, however, we actually have a very narrow spectrum of antifungals at our disposal as treatment options. The discovery and development of new therapeutic agents to use against fungal infections is even more challenging than against bacterial pathogens due to the fact that fungi being eukaryotes like their human hosts share multiple cellular and molecular features with

* Corresponding Author: vijay.kothari@nirmauni.ac.in

DOI: 10.1201/9781032642864-2

humans (Chen et al., 2023; Kumar et al., 2023). This makes it difficult for many agents showing antifungal activity in preliminary screening assays to pass the criteria of 'selective toxicity,' i.e., being toxic to the fungi without harming the host. Additionally, the antibiotic susceptibility assays with fungi (particularly the filamentous ones, i.e., molds) are not as straightforward as those of many bacterial pathogens.

This chapter reviews the recent literature regarding the growing global prevalence of fungal infections the potential risk from certain emergent fungal pathogens, antimicrobial resistance among fungi, and potential antifungals in the development pipeline.

2.2 INCREASING PREVALENCE OF FUNGAL INFECTIONS AND ASSOCIATED RISK FACTORS

Since the 1970s, there has been a consistent rise in the occurrence of severe secondary systemic fungal infections (Friedman and Schwartz, 2019). One contributing factor to the proliferation of fungal diseases has been the widespread utilization of broad-spectrum antibiotics, which either eliminate or reduce the populations of non-pathogenic bacteria that typically compete with fungi (de Nies et al., 2023). Another factor is the growing number of individuals with compromised immune responses due to factors like acquired immunodeficiency syndrome (AIDS), the use of immunosuppressant drugs, or cancer chemotherapy. This has led to an increased prevalence of opportunistic infections, where fungi that seldom cause illness in healthy individuals now pose a greater threat.

The increasing number of diabetic patients globally presents a large population for fungal pathogens to attack. Fungal infections of nails and wounds in diabetic patients lasts for long periods of time and are often difficult to cure completely. Hyperglycemia or diabetes mellitus (DM) weakens the immune system, increasing infection vulnerability, particularly to fungal infections (Figure 2.1). While mucormycosis is typically rare and benign in healthy individuals, it takes on a severe and potentially fatal form in the context of diabetes, demanding urgent and effective treatment. Diabetes

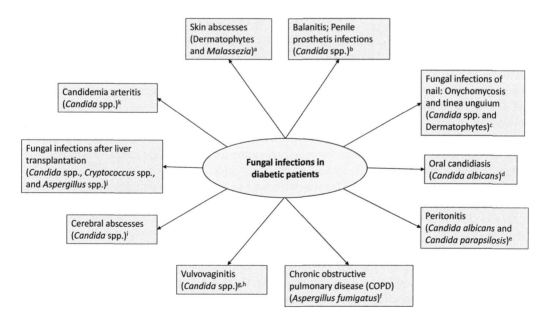

FIGURE 2.1 Various fungal infections observed in diabetic patients at higher frequency. [a]Kakeya et al., 2014; [b]Talapko et al., 2022; [c]Aragón-Sánchez et al., 2023; [d]Mussi et al., 2022; [e]Yuvaraj et al., 2014; [f]Lao et al., 2020; [g]Kendirci et al., 2004; [h]Malazy et al., 2007; [i]Gupta et al., 2022; [j]Araujo et al., 2013; [k]Shazniza Shaaya et al., 2021.

elevates the risk of fungal infections through compromised immunity from high blood sugar, often leading to hospitalizations for soft tissue and lower extremity bone infections due to impaired white blood cell function, hypoxia, and poor circulation (Bansal et al., 2008).

Though *Candida* infections are not recognized as the initial sign of diabetes, *Candida* spp., particularly *C. albicans*, are frequently isolated from diabetic foot ulcers, which can lead to severe inflammation (Rasoulpoor et al., 2021; Sundaravadivelu et al., 2022). Diabetic foot ulcers pose a serious medical concern, with non-albicans *Candida* species, known for their biofilm-forming capabilities, emerging as a prominent cause of this condition. A recent study revealed the prevalence of different *Candida* species in diabetic patients, with *Candida tropicalis* being the most common (34.6%) (Kumar et al., 2016). According to a study conducted by Raiesi et al. (2017), 24.5% of diabetic patients were found to have fungal infections, more frequently those with skin and nail lesions (28%) than those with foot ulcers (19.1%). Notably, a high incidence of fluconazole-resistant *Candida* species was observed, particularly in diabetic foot ulcers. These findings emphasize the need for effective prevention, early detection, and treatment strategies to address the growing challenge of fungal infections in diabetic patients, particularly those with foot ulcers. Numerous other dermatological studies have consistently reported diabetes as the most prevalent systemic condition associated with a spectrum of cutaneous complications (Khoharo et al., 2009). For instance, a study by Rasoulpoor et al. (2021) reported that fungal infections accounted for a significant 37.9% of the cases of cutaneous infections, which were the most common skin problem among diabetic patients. Darjani et al.'s (2002) investigation found that 28.6% of skin lesions in the diabetic population were infectious, with 24% attributed to fungal etiology. Similarly, Phulari and Kaushik (2018) reported infectious lesions in 61% of their diabetic cohort, of which 39.5% were associated with fungal infections. These findings underscore the common occurrence of skin complications, particularly fungal infections, in individuals with diabetes.

The delayed and disrupted healing of wounds in individuals with DM is a global issue that causes substantial morbidity and economic stress (Mirpour et al., 2020). This problem mainly arises from the difficulties in finding effective treatments for wound healing. Due to their immunosuppressed state, individuals with DM are more susceptible to infections, especially fungal infections. Factors such as elevated glucose levels, the release of destructive enzymes, and the immunosuppressed state of the patient can disrupt the delicate balance between the host and the yeast, ultimately leading to the transition of the typically harmless *C. albicans* into a pathogenic form that causes infection (Wilson and Reeves, 1986; Rodrigues et al., 2019).

Chronic wounds are difficult to heal due to disrupted healing mechanisms, and fungi play a substantial role in these wounds, with *C. albicans* and *Trichophyton rubrum* being the main fungal species involved. Fungal infections in chronic wounds are often underestimated and they can lead to invasive fungal illness, local infections, and fungemia, contributing notably to the microbial bioburden of wounds and ulcers (Ge and Wang, 2023). They are important contributors to biofilm production and polymicrobial wound infections in chronic wounds (Durand et al., 2022). Additional studies that acknowledge the fungal component of interkingdom biofilms are required to develop more effective treatment strategies for chronic wounds.

Another example of fungal infection on the rise, mucormycosis, also known as zygomycosis, is a serious fungal infection caused by fungi belonging to the order Mucorales. These fungi are commonly found in the environment as saprophytes, feeding on dead or decaying organic matter. The prevalence of mucormycosis or invasive zygomycosis in India is significantly higher when compared to global data. In fact, it is estimated to be around 70 times higher in India (Chakrabarti et al., 2001, 2009). This higher prevalence may be attributed to various factors, including environmental conditions and the population's susceptibility. The most common risk factor in India is diabetes mellitus. Other risk factors include hematological malignancies (such as blood cancers), solid-organ transplant, a history of postpulmonary tuberculosis, and chronic kidney disease (Chakrabarti et al., 2006). The incidence of mucormycosis in India has shown an increasing trend over the years. The yearly incidence rates have been documented as 12.9, 35.6, and 50 cases per year during the periods of 1990–1999, 2000–2004, and 2006–2007, respectively (Chakrabarti et al., 2009). There has been a significant overall increase in the number of cases, with a shift from 25 cases per year (1990–2007) to 89 cases per year (2013–2015). This indicates a growing burden of mucormycosis within the country.

2.3 FUNGAL PATHOGENS OF HIGH CLINICAL RELEVANCE

Invasive fungal diseases (IFDs) are following an ascending trend, particularly among immuno-compromised populations. This rise in IFDs is compounded by inadequate access to effective diagnostics and treatments, and further exacerbated by the emergence of antifungal resistance across multiple regions. This complex landscape poses a formidable challenge to global health, forcing the public health agencies to respond. For example, the Centers for Disease Control and Prevention (CDC) has categorized fungal diseases based on their prevalence (Box 2.1). The World Health Organization (WHO, 2022) has taken an important step by articulating the first Fungal Priority Pathogens List (WHO FPPL) to guide research, development and public health action (https://www.who.int/publications/i/item/9789240060241). This list has placed various pathogenic fungi into 'critical', 'high', or 'medium' priority groups. A brief description of four pathogenic fungi placed in the critical priority group, i.e., *Candida auris*, *Cryptococcus neoformans*, *Aspergillus fumigatus*, and *Candida albicans* is presented in Box 2.1.

2.3.1 *CANDIDA AURIS*

This globally distributed pathogenic yeast can cause invasive candidiasis of the blood (candidemia), eyes, central nervous system, heart, bones and internal organs. Invasive candidiasis is a serious noso-comial infection, particularly affecting immunocompromised and critically ill patients, e.g., those suffering from cancer, or having undergone bone marrow and organ transplantation. The overall mortality of invasive candidiasis with *C. auris* is reported to range from 29% to 53% (Ortiz-Roa et al., 2023). As per the CDC fact sheet, *C. auris*, an emergent and extensively drug-resistant yeast species,

---

BOX 2.1 TYPES OF FUNGAL DISEASES LISTED BY THE CENTERS FOR DISEASE CONTROL AND PREVENTION (CDC)

Most common fungal diseases
1. Fungal nail infections
2. Vaginal candidiasis
3. Ringworm
4. *Candida* infections of the mouth, throat, and esophagus

Fungal diseases that affect people with weakened immune systems
1. Aspergillosis
2. *Candida auris* infection
3. Invasive candidiasis
4. Pneumocystis pneumonia (PCP)
5. Candidiasis
6. *Cryptococcus neoformans* infection
7. Mucormycosis
8. Talaromycosis

Fungal diseases that affect people who live in or travel to certain areas
1. Blastomycosis
2. *Cryptococcus gattii* infection
3. Paracoccidioidomycosis
4. Coccidioidomycosis (Valley Fever)
5. Histoplasmosis

Other diseases and health problems caused by fungi
1. Fungal eye infections
2. Sporotrichosis
3. Mycetoma
4. Healthcare-associated fungal meningitis

Source: https://www.cdc.gov/fungal/diseases/index.html.

represents a substantial menace owing to its capacity to cause severe infections and facile transmission within healthcare institutions, particularly hospitals and nursing homes. Initially reported in 2009 in Asia, *C. auris* has rapidly intensified into a prominent causative agent of grave infections on a global scale. Within the United States, hospitalized patients are subjected to approximately 400 infections caused by antifungal-resistant *C. auris* annually, while other variants of antifungal-resistant *Candida* account for roughly 35,000 infections. *C. auris* started spreading in the United States in 2015, with an increase in reported cases by 318% in 2018 compared to the average number between 2015 and 2017 (https://www.cdc.gov/drugresistance/pdf/threats-report/candida-auris-508.pdf).

2.3.2 *Cryptococcus neoformans*

This widely prevalent yeast can infect humans following inhalation of fungal cells from the environment, primarily affecting the lungs but also having the potential of spreading to the central nervous system and blood. Risk factors include autoimmune disease, HIV infection, liver cirrhosis, and immunosuppression. Cerebral cryptococcosis is a life-threatening disease with high mortality despite antifungal therapy, with mortality ranging from 41%–61%, especially in HIV-infected patients (Zhao et al., 2023). Median duration of hospital stay of patients with *C. neoformans* infection was reported to range from 18–39 days (Yoon et al., 2020).

2.3.3 *Aspergillus fumigatus*

This ubiquitous environmental mold can infect humans and cause aspergillosis following inhalation from the environment. Though predominantly causing pulmonary disease, it can disseminate to other sites too, such as the brain. Risk factors for aspergillosis susceptibility include conditions like transplantation, hematological malignancy, chronic liver or lung disease, neutropenia, and corticosteroid therapy (Latgé and Chamilos, 2019). The prevalence of aspergillosis varies geographically from less than 1% to 5%–10%, and annual incidence rates are observed to fluctuate. Mortality in patients with azole-resistant *A. fumigatus* infection is high (47%–88%) and has been reported to reach 100% in some studies. Data on overall length of hospital stay related to invasive aspergillosis are limited and range widely from 21–532 days (Arastehfar et al., 2021).

Allergic bronchopulmonary aspergillosis (ABPA) is a hypersensitivity or allergic reaction triggered by exposure to the fungus *A. fumigatus*. It is estimated to affect between 1 and 15% of cystic fibrosis patients (Stevens et al., 2003). One study calculated that 2.5% of adults who have asthma also have ABPA, which is approximately 4.8 million people worldwide. Of these 4.8 million people who have ABPA, an estimated 400,000 also have chronic pulmonary aspergillosis (CPA) (Denning et al., 2013a). Another 1.2 million people were estimated to have CPA after having tuberculosis (Denning et al., 2011), and over 70,000 people were estimated to have CPA as a complication of sarcoidosis (Denning et al., 2013b).

2.3.4 *Candida albicans*

This fungus can be considered part of the healthy human microbiome, but may also cause infections of the mucosae or invasive candidiasis. According to conservative estimates, invasive candidiasis affects more than 250,000 people worldwide every year and is the cause of more than 50,000 deaths (Arendrup, 2010). Incidence rates of candidemia have been reported to be between 2–14 cases per 100,000 people in population-based studies (Cleveland et al., 2015).

C. albicans, and to a lesser extent, other *Candida* species, are commonly found in the oral cavity of as many as 75% of the population (Vila et al., 2020). In individuals who are generally healthy, this colonization typically remains harmless. However, individuals with mild immune system compromises are more likely to experience persistent oral infections. These infections caused by *Candida* species are referred to as "oral candidiasis" (OC) (Talapko et al., 2021). OC is most commonly attributed to *C. albicans* and can affect the oropharynx and/or the esophagus in individuals with impaired adaptive immune systems. Notably, HIV infection is a significant risk factor for developing OC. Additionally, wearing dentures and extreme age are also recognized as risk factors for the development of OC (Erfaninejad et al., 2022).

2.4 ANTIMICROBIAL RESISTANCE AMONG MAJOR FUNGAL PATHOGENS

In clinical practice, treatment of systemic fungal infections largely relies on antifungals belonging to four main categories: Azoles, echinocandins, polyenes, and pyrimidines (Campoy and Adrio, 2017). In the 1950s, a category of antifungal agents known as polyenes was discovered (Carolus et al., 2020). Similar to some other antifungals, e.g., azoles, these compounds selectively interact with the distinctive fungal membrane constituent ergosterol. They accomplish this through direct sterol binding, which induces a membrane potential disruption and alteration in membrane permeability (Odds et al., 2003; Ivanov et al., 2022). Since the late 1960s, azole antifungal drugs have been known to inhibit the synthesis of ergosterol, an essential component of fungal cell membranes. Disruption of cell membrane integrity caused by this group of antifungals ultimately leads to death of the fungi. Aspergillosis, candidiasis, and dermatophytosis are among the various fungal infections that azoles are employed to treat (Assress et al., 2021). In the 1980s and 1990s, echinocandins, another class of antifungal agents, were discovered. They function by selectively targeting β-(1,3)-glucan, a crucial constituent of the fungal cell wall that is vital for maintaining its structural integrity. Due to its distinctive composition, which consists of mannan, chitin, α- and β-glucans, and other compounds, the fungal cell wall is a prime target for antifungal agents. Due to the fact that these components are unique to fungi and do not exist in other organisms, they are a compelling target for drugs that aim to eliminate fungal cells selectively while limiting harm to host cells (Hüttel, 2021; Szymański et al., 2022). Pyrimidine analogues constitute a different group of antifungal agents that function through the inhibition of DNA and RNA synthesis (Basha and Goudgaon, 2021). As a pyrimidine analogue, flucytosine (5FC) is the most frequently employed drug in the treatment of fungal infections. It has been demonstrated that the 5FC inhibits the development of numerous yeasts, such as *Candida* and *Cryptococcus neoformans*. Due to the high prevalence of primary resistance in numerous fungal species, its initial potential as an antifungal agent has been weakened (Lotfali et al., 2021).

Target modification is one of the several strategies employed by the pathogens, including fungi to develop resistance against antimicrobial drugs (Bhattacharya et al., 2020). Genetic alterations in target binding locations, like lanosterol demethylase genes for azoles or β-glucan synthase genes for echinocandins, can cause resistance (Jha and Kumar, 2019). Increased target availability (Palmer and Kishony, 2014), higher drug efflux activity for drugs such as azoles (Cannon et al., 2009; Jangir et al., 2023), or flucytosine prodrug activation inhibition (Hossain et al., 2022) can also cause resistance. Certain phenotypes of the pathogens fall between 'sensitive' and 'resistant', and they are described as 'tolerant.' These tolerant cells may not be able to reproduce in presence of inhibitory concentrations of antifungal agents, but they can still survive. Berman and Krysan (2020) mentions important role of stress-response machinery and epigenetic pathways toward development of such tolerant-but not resistant phenotypes. Drug tolerance can impact treatment response with fluconazole-tolerant *Candida albicans* (Levinson et al., 2021). Fungistatic drugs may promote tolerance (defined as the fraction of growth above the minimum inhibitory concentration [MIC]) more than the fungicidal drugs. While the latter may promote development of resistance more.

Judicious use of available effective antimicrobial agents is necessary to contain development of resistant phenotypes. Overuse and misuse of antimicrobial agents can be reduced by prescribing the drugs after knowing the antimicrobial susceptibility pattern (i.e., antibiogram) of the infectious agent. Antifungal susceptibility tests are of crucial importance in ensuring that satisfactory and proper antimicrobial therapy is prescribed to optimize treatment outcomes (Wanger, 2009).

However, susceptibility testing with fungi is more complicated than with many non-fastidious bacteria. Some of the challenges encountered while interpreting fungal susceptibility results are listed below (Petrikkou et al., 2001; Rex et al., 2001; CLSI, 2017):

1. *In vitro* susceptibility does not consistently predict clinical efficacy *in vivo*.
2. Establishing susceptibility breakpoints for drugs against susceptible fungal organisms is a complex process.
3. Many pathogenic fungi exhibit dimorphism. They grow as unicellular yeast in the human host, but as filamentous mold outside the host (Playfair and Bancroft, 2013). Response of the same species grown in mold form in lab for antifungal susceptibility test may vary from that of its unicellular form inside host.
4. Although RPMI is a recommended medium for antifungal susceptibility testing (AFST), fastidious organisms (e.g., lipophilic yeasts like *Malassezia*) cannot grow in this medium due to deficiency of specific lipids in the medium.
5. Longer incubation periods (48 hours) are required for fungal AFST.
6. The CLSI method proposes inoculum preparation using a spectrophotometric approach. However, this approach may not be applicable equally to spores vs. filaments. Inoculum of filamentous organisms do suffer from some degree of heterogeneity; in such cases, the relationship between biomass and optical density is not always linear. Inoculum size is known to significantly affect minimum inhibitory concentrations (MICs) for filamentous fungi.
7. Additional factors such as spore color can influence optical density measurements.
8. Antifungal susceptibility testing lacks the standardization achieved in such testing with bacteria.
9. Prolonged incubation times can lead to evaporation-related issues.

2.5 DISCOVERY AND DEVELOPMENT OF NOVEL ANTIFUNGALS, AND THE ASSOCIATED CHALLENGES

Why is the antifungal pipeline dry? There are multiple reasons why we lack a sufficient number of safe and potent antifungals in our arsenal for clinical applications. These include toxicity exerted by antifungal compounds toward the eukaryotic host; slow pace of action associated with most antifungals requiring longer period of therapy and reducing patient compliance; rapid development of resistance in the pathogenic strains, and so on.

Identification of valid cellular and molecular targets in pathogens is one of the crucial requirements for progressing toward the discovery of novel antimicrobial agents. For these novel antimicrobials to satisfy the criteria of selective toxicity, it is necessary that the molecular target attacked by them is absent from the human host, and present exclusively in the pathogen. While in the case of bacteria, there are some such targets like peptidoglycan, 70S ribosome, etc., which are absent from their human host; fungi, being eukaryotes, share a much higher similarity with human cells. Both the fungal pathogen and the human host possess 80S ribosomes, membrane-bound organelles like mitochondria, whose targeting by any potent antifungal will also prove toxic to the human cells. This leaves only a few targets like chitin in fungal cell walls or ergosterol in fungal cell membranes available to explore by antifungal discovery programs (Figure 2.2). In addition, the calcineurin pathway, sphingolipid synthesis, and trehalose pathway are also considered viable targets for the next generation of antifungals (Perfect, 2017). A brief description of some of these potential targets in fungal pathogens is presented below:

a. *Chitin in fungal cell walls*: The fungal cell wall is mainly composed of chitin, β-glucan, and glycoproteins, such as mannose and hydrophobins. The outer surface is rich in glucans acting as mucilage, while the inner layer contains chitin microfibrils, which are

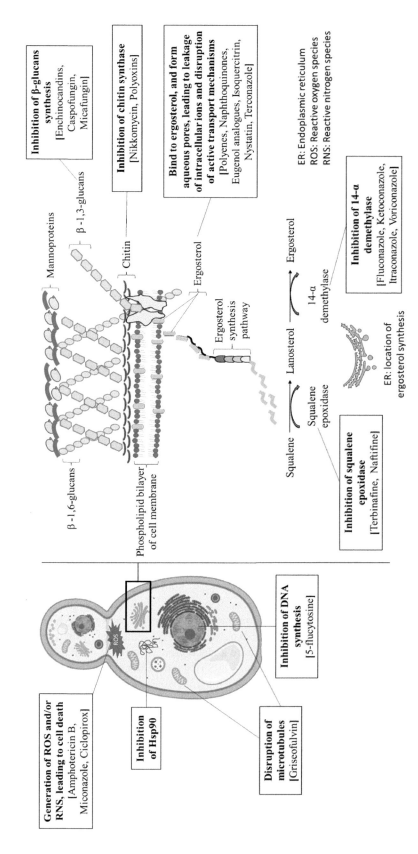

FIGURE 2.2 Important antifungal mechanisms exploited by different antifungal agents, and their cellular or molecular targets. (From Carrillo-Munoz et al., 2006.)

covalently cross-linked with other polysaccharides such as glucans (Ruiz-Herrera and Ortiz-Castellanos, 2019). Chitin, a polymer of N-acetylglucosamine, is essential for cell wall integrity. The amount and localization of chitin are species-specific. Chitin deacetylase (CDA) enzymes involved in chitin synthesis are found in both the cell membrane and cell wall. Besides maintaining the integrity of the cell wall, chitin is also responsible for the linkage between the cell wall and capsule for epithelial adhesion (Goldman and Vicencio, 2012; Free, 2013). The composition and localization of cell wall polysaccharides remains fluidic enough to adapt to the varying environmental conditions, to protect the fungi against the harsh environment. By specifically focusing on inhibiting chitin synthesis or its integration into the fungal cell wall (a trait absent in humans), we can undermine the structural stability of the fungal cell. Chitin-targeting agents can be expected to trigger fungal cell lysis and guarantee targeted toxicity.

b. *Ergosterol in fungal cell membranes*: Ergosterol is the major sterol component of yeast and filamentous fungal cell membranes, which is vital for maintaining normal growth, viability, cell integrity, and function (Mukhopadhyay et al., 2004). Considering the absence of ergosterol in animals, it is an appropriate target for antifungal drugs (Konuk and Ergüden, 2022). Targeting ergosterol leads to membrane disruption, ultimately causing fungal cell death. Squalene epoxidase is a key flavin adenine dinucleotide (FAD)-dependent enzyme of ergosterol and cholesterol biosynthetic pathways and is viewed as a plausible target with respect to the development of novel inhibitors of the growth of pathogenic fungi (Nowosielski et al., 2011). However, such antifungals may also interfere with cholesterol synthesis in humans, though they may also be of increased value in treating fungal infections in people suffering from hypercholesterolemia. Disrupting ergosterol synthesis or function, which differs from cholesterol found in human systems, can compromise the integrity and fluidity of the fungal cell membrane, affecting vital cellular processes and establishing selective toxicity (Vishwakarma et al., 2023).

c. *Fungal-specific signaling pathways*: Fungi often have unique signaling pathways that regulate essential processes such as cell growth, differentiation, and virulence (Ikeda et al., 2019). Several fungal-specific signaling pathways have been identified and they play crucial roles in regulating various aspects of fungal biology, including Cyclic AMP (cAMP) pathway, Mitogen-Activated Protein Kinase (MAPK) pathway, calcineurin pathway, Hog1 pathway, and Ras pathway (Jung and Bahn, 2009; Albataineh and Kadosh, 2015; Earle et al., 2023). Important traits regulated by each of these pathways are listed in Table 2.1. Inhibiting fungal signaling pathways, which disrupt critical cellular functions and impede pathogen survival and proliferation, achieves selective toxicity by specifically targeting pathways with fungal-specific components, minimizing the risk of affecting human cellular processes.

d. *Transporters and pumps*: These play essential roles in nutrient uptake, maintaining cellular homeostasis, and efflux of toxic compounds (Robinson et al., 2021). These proteins are responsible for the movement of small molecules, ions, and other substrates across the fungal cell membrane. Targeting these transporters and pumps is a strategy for developing antifungal agents with selective toxicity, as many of these proteins have fungal-specific characteristics. They are either absent from human system or present in significantly different form than in fungi. Targeting fungal-specific transporters and pumps, including Major Facilitator Superfamily (MFS) transporters, ATP-binding cassette (ABC) transporters, drug efflux pumps, P-Type ATPases, sugar transporters, and amino acid transporters, presents a strategy for developing antifungal agents with selective toxicity, as these proteins are involved in essential fungal cellular processes absent or significantly different in human cells.

e. *Metal homeostasis*: Inorganic metal ions are important to all life forms as micronutrients or cofactors for different enzymes. The same metals at high concentrations can prove toxic. Hence, biological systems are required to maintain a proper metal homeostasis. Disturbance of this metal homeostasis can trigger dysregulation of multiple

TABLE 2.1

Some Signaling Pathways in Fungi Viewed as Potential Targets for Novel Antifungals

Sr. No.	Pathway	Traits Regulated	Remarks
1.	Cyclic AMP (cAMP) Pathway	Cellular morphogenesis and virulence factors	The cAMP pathway regulates hyphal growth, thereby influencing the development of filamentous structures important for fungal colonization and invasion (Hu et al., 2014). This pathway also regulates secretion of virulence factors, including toxins, which are of critical importance for the pathogenicity of fungi (D'Souza et al., 2001).
2.	Mitogen-Activated Protein Kinase (MAPK) Pathway	Cell differentiation and stress response	The MAPK regulates conidiation, i.e., formation and release of asexual spores, thereby contributing toward the dissemination. This pathway also participates in fungal response to heat shock and other environmental stresses, allowing fungi to adapt and survive in diverse conditions (Ryter et al., 2004; May et al., 2005; Zhao et al., 2007; González-Rubio et al., 2019)
3.	Calcineurin Pathway	Cell wall integrity and virulence	The calcineurin pathway regulates synthesis of chitin- a key component of the fungal cell wall, essential for maintaining structural integrity. This pathway also regulates evasion of the host immune system, and thus plays an important role toward the establishment of infection (Park et al., 2016; Matsumoto et al., 2022; Wang et al., 2023).
4.	Hog1 Pathway	Osmotic stress response and cell cycle regulation	The Hog1 pathway regulates fungal response to osmotic stress, enabling fungi to survive in high-salt environments. In stressful situations, it triggers cell cycle arrest, allowing fungi to divert resources to stress response mechanisms (Matsumoto et al., 2023; Zhang et al., 2023).
5.	Ras Pathway	Cell proliferation and virulence	The Ras pathway regulates mitotic signaling, influencing the rate of cell division and proliferation. Components of this pathway may influence adhesion and invasion processes, supporting the ability of fungi to establish infections in host tissues (Zong et al., 2023).

enzyme-catalyzed biochemical reactions. If metal homeostasis in a pathogen can be disturbed selectively without affecting the host, then it can be a useful target in bacterial (Joshi et al., 2019; Gajera et al., 2023) as well as fungal (Hameed et al., 2020) pathogens. Depriving the fungal pathogens of the essential micronutrient metals can compromise their fitness inside the host. Hosts compete with the pathogens for the same micronutrients like iron, zinc, etc., which be viewed as a non-specific defense strategy on the part of the host. Micronutrient deprivation has been considered a worthy strategy against *C. albicans*. Hans et al. (2022) demonstrated Mg availability to be indispensable for the evasion of host immunity by *C. albicans*, and that specific Mg-dependent pathways could be targeted for therapeutic purposes. Availability of sufficient Zn and iron was reported to have contributed toward success of *Mucor* species in causing mucormycosis in patients infected with the coronavirus (Dogra et al., 2022; Kumar et al., 2022).

Despite theoretical knowledge and some experimental demonstrations regarding probable drugability of the abovementioned targets in fungi, in actuality, the discovery pipeline of effective and safe antifungals is not rich. Readers may refer to McCarty and Pappas (2021) and Jacobs et al. (2022) for an update on global efforts to find new antifungals. Some of the recent efforts from India are tabulated in Table 2.2. In addition to the conventional efforts for discovery of new antifungal antibiotics,

TABLE 2.2

Some Antifungal Research Projects Listed on Clinical Trial Registries (India)

Sr. No.	Clinical Trial ID	Test Formulation	Title of Study
1.	CTRI/2020/03/023756	*Abhayarishtadi Churna* and *Durvanishadi Lepa*	Randomized clinical trial to evaluate combine efficacy of *Abhayarishtadi Churna* and *Durvanishadi Lepa* in Dadru
2.	CTRI/2020/05/025361	*Ushq* and *Sirka*	Clinical study to compare safety and efficacy of unani formulation, *Ushq* (*Dorema ammoniacum*) & *Sirka* (vinegar) with Terbinafine in patients of Quba (dermatophytosis, ringworm)
3.	CTRI/2021/04/032728	B C Caps *Nimba Beeja Taila*	An open label clinical study to evaluate the efficacy and safety of B C Caps manufactured by Nutra Grace in abnormal vaginal discharge due to microbial infections
4.	CTRI/2022/11/047087	*Edagajadi lepa*	A clinical study to evaluate the therapeutic effect of *Virechanottara Edagajadi lepa* in Dadrukushta with special reference to Tinea infection
5.	CTRI/2020/04/024864	Griseofulvin	A clinico-epidemiological, microbiological, histopathological and therapeutic study of recurrent dermatophytosis
6.	CTRI/2021/06/033994	Accelerated corneal collagen cross with standard antifungal, i.e., natamycin 5% eyedrops	Assessment of accelerated corneal collagen cross-linking as an adjuvant therapy in moderate to severe fungal keratitis
7.	CTRI/2021/05/033812	Topical oxiconazole nitrate 1% lotion in treatment of pityriasis versicolor	A randomized double-blind study comparing topical ketoconazole 2% lotion and topical oxiconazole nitrate 1% lotion in treatment of pityriasis versicolor
8.	CTRI/2020/06/026097	Doxycycline 100 mg	A randomized controlled trial to evaluate the role of oral doxycycline in the treatment of fungal keratitis
9.	CTRI/2021/05/033591	Topical posaconazole	Efficacy of topical posaconazole in recalcitrant *Aspergillus flavus* keratitis
10.	CTRI/2022/04/042226	Peribulbar injection of amphotericin b	Efficacy of using peribulbar injection of amphotericin b for invasive rhino-orbital mucor-mycosis – a pilot study
11.	CTRI/2022/05/042593	Dupilumab	Study assessing the efficacy and safety of dupilumab in patients with Allergic Fungal Rhinosinusitis (AFRS)

Source: https://ctri.nic.in/Clinicaltrials/login.php.

alternative approaches are also being pursued. Among them are the efforts for finding antifungal leads from traditional medicine (e.g., some entries in Table 2.2; Joshi and Vaidya, 2019) and phytocompounds (De Filippis et al., 2019), developing monoclonal antibodies against specific fungal antigens, vaccine development (Nicola et al., 2019), drug repurposing, developing new delivery systems or formulations of the known antifungals, and myconanotechnological approaches (Sousa et al., 2020). Although these approaches hitherto have not yielded any major success, they certainly hold promise and are worth pursuing. This review does not attempt to provide any detailed overview of these non-antibiotic approaches, but readers may refer to the above-cited articles for details.

2.6 CONCLUSION

Owing to the increasing number of immunocompromised people in the modern world, the clinical relevance of fungal pathogens is on the rise. Invasive and systemic fungal infections usually prove difficult to treat. The requirement for longer duration of treatment with antifungal agents may lead to issues of non-compliance and side effects. The development of antifungal resistance among clinically important fungi further complicates the situation. In the context of the limited number of non-toxic antifungals available currently, there is an urgent need for discovery and development of novel antifungal agents and strategies. Besides pursuing the conventional efforts in the search of new antifungal antibiotics, alternative strategies like mining traditional medicine wisdom and natural products for identifying novel fungicidal and antivirulence (non-antibiotic) leads, vaccine development against important fungal antigens, and so on, also need to be pursued. Combinatorial chemistry, and *in silico* screening of large libraries of natural and synthetic compounds followed by *in vivo* evaluation of predicted 'hits' also hold promise.

REFERENCES

Albataineh MT, Kadosh D. Regulatory roles of phosphorylation in model and pathogenic fungi. *Medical Mycology.* 2015;54(4):333–52. https://doi.org/10.1093/mmy/myv098

Aragón-Sánchez J, López-Valverde ME, Víquez-Molina G, Milagro-Beamonte A, Torres-Sopena L. Onychomycosis and tinea pedis in the feet of patients with diabetes. *International Journal of Lower Extremity Wounds.* 2023;22(2):321–7. https://doi.org/10.1177/15347346211009409

Arastehfar A, Carvalho A, Houbraken J, Lombardi L, Garcia-Rubio R, Jenks JD, Rivero-Menendez O, Aljohani R, Jacobsen ID, Berman J, Osherov N. *Aspergillus fumigatus* and aspergillosis: From basics to clinics. *Studies in Mycology.* 2021;100(1):100115. https://doi.org/10.1016/j.simyco.2021.100115

Araujo PS, Medeiros Z, Melo FL, Maciel MA, Melo HR. *Candida famata*-induced fulminating cholecystitis. *Revista da Sociedade Brasileira de Medicina Tropical.* 2013;46:795–6. https://doi.org/10.1590/0037-8682-0162-2013

Arendrup MC. Epidemiology of invasive candidiasis. *Current Opinion in Critical Care.* 2010;16(5):445–52. https://doi.org/10.1097/MCC.0b013e32833e84d2

Assress HA, Selvarajan R, Nyoni H, Mamba BB, Msagati TA. Antifungal azoles and azole resistance in the environment: Current status and future perspectives—A review. *Reviews in Environmental Science and Bio/Technology.* 2021:1–31. https://doi.org/10.1007/s11157-021-09594-w

Bansal E, Garg A, Bhatia S, Attri AK, Chander J. Spectrum of microbial flora in diabetic foot ulcers. *Indian Journal of Pathology and Microbiology.* 2008;51(2):204. https://doi.org/10.4103/0377-4929.41685

Barin F. HIV/AIDS as a model for emerging infectious disease: Origin, dating and circumstances of an emblematic epidemiological success. *La Presse Médicale.* 2022;51(3):104128. https://doi.org/10.1016/j.lpm.2022.104128

Basha J, Goudgaon NM. A comprehensive review on pyrimidine analogs-versatile scaffold with medicinal and biological potential. *Journal of Molecular Structure.* 2021;1246:131168. https://doi.org/10.1016/j.molstruc.2021.131168

Berman J, Krysan DJ. Drug resistance and tolerance in fungi. *Nature Reviews Microbiology.* 2020;18(6):319–31. https://doi.org/10.1038/s41579-019-0322-2

Bhattacharya S, Sae-Tia S, Fries BC. Candidiasis and mechanisms of antifungal resistance. *Antibiotics.* 2020;9(6):312. https://doi.org/10.3390/antibiotics9060312

Campoy S, Adrio JL. Antifungals. *Biochemical Pharmacology.* 2017;133:86–96. https://doi.org/10.1016/j.bcp.2016.11.019

Cannon RD, Lamping E, Holmes AR, Niimi K, Baret PV, Keniya MV, Tanabe K, Niimi M, Goffeau A, Monk BC. Efflux-mediated antifungal drug resistance. *Clinical Microbiology Reviews.* 2009;22(2):291–321. https://doi.org/10.1128/cmr.00051-08

Carolus H, Pierson S, Lagrou K, Van Dijck P. Amphotericin B and other polyenes—Discovery, clinical use, mode of action and drug resistance. *Journal of Fungi.* 2020;27;6(4):321. https://doi.org/10.3390/jof6040321

Carrillo-Munoz AJ, Giusiano G, Ezkurra PA, Quindós G. Antifungal agents: Mode of action in yeast cells. *Revista española de quimioterapia.* 2006;19(2):130–9.

Chakrabarti A, Chatterjee SS, Das A, Panda N, Shivaprakash MR, Kaur A, Varma SC, Singhi S, Bhansali A, Sakhuja V. Invasive zygomycosis in India: Experience in a tertiary care hospital. *Postgraduate Medical Journal.* 2009;85(1009):573–81. https://doi.org/10.1136/pgmj.2008.076463

Chakrabarti A, Das A, Mandal J, Shivaprakash MR, George VK, Tarai B, Rao P, Panda N, Verma SC, Sakhuja V. The rising trend of invasive zygomycosis in patients with uncontrolled diabetes mellitus. *Medical Mycology.* 2006;44(4):335–42. https://doi.org/10.1080/13693780500464930

Chakrabarti A, Das A, Sharma A, Panda N, Das S, Gupta KL, Sakhuja V. Ten years' experience in zygomycosis at a tertiary care centre in India. *Journal of Infection.* 2001;42(4):261–6. https://doi.org/10.1053/jinf.2001.0831

Chen L, Zhang L, Xie Y, Wang Y, Tian X, Fang W, Xue X, Wang L. Confronting antifungal resistance, tolerance, and persistence: Advances in drug target discovery and delivery systems. *Advanced Drug Delivery Reviews*. 2023:115007. https://doi.org/10.1016/j.addr.2023.115007

Cleveland AA, Harrison LH, Farley MM, Hollick R, Stein B, Chiller TM, Lockhart SR, Park BJ. Declining incidence of candidemia and the shifting epidemiology of *Candida* resistance in two US metropolitan areas, 2008–2013: Results from population-based surveillance. *PLOS One*. 2015;10(3):e0120452. https://doi.org/10.1371/journal.pone.0120452

Clinical and Laboratory Standards Institute. *Reference Method for Broth Dilution Antifungal Susceptibility Testing of Filamentous Fungi*, 3rd ed. Wayne, PA: Clinical and Laboratory Standards Institute, 2017.

D'Souza CA, Alspaugh JA, Yue C, Harashima T, Cox GM, Perfect JR, Heitman J. Cyclic AMP-dependent protein kinase controls virulence of the fungal pathogen *Cryptococcus neoformans*. *Molecular and Cellular Biology*. 2001;21(9):3179–91. https://doi.org/10.1128/MCB.21.9.3179-3191.2001

Darjani A, Farrokhian K, Sobhani A, Kalantari S. Prevalence of skin lesions in diabetic patients. *Journal of Guilan University of Medical Sciences*. 2002;11(43):60–6.

De Filippis B, Ammazzalorso A, Amoroso R, Giampietro L. Stilbene derivatives as new perspective in antifungal medicinal chemistry. *Drug Development Research*. 2019;80(3):285–93. https://doi.org/10.1002/ddr.21525

de Nies L, Kobras CM, Stracy M. Antibiotic-induced collateral damage to the microbiota and associated infections. *Nature Reviews Microbiology*. 2023:1–6. https://doi.org/10.1038/s41579-023-00936-9

Denning DW, Pleuvry A, Cole DC. Global burden of chronic pulmonary aspergillosis as a sequel to pulmonary tuberculosis. *Bulletin of the World Health Organization*. 2011;89(12):864–72.

Denning DW, Pleuvry A, Cole DC. Global burden of allergic bronchopulmonary aspergillosis with asthma and its complication chronic pulmonary aspergillosis in adults. *Medical Mycology*. 2013a;51(4):361–70. https://doi.org/10.3109/13693786.2012.738312

Denning DW, Pleuvry A, Cole DC. Global burden of chronic pulmonary aspergillosis complicating sarcoidosis. *European Respiratory Journal*. 2013b;41(3):621–6. https://doi.org/10.1183/09031936.00226911

Dogra S, Arora A, Aggarwal A, Passi G, Sharma A, Singh G, Barnwal RP. Mucormycosis amid COVID-19 crisis: Pathogenesis, diagnosis, and novel treatment strategies to combat the spread. *Frontiers in Microbiology*. 2022;12:794176. https://doi.org/10.3389/fmicb.2021.794176

Durand BA, Pouget C, Magnan C, Molle V, Lavigne JP, Dunyach-Remy C. Bacterial interactions in the context of chronic wound biofilm: A review. *Microorganisms*. 2022;10(8):1500. https://doi.org/10.3390/microorganisms10081500

Earle K, Valero C, Conn DP, Vere G, Cook PC, Bromley MJ, Bowyer P, Gago S. Pathogenicity and virulence of *Aspergillus fumigatus*. *Virulence*. 2023;14(1):2172264. https://doi.org/10.1080/21505594.2023.2172264

Erfaninejad M, Zarei Mahmoudabadi A, Maraghi E, Hashemzadeh M, Fatahinia M. Epidemiology, prevalence, and associated factors of oral candidiasis in HIV patients from southwest Iran in post-highly active antiretroviral therapy era. *Frontiers in Microbiology*. 2022;13:983348. https://doi.org/10.3389/fmicb.2022.983348

Fisher MC, Alastruey-Izquierdo A, Berman J, Bicanic T, Bignell EM, Bowyer P, Bromley M, Brüggemann R, Garber G, Cornely OA, Gurr SJ. Tackling the emerging threat of antifungal resistance to human health. *Nature Reviews Microbiology*. 2022;20(9):557–71. https://doi.org/10.1038/s41579-022-00720-1

Free SJ. Fungal cell wall organization and biosynthesis. *Advances in Genetics*. 2013;81:33–82. https://doi.org/10.1016/B978-0-12-407677-8.00002-6

Friedman DZ, Schwartz IS. Emerging fungal infections: New patients, new patterns, and new pathogens. *Journal of Fungi*. 2019;5(3):67. https://doi.org/10.3390/jof5030067

Gajera G, Thakkar N, Godse C, DeSouza A, Mehta D, Kothari V. Sub-lethal concentration of a colloidal nanosilver formulation (Silversol®) triggers dysregulation of iron homeostasis and nitrogen metabolism in multidrug resistant *Pseudomonas aeruginosa*. *BMC Microbiology*. 2023;23(1):303. https://doi.org/10.1186/s12866-023-03062-x

Ge Y, Wang Q. Current research on fungi in chronic wounds. *Frontiers in Molecular Biosciences*. 2023;9:1057766. https://doi.org/10.3389/fmolb.2022.1057766

Gold JA, Adjei S, Gundlapalli AV, Huang YL, Chiller T, Benedict K, Toda M. Increased hospitalizations involving fungal infections during COVID-19 pandemic, United States, January 2020–December 2021. *Emerging Infectious Diseases*. 2023;29(7):1433. https://doi.org/10.3201/eid2907.221771

Goldman DL, Vicencio AG. The chitin connection. *MBio*. 2012;3(2):10–128. https://doi.org/10.1128/mbio.00056-12

González-Rubio G, Fernández-Acero T, Martín H, Molina M. Mitogen-activated protein kinase phosphatases (MKPs) in fungal signaling: Conservation, function, and regulation. *International Journal of Molecular Sciences*. 2019;20(7):1709. https://doi.org/10.3390/ijms20071709

Gunathilaka MG, Senevirathna RM, Illappereruma SC, Keragala KA, Hathagoda KL, Bandara HM. Mucormycosis-causing fungi in humans: A meta-analysis establishing the phylogenetic relationships using internal transcribed spacer (ITS) sequences. *Journal of Medical Microbiology*. 2023;72(2):001651. https://doi.org/10.1099/jmm.0.001651

Gupta P, Kaur H, Mohindra S, Verma S, Rudramurthy S. Recurrent fatal brain abscess by *Cladophialophora bantiana* in a diabetic patient. *Indian Journal of Medical Microbiology*. 2022;40(1):152–5. https://doi.org/10.1016/j.ijmmb.2021.11.011

Hameed S, Hans S, Singh S, Fatima Z. Harnessing metal homeostasis offers novel and promising targets against *Candida albicans*. *Current Drug Discovery Technologies*. 2020;17(4):415–29. https://doi.org/10.2174/1570163816666190227231437

Hans S, Fatima Z, Ahmad A, Hameed S. Magnesium impairs *Candida albicans* immune evasion by reduced hyphal damage, enhanced β-glucan exposure and altered vacuole homeostasis. *PLOS One*. 2022;17(7):e0270676. https://doi.org/10.1371/journal.pone.0270676

Hossain CM, Ryan LK, Gera M, Choudhuri S, Lyle N, Ali KA, Diamond G. Antifungals and drug resistance. *Encyclopedia*. 2022;2(4):1722–37. https://doi.org/10.3390/encyclopedia2040118

Hüttel W. Echinocandins: Structural diversity, biosynthesis, and development of antimycotics. *Applied Microbiology and Biotechnology*. 2021;105:55–66. https://doi.org/10.1007/s00253-020-11022-y

Hu S, Zhou X, Gu X, Cao S, Wang C, Xu JR. The cAMP-PKA pathway regulates growth, sexual and asexual differentiation, and pathogenesis in *Fusarium graminearum*. *Molecular Plant-Microbe Interactions*. 2014;27(6):557–66. https://doi.org/10.1094/MPMI-10-13-0306-R

Ikeda K, Park P, Nakayashiki H. Cell biology in phytopathogenic fungi during host infection: Commonalities and differences. *Journal of General Plant Pathology*. 2019;85:163–73. https://doi.org/10.1007/s10327-019-00846-w

Ivanov M, Ćirić A, Stojković D. Emerging antifungal targets and strategies. *International Journal of Molecular Sciences*. 2022;23(5):2756. https://doi.org/10.3390/ijms23052756

Jacobs SE, Zagaliotis P, Walsh TJ. Novel antifungal agents in clinical trials. *F1000Research*. 2022;10. https://doi.org/10.12688/f1000research.28327.2

Jampilek J. Novel avenues for identification of new antifungal drugs and current challenges. *Expert Opinion on Drug Discovery*. 2022;17(9):949–68. https://doi.org/10.1080/17460441.2022.2097659

Jangir P, Kalra S, Tanwar S, Bari VK. Azole resistance in *Candida auris*: Mechanisms and combinatorial therapy. *Journal of Pathology, Microbiology and Immunology*. 2023. https://doi.org/10.1111/apm.13336

Jha A, Kumar A. Anticandidal agent for multiple targets: The next paradigm in the discovery of proficient therapeutics/overcoming drug resistance. *Future Medicinal Chemistry*. 2019;11(22):2955–74. https://doi.org/10.4155/fmc-2018-0479

Joshi C, Patel P, Palep H, Kothari V. Validation of the anti-infective potential of a polyherbal 'Panchvalkal' preparation, and elucidation of the molecular basis underlining its efficacy against *Pseudomonas aeruginosa*. *BMC Complementary and Alternative Medicine*. 2019;19(1):1–5. https://doi.org/10.1186/s12906-019-2428-5

Joshi J, Vaidya R A monograph on Panchvalkal (Modified) for the Treatment of Leucorrhoea. 2019. https://ayushportal.nic.in/EMR/CLINICAL_FINAL_REPORT-9.pdf

Jung KW, Bahn YS. The stress-activated signaling (SAS) pathways of a human fungal pathogen, *Cryptococcus neoformans*. *Mycobiology*. 2009;37(3):161–70. https://doi.org/10.4489/MYCO.2009.37.3.161

Kakeya H, Izumikawa K, Yamada K, Narita Y, Nishino T, Obata Y, Takazono T, Kurihara S, Kosai K, Morinaga Y, Nakamura S. Concurrent subcutaneous candidal abscesses and pulmonary cryptococcosis in a patient with diabetes mellitus and a history of corticosteroid therapy. *Internal Medicine*. 2014;53(12):1385–90. https://doi.org/10.2169/internalmedicine.53.1409

Kendirci M, Kurtoglu S, Keskin M, Kuyucu T. Vulvovaginal candidiasis in children and adolescents with Type I diabetes mellitus. *Journal of Pediatric Endocrinology and Metabolism*. 2004;17(11):1545–50. https://doi.org/10.1515/JPEM.2004.17.11.1545

Khoharo HK, Ansari S, Qureshi F. Frequency of skin manifestations in 120 type 2 diabetics presenting at tertiary care hospital. *Journal of Liaquat University of Medical and Health Sciences*. 2009;8:12–5.

Konuk HB, Ergüden B. Investigation of antifungal activity mechanisms of alpha-pinene, eugenol, and limonene. *Journal of Advances in VetBio Science and Techniques*. 2022;7(3):385–90. https://doi.org/10.31797/vetbio.1173455

Kumar A, Bansal P, Katiyar D, Prakash S, Rao NG. Molecular targeting and novel therapeutic approaches against fungal infections. *Current Molecular Medicine*. 2023. https://doi.org/10.2174/1566524023666230302123310

Kumar D, Banerjee T, Chakravarty J, Singh SK, Dwivedi A, Tilak R. Identification, antifungal resistance profile, *in vitro* biofilm formation and ultrastructural characteristics of *Candida* species isolated from diabetic foot patients in Northern India. *Indian Journal of Medical Microbiology.* 2016;34(3):308–14. https://doi.org/10.4103/0255-0857.188320

Kumar S, Acharya S, Jain S, Shukla S, Talwar D, Shah D, Hulkoti V, Parveen S, Patel M, Patel S, Hulkoti V. III Role of zinc and clinicopathological factors for COVID-19-associated Mucormycosis (CAM) in a rural hospital of central India: A case-control study. *Cureus.* 2022;14(2). https://doi.org/10.7759/cureus.22528

Lao M, Li C, Li J, Chen D, Ding M, Gong Y. Opportunistic invasive fungal disease in patients with type 2 diabetes mellitus from Southern China: Clinical features and associated factors. *Journal of Diabetes Investigation.* 2020;11(3):731–44. https://doi.org/10.1111/jdi.13183

Latgé JP, Chamilos G. *Aspergillus fumigatus* and Aspergillosis in 2019. *Clinical Microbiology Reviews.* 2019;33(1):10–128. https://doi.org/10.1128/cmr.00140-18

Levinson T, Dahan A, Novikov A, Paran Y, Berman J, Ben-Ami R. Impact of tolerance to fluconazole on treatment response in *Candida albicans* bloodstream infection. *Mycoses.* 2021;64(1):78–85. https://doi.org/10.1111/myc.13191

Lotfali E, Fattahi A, Sayyahfar S, Ghasemi R, Rabiei MM, Fathi M, Vakili K, Deravi N, Soheili A, Toreyhi H, Shirvani F. A review on molecular mechanisms of antifungal resistance in *Candida glabrata*: Update and recent advances. *Microbial Drug Resistance.* 2021;27(10):1371–88. https://doi.org/10.1089/mdr.2020.0235

Malazy OT, Shariat M, Heshmat R, Majlesi F, Alimohammadian M, Tabari NK, Larijani B. Vulvovaginal candidiasis and its related factors in diabetic women. *Taiwanese Journal of Obstetrics and Gynecology.* 2007;46(4):399–404. https://doi.org/10.1016/S1028-4559(08)60010-8

Matsumoto Y, Sugiyama Y, Nagamachi T, Yoshikawa A, Sugita T. Hog1-mediated stress tolerance in the pathogenic fungus *Trichosporon asahii. Scientific Reports.* 2023;13(1):13539. https://doi.org/10.1038/s41598-023-40825-y

Matsumoto Y, Yoshikawa A, Nagamachi T, Sugiyama Y, Yamada T, Sugita T. A critical role of calcineurin in stress responses, hyphal formation, and virulence of the pathogenic fungus *Trichosporon asahii. Scientific Reports.* 2022;12(1):16126. https://doi.org/10.1038/s41598-022-20507-x

May GS, Xue T, Kontoyiannis DP, Gustin MC. Mitogen activated protein kinases of *Aspergillus fumigatus. Medical Mycology.* 2005;43(supl):83–6. https://doi.org/10.1080/13693780400024784

McCarty TP, Pappas PG. Antifungal pipeline. *Frontiers in Cellular and Infection Microbiology.* 2021;11:732223. https://doi.org/10.3389/fcimb.2021.732223

Mirpour S, Fathollah S, Mansouri P, Larijani B, Ghoranneviss M, Mohajeri Tehrani M, Amini MR. Cold atmospheric plasma as an effective method to treat diabetic foot ulcers: A randomized clinical trial. *Scientific Reports.* 2020;10(1):10440. https://doi.org/10.1038/s41598-020-67232-x

Morton CO, Griffiths JS, Loeffler J, Orr S, White PL. Defective antifungal immunity in patients with COVID-19. *Frontiers in Immunology.* 2022;13. https://doi.org/10.3389/fimmu.2022.1080822

Mukhopadhyay K, Prasad T, Saini P, Pucadyil TJ, Chattopadhyay A, Prasad R. Membrane sphingolipid-ergosterol interactions are important determinants of multidrug resistance in *Candida albicans. Antimicrobial Agents and Chemotherapy.* 2004;48(5):1778–87. https://doi.org/10.1128/aac.48.5.1778-1787.2004

Mussi MC, Fernandes KS, Gallottini MH. A call for further research on the relation between type 2 diabetes and oral candidiasis. *Oral Surgery, Oral Medicine, Oral Pathology and Oral Radiology.* 2022;134(2):206–12. https://doi.org/10.1016/j.oooo.2022.02.009

Ngan NT, Flower B, Day JN. Treatment of cryptococcal meningitis: How have we got here and where are we going? *Drugs.* 2022;82(12):1237–49. https://doi.org/10.1007/s40265-022-01757-5

Nicola AM, Albuquerque P, Paes HC, Fernandes L, Costa FF, Kioshima ES, Abadio AK, Bocca AL, Felipe MS. Antifungal drugs: New insights in research & development. *Pharmacology & Therapeutics.* 2019;195:21–38. https://doi.org/10.1016/j.pharmthera.2018.10.008

Nowosielski M, Hoffmann M, Wyrwicz LS, Stepniak P, Plewczynski DM, Lazniewski M, Ginalski K, Rychlewski L. Detailed mechanism of squalene epoxidase inhibition by terbinafine. *Journal of Chemical Information and Modeling.* 2011;51(2):455–62. https://doi.org/10.1021/ci100403b

Odds FC, Brown AJ, Gow NA. Antifungal agents: Mechanisms of action. *Trends in Microbiology.* 2003;11(6):272–9. https://doi.org/10.1016/S0966-842X(03)00117-3

Ortiz-Roa C, Valderrama-Rios MC, Sierra-Umaña SF, Rodríguez JY, Muñetón-López GA, Solórzano-Ramos CA, Escandón P, Alvarez-Moreno CA, Cortés JA. Mortality caused by *Candida auris* bloodstream infections in comparison with other *Candida* species, a multicentre retrospective cohort. *Journal of Fungi.* 2023;9(7):715. https://doi.org/10.3390/jof9070715

Palmer AC, Kishony R. Opposing effects of target overexpression reveal drug mechanisms. *Nature Communications.* 2014;5(1):4296. https://doi.org/10.1038/ncomms5296

Park HS, Chow EW, Fu C, Soderblom EJ, Moseley MA, Heitman J, Cardenas ME. Calcineurin targets involved in stress survival and fungal virulence. *PLOS Pathogens.* 2016;12(9):e1005873. https://doi.org/10.1371/journal.ppat.1005873

Perfect JR. The antifungal pipeline: A reality check. *Nature Reviews Drug Discovery.* 2017;16(9):603–16. https://doi.org/10.1038/nrd.2017.46

Petrikkou E, Rodríguez-Tudela JL, Cuenca-Estrella M, Gómez A, Molleja A, Mellado E. Inoculum standardization for antifungal susceptibility testing of filamentous fungi pathogenic for humans. *Journal of Clinical Microbiology.* 2001;39(4):1345–7. https://doi.org/10.1128/jcm.39.4.1345-1347.2001

Phulari YJ, Kaushik V. Study of cutaneous manifestations of type 2 diabetes mellitus. *International Journal of Research in Dermatology.* 2018;4(1):8–13.

Playfair J, Bancroft G. Fungi. In: *Infection and Immunity,* 4th ed. p. 31–34. Oxford University Press, Oxford, 2013.

Raiesi O, Siavash M, Mohammadi F, Chabavizadeh J, Mahaki B, Maherolnaghsh M, Dehghan P. Frequency of cutaneous fungal infections and azole resistance of the isolates in patients with diabetes mellitus. *Advanced Biomedical Research.* 2017;6. doi: 10.4103/2277-9175.191003

Rasoulpoor S, Shohaimi S, Salari N, Vaisi-Raygani A, Rasoulpoor S, Shabani S, Jalali R, Mohammadi M. *Candida albicans* skin infection in patients with type 2 diabetes: A systematic review and meta-analysis. *Journal of Diabetes and Metabolic Disorders.* 2021;20:665–72. doi: 10.1007/s40200-021-00797-0

Rex JH, Pfaller MA, Walsh TJ, Chaturvedi V, Espinel-Ingroff A, Ghannoum MA, Gosey LL, Odds FC, Rinaldi MG, Sheehan DJ, Warnock DW. Antifungal susceptibility testing: Practical aspects and current challenges. *Clinical Microbiology Reviews.* 2001;14(4):643–58. https://doi.org/10.1128/cmr.14.4.643-658.2001

Robinson JR, Isikhuemhen OS, Anike FN. Fungal–metal interactions: A review of toxicity and homeostasis. *Journal of Fungi.* 2021;7(3):225. https://doi.org/10.3390/jof7030225

Rodrigues CF, Rodrigues ME, Henriques M. *Candida sp.* infections in patients with diabetes mellitus. *Journal of Clinical Medicine.* 2019;8(1):76. https://doi.org/10.3390/jcm8010076

Ruiz-Herrera J, Ortiz-Castellanos L. Cell wall glucans of fungi. A review. *Cell Surface.* 2019;5:100022. https://doi.org/10.1016/j.tcsw.2019.100022

Ryter SW, Xi S, Hartsfield CL, Choi AM. Mitogen activated protein kinase (MAPK) pathway regulates heme oxygenase-1 gene expression by hypoxia in vascular cells. *Antioxidants and Redox Signaling.* 2004;4(4):587–92. https://doi.org/10.1089/15230860260220085

Shazniza Shaaya E, Halim SA, Leong KW, Ku KB, Lim PS, Tan GC, Wong YP. *Candida chorioamnionitis* in mothers with gestational diabetes mellitus: A report of two cases. *International Journal of Environmental Research and Public Health.* 2021;18(14):7450. https://doi.org/10.3390/ijerph18147450

Sousa F, Ferreira D, Reis S, Costa P. Current insights on antifungal therapy: Novel nanotechnology approaches for drug delivery systems and new drugs from natural sources. *Pharmaceuticals.* 2020;13(9):248. https://doi.org/10.3390/ph13090248

Stevens DA, Moss RB, Kurup VP, Knutsen AP, Greenberger P, Judson MA, Denning DW, Crameri R, Brody AS, Light M, Skov M. Allergic bronchopulmonary aspergillosis in cystic fibrosis – State of the art: Cystic Fibrosis Foundation consensus conference. *Clinical Infectious Diseases.* 2003;37(Supplement_3): S225–64. https://doi.org/10.1086/376525

Sundaravadivelu S, Kumar S, Sripathi SK. Antimicrobial and anti-inflammatory response by two formulations of Jatyadithailam in healing diabetic foot ulcers. *Indian Journal of Traditional Knowledge.* 2022;21(3):489–98. https://doi.org/10.56042/ijtk.v21i3.50678

Szymański M, Chmielewska S, Czyżewska U, Malinowska M, Tylicki A. Echinocandins–structure, mechanism of action and use in antifungal therapy. *Journal of Enzyme Inhibition and Medicinal Chemistry.* 2022;37(1):876–94. https://doi.org/10.1080/14756366.2022.2050224

Talapko J, Juzbašić M, Matijević T, Pustijanac E, Bekić S, Kotris I, Škrlec I. *Candida albicans* – The virulence factors and clinical manifestations of infection. *Journal of Fungi.* 2021;7(2):79. https://doi.org/10.3390/jof7020079

Talapko J, Meštrović T, Škrlec I. Growing importance of urogenital candidiasis in individuals with diabetes: A narrative review. *World Journal of Diabetes.* 2022;13(10):809.

Vila T, Sultan AS, Montelongo-Jauregui D, Jabra-Rizk MA. Oral Candidiasis: A disease of opportunity. *Journal of Fungi.* 2020;6(1):15. https://doi.org/10.3390/jof6010015

Vishwakarma M, Haider T, Soni V. Update on fungal lipid biosynthesis inhibitors as antifungal agents. *Microbiological Research*. 2023:127517. https://doi.org/10.1016/j.micres.2023.127517

Wang ZH, Shen ZF, Wang JY, Cai YY, Li L, Liao J, Lu JP, Zhu XM, Lin FC, Liu XH. MoCbp7, a novel calcineurin B subunit-binding protein, is involved in the calcium signaling pathway and regulates fungal development, virulence, and ER homeostasis in *Magnaporthe oryzae*. *International Journal of Molecular Sciences*. 2023;24(11):9297. https://doi.org/10.3390/ijms24119297

Wanger A. Antibiotic Susceptibility Testing, In: Goldman E, Green LH, eds., *Practical Handbook of Microbiology*, 2nd ed. p. 149–158. Boca Raton: CRC Press, 2009.

Wilson RM, Reeves WG. Neutrophil phagocytosis and killing in insulin-dependent diabetes. *Clinical and Experimental Immunology*. 1986;63(2):478.

World Health Organization. WHO Fungal Priority Pathogens List to Guide Research, Development and Public Health Action. Geneva: World Health Organization, 2022. Licence: CC BY-NC-SA 3.0 IGO.

Xin H, Rosario-Colon J, Eberle K. Novel intravenous immunoglobulin therapy for the prevention and treatment of *Candida auris* and *Candida albicans* disseminated candidiasis. *mSphere*. 2023;8. https://doi.org/10.1128/msphere.00584-22.

Yoon HA, Felsen U, Wang T, Pirofski LA. *Cryptococcus neoformans* infection in Human Immunodeficiency Virus (HIV)-infected and HIV-uninfected patients at an inner-city tertiary care hospital in the Bronx. *Medical Mycology*. 2020;58(4):434–43. https://doi.org/10.1093/mmy/myz082

Yuvaraj A, Rohit A, Koshy PJ, Nagarajan P, Nair S, Abraham G. Rare occurrence of fatal *Candida haemulonii* peritonitis in a diabetic CAPD patient. *Renal Failure*. 2014;36(9):1466–7. https://doi.org/10.3109/0886022X.2014.944067

Zhang Y, Zhu M, Wang H, Yu G, Guo A, Ren W, Li B, Liu N. The mitogen-activated protein kinase Hog1 regulates fungal development, pathogenicity and stress response in *Botryosphaeria dothidea*. *Phytopathology*. 2023. https://doi.org/10.1094/PHYTO-07-23-0260-R

Zhao X, Mehrabi R, Xu JR. Mitogen-activated protein kinase pathways and fungal pathogenesis. *Eukaryotic Cell*. 2007;6(10):1701–14. https://doi.org/10.1128/ec.00216-07

Zhao Y, Ye L, Zhao F, Zhang L, Lu Z, Chu T, Wang S, Liu Z, Sun Y, Chen M, Liao G. *Cryptococcus neoformans*, a global threat to human health. *Infectious Diseases of Poverty*. 2023;12(1):20. https://doi.org/10.1186/s40249-023-01073-4

Zong Y, Zhang X, Gong D, Zhang F, Yu L, Bi Y, Sionov E, Prusky D. Small GTPases RasA and RasB regulate development, patulin production, and virulence of *Penicillium expansum*. *Postharvest Biology and Technology*. 2023;197:112192. https://doi.org/10.1016/j.postharvbio.2022.112192

3 Understanding *Trichophyton*-Mediated Dermatophytosis
Key Factors and Treatment Insights

*Shikha Chandra, Khushboo Arya, Sana Akhtar Usmani,
Saumya Chaturvedi, Sudhir Mehrotra, Nitin Bhardwaj,
and Ashutosh Singh**

3.1 INTRODUCTION

A specialized group of filamentous fungi, known as dermatophytes, is the most prevalent agent for superficial infections affecting humans and animals. Dermatophytosis alone infects 900 million people worldwide. Dermatophyte infection encompasses a broad spectrum of disorders affecting the keratin rich organ structures like nails, hair, and skin specifically. Dermatophytosis primarily results from the infection of the epidermal layer of the skin and is caused by several species of dermatophytes.

Dermatophytes are a distinct category of fungus classified under seven genera: *Trichophyton, Epidermophyton, Nannizzia, Paraphyton, Lophophyton, Microsporum,* and *Arthroderma* (Moskaluk and VandeWoude, 2022). Dermatophytosis in humans is frequently connected with the genus *Trichophyton*.

3.1.1 COMMONLY OCCURRING *TRICHOPHYTON* SPECIES

Trichophyton from the phylum Ascomycota is the most frequent causing agent of dermatophytosis. The transmission of *Trichophyton* can be categorized into three types: anthrophilic, zoophilic, and geophilic. Anthrophilic transmission occurs when the fungus is transmitted through direct contact with an infected person or contaminated items such as towels, clothing, and shared shower stalls. Zoophilic transmission occurs when the fungus is transmitted from animals to humans while geophilic transmission occurs when the fungus is transmitted from soil (Segal and Elad, 2021). Table 3.1 shows the examples of known *Trichophyton* species and their mode of transmission. *T. rubrum, T. mentagrophyte, T. interdigitale, T. tonsurans, T. indotineae,* and *T. violaceum* are the most common *Trichophyton* species that cause tinea infections.

3.1.2 TYPES OF TINEA INFECTIONS

Tinea infections are categorized according to their site of infection. Different *Trichophyton* species are responsible for different kinds of tinea infections. For example, tinea corporis is mainly caused by *T. rubrum* and tinea capitis is mainly caused by *T. tonsurans.*

Figures 3.1A and 3.1B show different types of tinea infections, their site of infection, and causal agents.

---

* Corresponding Author: singh-ashutosh@lkouniv.ac.in

DOI: 10.1201/9781032642864-3

TABLE 3.1

Examples of *Trichophyton* Species and Their Mode of Transmission

	Mode of Transmission		
Trichophyton	**Anthropophilic**	**Zoophilic**	**Geophilic**
	T. rubrum	*T. verrucosum*	*T. ajelloi*
	T. tonsurans	*T. simii*	*T. terrestre*
	T. interdigitale	*T. equinum*	*T. vanbreuseghemii*
	T. conentricum	*T. erinacei*	*T. phaseoliforme*
	T. schoenicinii	*T. benhamiae*	*T. gloriae*
	T. soudanese	*T. bullosum*	*T. flavescens*
	T. violaceum	*T. mentagrophytes*	

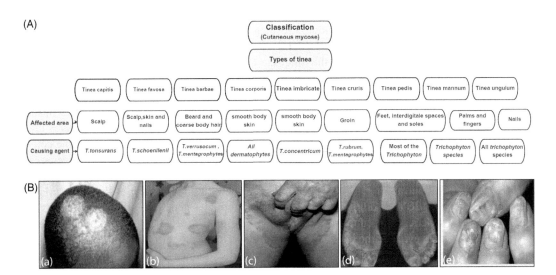

FIGURE 3.1   (A) Types of tinea infections, respective sites of infections, and associated causal organisms. (B) Panels (a), (b), (c), (d), and (e) show tinea capitis infecting the scalp, tinea corporis infecting the body, tinea cruris infecting the groin area, tinea pedis infecting the feet skin, and tinea unguium infecting nails, respectively.

3.2   CHALLENGES IN *TRICHOPHYTON*-MEDIATED DERMATOPHYTOSIS

Among dermatophytes, *Trichophyton* species is the most commonly reported in clinics. tinea pedis and tinea unguium are the predominant types of dermatophytosis in Europe, the Middle East, North and Central America, and Japan. *T. rubrum* is the primary pathogen responsible for causing tinea pedis and tinea unguium. Tinea corporis is more widespread in Lebanon, Saudi Arabia, and Northern Iran, where *T. mentagrophytes* and *T. verrucosum* are the major dermatophytic agents (Havlickova et al., 2008). In India, these dermatophytic infections are primarily caused by the *T. mentagrophytes* complex. Tinea capitis, caused by *T. tonsurans*, is the most common scalp fungal infection in the Caribbean and Africa. Literature indicates that *Trichophyton*-mediated dermatophytosis is highly prevalent worldwide. As far as the environmental factors are concerned, the fungus thrives in warm and humid conditions and can survive on contaminated objects such as towels and clothing for weeks. Its ability to colonize and persist on inanimate surfaces makes it a significant public health concern. Despite several treatment options, *Trichophyton* infections can be challenging to eradicate, particularly chronic or recurrent. Its high prevalence can be attributed to several biological and environmental factors that support its growth and make it difficult to treat. Several factors contribute to making the treatment of dermatophytosis challenging as described below.

3.2.1 Biological Factors

3.2.1.1 High Prevalence

The contagious nature of dermatophytosis makes it easy to spread from person to person or from animals to humans. The prevalence of fungal infections has markedly grown over the past 20 years. Around 100,000 fungal species out of the millions on Earth can infect humans and animals, particularly in tropical and temperate regions. There are a minimum of 40 distinct species that possess various morphological characteristics. About 20%–25% of the global population is affected by dermatophytosis (Zhan et al., 2021). *T. rubrum* is the most frequent dermatophyte found in humans, followed by *T. mentagrophytes*. The high prevalence of *T. rubrum* infection was well documented in Europe. Its broad spectrum of survival mechanisms utilizing several proteolytic enzymes, invasive ability in various ecological niches, and manageable human and animal body temperature growth make it robust and highly prevalent (Sypka et al., 2021). The occurrence of dermatophytosis is significantly influenced by travel and migration, which can lead to the spread of previously unidentified species or strains in any region of the world. The infection follows a specific pattern that varies depending on the geographic location. Prevalence and frequency of dermatophytosis can be increased by the ability of the fungus to adapt and remain dormant as spores in home dust for many years. The dust serves as a reservoir for the fungus.

3.2.1.2 Adaptation in Hosts – Immune System

The *Trichophyton* species use several mechanisms to survive and cause dermatophytosis in the host system. The HypA protein, associated with the cell wall of *T. mentagrophytes*, creates layers of rodlet structures on the surface to protect from being recognized by human immune cells. They produce several proteolytic enzymes, including proteases, lipases, elastases, collagenases, phosphatases, and esterases, which break down the protein components of the cell wall, allowing for adherence and penetration of the dermatophytes in the stratum corneum of the epidermis in the skin (Zahur et al., 2014). Additionally, specific dermatophytes generate enduring infections by employing various tactics to circumvent the microbicidal effects of leukocytes and immunological identification, such as evading the host defense through pattern recognition receptor (PRR)-based mechanisms. Cell wall glycoproteins, endoproteases, and exoproteases derived from *T. rubrum* are virulence factors that influence the host immunological response. Dermatophytes can trigger potent anti-inflammatory cytokines in specific circumstances to evade the host's immune response. In other instances, the Lysine motif (LysM)-binding domains, which are primarily found in dermatophytes can invade the host's innate immunity by adhering to pieces of chitin. *T. verrucosum* possesses a set of nine genes that encode for LysM domains. Furthermore, *T. rubrum* can inhibit the production of toll-like receptors (TLR) in keratinocytes and Langerhans cells in the dermis and epidermis. These receptors are essential for the activation of Th1-type cell response, which in turn would boost into Th2-type immune responses. Obviously, the reduced TLR response renders the immune system insufficient in combating these fungal infections. It is observed that nonsynonymous mutations present in the genes TERG_06754 and TERG_03373 had a significant role in enabling the adaption and invasion of *T. rubrum*, resulting in the formation of the specific kind of granuloma (Martínez-Herrera et al., 2023). Also, certain dermatophytes secrete glycopeptides that can reversibly hinder the proliferation of T cells in a laboratory setting, influencing the host's immune response.

3.2.1.3 Emergence of New *Trichophyton* Species

Trichophyton species show flexibility toward biological and environmental changes. This ability allows them to mutate and eventually develop as new strains with an enhanced range of tolerance to adverse conditions. Recently, *T. mentagrophytes* Type VIII, recognized as an Indian genotype, has emerged (Nenoff et al., 2019). This genotype has been observed in human populations and documented in several Asian nations. Currently, there have been approximately 30 species of dermatophytes identified as human pathogens. The occurrence of dermatophyte is documented in regions characterized by tropical and subtropical climates characterized by high temperatures and humidity levels. Furthermore, in 2020, the emergence of an independent new species, *T. indotinee*, separate from *T. interdigitale* and *T. mentagrophytes*, was reported. Such classifications are based on the analysis of ITS (Internal Transcribed Spacers) from highly terbinafine-resistant strains discovered in different locations of the world. There has been a concerning rise in superficial fungal infections due to this species observed globally. Most *T. indotinee* strains display a modification in the sequence of the squalene epoxidase (SQLE) gene, resulting in a high terbinafine resistance (Chowdhary et al., 2022).

3.2.1.4 Recurrence of Dermatophytosis

The majority of dermatophyte infections exhibit recurrence within a few weeks following the completion of treatment. Research findings indicate that 12.5% of patients who received an entire course of terbinafine treatment saw this observation materialize by the sixth week. 32% of treated cases with continuous terbinafine for managing onychomycosis have experienced dermatophytosis recurrence (Gupta and Cooper, 2008). Following an extended period of monitoring, researchers have concluded that the use of systemic terbinafine and itraconazole in the treatment of onychomycosis results in a higher likelihood of its recurrence with itraconazole therapy, compared to terbinafine therapy. Dermatophytes in deep-seated skin lesions might lead to chronic or recurrent infection. Deep dermatophytosis is a severe and potentially life-threatening fungal illness caused by dermatophytes. *T. rubrum* and *T. verrucosum* were found to be responsible for the development of deep-seated hyphae in a patient undergoing immunosuppressant treatment for non-Hodgkin's lymphoma. Spores detectable in skin biopsy tissues. The incomplete course and the improper administration of antifungal medications are regarded as significant contributors to the emergence of the reoccurrence of infections, which have the potential to become prevalent among populations. Patients may experience a recurrence of dermatophytosis within 1–4 weeks of treatment with clotrimazole cream. Nevertheless, the recurrence of dermatophytosis infection has been attributed to lesions and socioeconomic circumstances. The prevalence of recurrent dermatophytosis is high among Indian patients with tinea pedis; higher occurrences of relapse tinea cruris and tinea corporis have been recorded as well.

3.2.1.5 Physiological Similarities between Human and Fungi Systems

The similarities in cell structure between humans and fungi are so significant that it is challenging to develop new antifungal drugs. Antifungals that target fungi can cause harm to humans and animals to an equal degree, leading to severe side effects that render the treatment ineffective. The primary targets of most antifungal agents are fungal cell membranes and their components. However, given the diversity of dermatophytes, treating all of them with the same antifungals is a complex task. To overcome this challenge, researchers must prioritize studying new targets by understanding the mechanism of action of antifungal medicines and fungal pathology. They need to focus on developing drugs that selectively target fungi without harming humans and animals, which requires a deep understanding of the genetic and molecular differences between fungal and human cells.

3.2.1.6 Zoophilic Nature of Dermatophytes

The *Trichophyton* genus exhibits a zoophilic and anthropophilic nature, meaning dermatophytosis can affect animals and humans. Although the clinical characteristics may vary, it is essential to note that the transfer of dermatophytosis from pets to humans via zoophilic dermatophyte agents is possible (Segal and Elad, 2021). Zoophilic lesions tend to be more advanced than those caused by anthropophilic agents, which are transferred directly between people. Furthermore, as observed in laboratory or other occupational settings, humans can serve as a reservoir of infection for wild animals. By being aware of these possibilities, we can take necessary precautions and measures to prevent and treat dermatophytosis effectively.

3.2.2 Environmental Factors

The prevalence of the disease is supported by various factors such as geographical features, health-care infrastructure, immigration patterns, climatic conditions (including temperature, humidity, and wind), population density, cultural practices related to environmental sanitation, age distribution, personal hygiene practices, and socioeconomic circumstances. *T. mentagrophytes* that can be classified as either zoophilic (*T. mentagrophytes*) or anthropophilic (*T. interdigitale*), is capable of causing human infections. Both species can survive in seawater and beach sand. This shows dermatophytes have a tremendous ability to survive different climatic conditions. Studies reported that more dermatophytosis cases occur in warmer temperatures with high humidity and rainy seasons (Jartarkar et al., 2021).

3.2.3 Limitations in Diagnoses and Treatments of Dermatophytosis

3.2.3.1 Incorrect Diagnosis

More than 40 species of the genus *Trichophyton* have been recognized and have the potential to infect humans and animals. All the species exhibit numerous morphological characteristics, which are crucial for diagnosing the particular species. Another genus like *Microsporum*, and *Epidermophyton* are also characterized and well known for their role in dermatophytosis. In some instances, a tinea lesion can be caused by a solitary dermatophyte species or by many species. Furthermore, a solitary dermatophyte species might give rise to several forms of tinea (Jena et al., 2018). The research-based characteristics of these fungi are crucial for distinguishing them from other clinical skin conditions. The clinical manifestations, including pruritus, maceration, discomfort, desquamation, vesicles, formation of crusts, and erythema, can range from mild to moderate severity. These symptoms might be mistaken for various conditions, such as contact dermatitis, bacterial folliculitis, psoriasis, and eczema. Subpar treatment, individuals with weakened immune systems, and inadequate duration of medication can all contribute to an inaccurate diagnosis. The concept of false negative culture is perplexing because it hinders the identification of new or recurring infections, which in turn restricts the appropriate delivery of medication.

3.2.3.2 Limited Availability of Antifungals

A limited number of antifungal drugs have been used to treat dermatophyte infections for more than a decade. Therefore, there is a need for new and enhanced antifungal medications that can be developed by making chemical or structural modifications to existing treatments. The primary targets of most antifungal agents are: The structures that are found in the fungal cell membrane, fungal nucleic acid, and components of the fungal cell wall. The two primary kinds of antifungal drugs commonly used to treat dermatophyte infections are azoles (including tioconazole, clotrimazole, oxiconazole, econazole, and miconazole) and allylamines (such as terbinafine and naftifine). Terbinafine is the predominant topical therapy for dermatophytosis. There is no evidence to suggest that one class of topical antifungal medication is superior to another in achieving a cure after treatment. Most of the common antifungals are based on the formula targeting the same factor (cell membrane components like ergosterol lanosterol); mutation in a single gene or factor can make several antifungals incompetent alone. So, new targets for antifungals are required to be elucidated (Scorzoni et al., 2017).

3.2.3.3 Side Effects Associated with Antifungals

Antifungal medications used to treat dermatophyte infections have various side effects, depending on the type and duration of treatment. Common side effects of these antifungal drugs include nausea, vomiting, diarrhea, abdominal pain, and headache. Sometimes, serious side effects such as liver damage, heart failure, and blood-related disorders may occur. Additionally, pregnant women, children, and the elderly are groups at higher risk of experiencing adverse effects and need to be closely monitored during treatment. The safety of systemic antifungal drugs during pregnancy is still inconclusive, and it is generally recommended to avoid their use. Systemic terbinafine is a relatively safe option for the aged as it lacks cardiac complications but it should be discontinued during lactation and pregnancy. In pregnant women, miconazole and clotrimazole are considered safe and are often the first-line choice, while econazole should be avoided during the first trimester. Although topical therapy is a generally safe alternative, it also has limitations in some cases.

3.2.3.4 Emerging Resistance for Common Antifungal Drugs

Clinical antifungal resistance is characterized by the continued presence of infection or symptoms despite using suitable antifungal medications, resulting in the inability to eradicate the fungal infection. The prevalence of antifungal drug resistance in dermatophytes and other pathogenic mycoses has significantly grown. It is generally seen due to the prolonged use of antifungal agents, mainly in HIV and immune-compromised patients. Consequently, there has been a rise in the occurrence of drug-resistant strains of pathogenic microorganisms, including dermatophytes. A study found that out of 36 dermatophytes, seven isolates showed drug resistance to one or more medications, including itraconazole, ketoconazole, voriconazole, and fluconazole. Reports also indicate that azole resistance in dermatophytes might reach up to 19% in some regions around the globe. The prevalence of resistance to fluconazole and terbinafine was highest, with rates of 61% and 48% respectively. All species of dermatophytes exhibited resistance to fluconazole, with terbinafine and clotrimazole showing subsequent resistance. Consequently, it is essential to rely on antifungal susceptibility testing when addressing the growing resistance to dermatophyte treatment. The primary mechanisms of drug resistance in dermatophytes are the upregulation of efflux pump proteins and the creation of biofilms. Additionally, resistance can arise from target enzyme alterations, increased gene expression encoding ATP-binding cassette transporters, and overexpression of stress-response-related proteins (Sacheli and Hayette, 2021) (Figure 3.2). The occurrence of terbinafine resistance was first documented in 2003. In addition, there have been reports of six single cases of *T. rubrum* resistant to terbinafine. A separate investigation demonstrated that *T. rubrum* resisted terbinafine

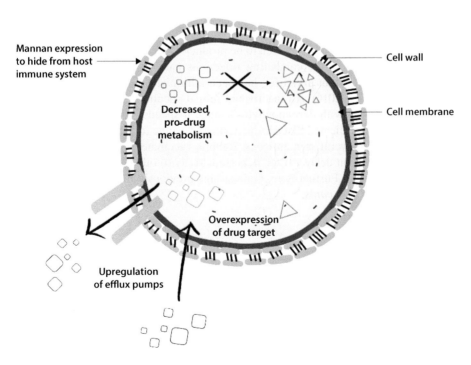

FIGURE 3.2 Different antifungal drug resistance mechanisms shown by dermatophytes.

with a minimum inhibitory concentration exceeding 4 μg/mL (Sacheli and Hayette, 2021). The resistance to antifungal agents can be strengthened and amplified by recurring dermatophyte infections, which result in the development of novel strains capable of withstanding traditional antifungal medications.

3.2.3.5   Long Duration of Treatment

Dermatophytosis therapy may necessitate an extended duration of several weeks to months, sometimes over a year. Prolonged treatment duration is regarded as a challenge linked to the utilization of antifungal medications. Moreover, several studies show that treating with suboptimum doses for a long time is critical in making the dermatophyte immune to that antifungal (Ghannoum, 2016). Also, the urgent need for rapid recovery from dermatophytosis has arisen due to the high cost and time-consuming nature of such long-term treatments. Therefore, developing antifungal agents with shorter treatment regimens is crucial to enhance patients' quality of life. Additionally, this approach is an essential preventive measure against the emergence of dermatophyte-resistant strains, which can occur due to prolonged exposure of antifungals to develop resistance (Scorzoni et al., 2017).

3.2.3.6   No Anti-Dermatophyte Vaccines for Humans

The superficial mycotic infections are prevalent worldwide, affecting approximately 20%–25% of the population. However, vaccinations are not available for humans to prevent these infections. Cell-mediated immune responses play a crucial role in resolving and protecting against dermatophytosis. Specific cytokines, including gamma interferon, interleukin-12, and interleukin-2, achieve this. No documented link exists between the antibody titer and the severity of dermatophyte infections. Although effective vaccines are available for specific animal species, there is a need for improvement and additional development in the vaccination against dermatophytosis in animals and humans. Several virulence factors of dermatophytes have been recently identified, with particular emphasis on the role of secreted proteases in invading the keratin network. Their specific functions in various stages of the infectious process and immunopathogenesis are currently under investigation, while comprehensive documentation of all elements of the host immune response to dermatophytes, including the innate response, is steadily rising. An inactivated *M. canis* vaccine has proven to be productive and secure in treating cats afflicted with dermatophytosis, effectively combating infections caused by *M. canis*. Furthermore, a deactivated vaccine called 'Fun-hikanifel' has been administered to dogs and cats for immunization (Nedosekov et al., 2016). An inactivated fungal vaccine has proven to be an effective treatment for cats infected with dermatophytosis since it speeds up the healing process of the lesions. The clinical trial indicated that administering a vaccine for a specific dermatophyte species can broadly protect several skin antigens. Rabbits vaccinated with a single dermatophyte, such as *M. canis*, can exhibit delayed-type hypersensitivity to other antigens, such as *T. mentagrophytes* and *M. gypseum*. Furthermore, Guinea pigs previously immunized with *T. equinum* showed immunity against *M. canis*.

3.3   DIAGNOSTICS AND TREATMENT APPROACHES

Species identification methods have come a long way, with newer techniques minimizing the drawbacks of traditional cultural and microscopic methods. The combined utilization of the E-test and disc diffusion method has demonstrated its appropriateness and dependability in assessing the antifungal susceptibility of dermatophytes (Namidi et al., 2021). Confocal laser scanning microscopy and advanced microscopic techniques have enabled real-time visualization of individual cells, and subcellular structures and have proven to be helpful screening tools for *Tinea incognito* diagnosis. PCR are advanced molecular techniques, with real-time PCR being less laborious and carrying a reduced contamination risk. Post-PCR strategies can increase the number of species identified, although this can also increase contamination risk and time. Genomic and proteomic approaches have also been explored, with MALDI-ToF mass spectroscopy being recognized as a fast and specific method for species differentiation (Kupsch et al., 2016). These advancements have enabled a more constructive approach to dermatophyte diagnosis.

Terbinafine, griseofulvin, and azoles are the primary antifungals used to treat dermatophytosis. Although some resistance cases of terbinafine emerge, these drugs effectively treat this infection when administered with proper doses and duration. The combinational effect of the common antifungals is also explored. In 2014, Tamura et al. conducted a study to examine the combined efficacy of six antifungal medications (amorolfine, terbinafine, butenafine, ketoconazole, itraconazole, and bifonazole) for the treatment of onychomycosis caused by both dermatophytes and non-dermatophytes (Tamura et al., 2014). In this study, a total of 44 strains of dermatophytes were utilized, including *T. rubrum*, *T. mentagrophytes*, *T. verrucosum*, *T. tonsurans*, *M. canis*, *M. gypseum*, and *E. floccosum*. Apart from the conventional antifungals and their combinational effects, multidrug theory, using novel preparation of topical agents, nano-formulation of antifungal drugs, use of anti-adhesives in conjunction with antifungal agents to prevent *Trichophyton* infection, use of polymeric films, photo remediation and exploring the natural sources of antifungals are being studied to treat dermatophytosis (Gnat et al., 2020).

3.4 CONCLUSION

Dermatophytes are blessed by nature; several environmental and biological factors enable them to survive well in harsh conditions. Their tremendous escape and defense mechanisms inside the host body make them *superorganisms* of nature. Understanding their biology thoroughly is the prerequisite, and developing new strategies and antifungals is an urgent need. While it is true that natural and chemical databases possess a considerable array of chemical compounds that promise to create potent pharmaceuticals, it is essential to acknowledge the hindrance posed by the absence of thorough investigations into their molecular targets and interaction mechanisms within fungi and mammalian cells. This lack of comprehensive *in vitro* and *in vivo* studies is a formidable obstacle to developing innovative antifungal medications for the inherent capacity of dermatophytes to manifest virulent traits.

ACKNOWLEDGMENTS

Ashutosh Singh, acknowledges the funding supported by the DBT, Govt. of India (BT/PR38505/MED/29/1513/2020), DST, Govt. of India (CRG/2022/001047; SR/PURSE Phase 2/29(C)), ICMR, Govt. of India (No.52/08/2019-BIO/BMS), UP Higher Education (No. 10/2021/281/-4-Sattar-2021-04(2)/2021), and the University of Lucknow.

Nitin Bhardwaj thanks ICMR No. 56/2/Hae/BMS and UGC start-up grant No. 30-496/2019 for providing support.

Shikha Chandra, Khushboo Arya, and Sana Akhtar Usmani are grateful for JRF fellowships from DST (CRG/2022/001047), ICMR (Ref. No 3/1/3/JRF2021/HRD(LS)) and UGC (NTA Ref. No. 22161023985), respectively.

REFERENCES

Chowdhary, A., Singh, A., Kaur, A., & Khurana, A. 2022. The emergence and worldwide spread of the species *Trichophyton indotineae* causing difficult-to-treat dermatophytosis: A new challenge in the management of dermatophytosis. PLOS Pathogens 18: e1010795.

Ghannoum, M. 2016. Azole resistance in dermatophytes: Prevalence and mechanism of action. Journal of the American Podiatric Medical Association 106: 79–86.

Gnat, S., Łagowski, D., & Nowakiewicz, A. 2020. Major challenges and perspectives in the diagnostics and treatment of dermatophyte infections. Journal of Applied Microbiology 129: 212–232.

Gupta, A. K., & Cooper, E. A. 2008. Update in antifungal therapy of dermatophytosis. Mycopathologia 166: 353–367.

Havlickova, B., Czaika, V. A., & Friedrich, M. 2008. Epidemiological trends in skin mycoses worldwide. Mycoses 51: 2–15.

Jartarkar, S. R., Patil, A., Goldust, Y., Cockerell, C. J., Schwartz, R. A., Grabbe, S., & Goldust, M. 2021. Pathogenesis, immunology and management of dermatophytosis. Journal of Fungi 8: 39.

Jena, A. K., Lenka, R. K., & Sahu, M. C. 2018. Dermatophytosis in a tertiary care teaching hospital of Odisha: A study of 100 cases of superficial fungal skin infection. Indian Journal of Public Health Research & Development 9: 7.

Kupsch, C., Ohst, T., Pankewitz, F., Nenoff, P., Uhrlaß, S., Winter, I., & Gräser, Y. 2016. The agony of choice in dermatophyte diagnostics—Performance of different molecular tests and culture in the detection of *Trichophyton rubrum* and *Trichophyton interdigitale*. Clinical Microbiology and Infection 22: 735–e11.

Martínez-Herrera, E., Moreno-Coutiño, G., Fuentes-Venado, C. E., Hernández-Castro, R., Arenas, R., Pinto-Almazán, R., & Rodríguez-Cerdeira, C. 2023. Main phenotypic virulence factors identified in *Trichophyton rubrum*. Journal of Biological Regulators and Homeostatic Agents 37(5): 2345–2356.

Moskaluk, A. E., & VandeWoude, S. 2022. Current topics in dermatophyte classification and clinical diagnosis. Pathogens 11: 957.

Namidi, M. H., Ananthnaraja, T., & Satyasai, B. 2021. Antifungal susceptibility testing of dermatophytes by ABDD and E-Test, a comparative study. Open Journal of Medical Microbiology 11: 129–143.

Nedosekov, V., Martynyuk, O., & Stetsiura, L. 2016. Evaluation of manufacturing specification of antifungal vaccines. Edukacja-Technika-Informatyka 1: 158–162.

Nenoff, P., Verma, S. B., Vasani, R., Burmester, A., Hipler, U. C., Wittig, F., & Uhrlaß, S. 2019. The current Indian epidemic of superficial dermatophytosis due to *Trichophyton mentagrophytes*—A molecular study. Mycoses 62: 336–356.

Sacheli, R., & Hayette, M. P. 2021. Antifungal resistance in dermatophytes: Genetic considerations, clinical presentations and alternative therapies. Journal of Fungi 7: 983.

Scorzoni, L., de Paula e Silva, A. C., Marcos, C. M., Assato, P. A., de Melo, W. C., de Oliveira, H. C., & Fusco-Almeida, A. M. 2017. Antifungal therapy: New advances in the understanding and treatment of mycosis. Frontiers in Microbiology 8: 36.

Segal, E., & Elad, D. 2021. Human and zoonotic dermatophytoses: Epidemiological aspects. Frontiers in Microbiology 12: 713532.

Sypka, M., Jodłowska, I., & Białkowska, A. M. 2021. Keratinases as versatile enzymatic tools for sustainable development. Biomolecules 11: 1900.

Tamura, T., Asahara, M., Yamamoto, M., Yamaura, M., Matsumura, M., Goto, K., & Makimura, K. 2014. In vitro susceptibility of dermatomycoses agents to six antifungal drugs and evaluation by fractional inhibitory concentration index of combined effects of amorolfine and itraconazole in dermatophytes. Microbiology and Immunology 58: 1–8.

Zahur, M., Afroz, A., Rashid, U., & Khaliq, S. 2014. Dermatomycoses: Challenges and human immune responses. Current Protein and Peptide Science 15: 437–444.

Zhan, P., Liang, G., & Liu, W. 2021. Dermatophytes and dermatophytic infections worldwide. In Dermatophytes and dermatophytosis, 15–40. Cham: Springer International Publishing.

4 Opportunistic Fungal Infections

*Nafis Raj, Parveen, Shabana Khatoon, and Nikhat Manzoor**

4.1 INTRODUCTION

4.1.1 OPPORTUNISTIC FUNGAL INFECTIONS

Invasive fungal infections have significantly increased due to advances in medical care, especially in the immunocompromised population. Approximately 7% (611,000 species) of all eukaryotic species on earth are fungi, widely distributed in diverse habitat. Interestingly, only around 600 fungal species are pathogenic to humans or cause some kind of ailment (Mora *et al.*, 2011; Brown *et al.*, 2012). Over the years, fungal illnesses have been neglected despite their serious consequences on human health. According to recent estimates, there are roughly 30 lakh cases of chronic pulmonary aspergillosis, over 2 lakh cases of cryptococcal meningitis, 7 lakh cases of invasive candidiasis, 5 lakh cases of pneumonia caused by *Pneumocystis jirovecii*, 2.5 lakh cases of invasive aspergillosis, 1 lakh cases of disseminated histoplasmosis, more than 1 million cases of fungal asthma, and 10 lakh cases of fungal keratitis reported worldwide each year (Bongomin *et al.*, 2017). More than ~1.5 million die each year due to fungal infections (Queiroz-Telles *et al.*, 2017).

Infections can spread through various routes: Ingestion of contaminated food or drink (gastro-intestinal disease), infiltration of the mucosa by commensals like *Candida albicans*, percutaneous inoculation in cutaneous and subcutaneous infections (dermatophytosis, Madura foot), and inhalation of spores (aspergillosis, cryptococcosis, histoplasmosis) (Cafarchia *et al.*, 2013). Mild infections that are limited to the skin surface or superficial tissues may be caused by dermatophytes and *Tinea versicolor* (*Pityriasis versicolor*). Infections that may result in serious, systemic diseases include mucormycosis, candidiasis, and aspergillosis. Depending on host immunity and physiological state, the clinical symptoms of the fungal infection can vary greatly. *Candida* species, for instance, can induce systemic infections, causing liver abscess and damaging renal, lung, and nervous system tissues; or it may invade local sites (mucocutaneous or cutaneous onychomycosis and candidiasis). Infections caused by *Aspergillus* species have also been associated with allergy symptoms (allergic bronchopulmonary aspergillosis) (Badiee *et al.*, 2012).

4.1.2 MAJOR CAUSATIVE AGENTS AND IMPACT ON PUBLIC HEALTH

Invasive fungal infections are very common in immunocompromised patients like the elderly and malnourished children, individuals on long-term drug therapy for chronic diseases like AIDS, cancer, and tuberculosis, as well as organ transplant cases undergoing immunosuppressive treatment (Fridkin and Jarvis, 1996; Groll *et al.*, 1996). The predominant opportunistic fungal pathogens

---

* Corresponding Author: nmanzoor@jmi.ac.in

DOI: 10.1201/9781032642864-4

include species of *Candida*, *Aspergillus*, *Mucorales*, *Fusarium*, and *Scedosporium*. (Hajjeh *et al.*, 2004). Medically important yeast, like *Candida* spp., cause bloodstream infections and are commonly found in both adult and paediatric intensive care units. The host gastrointestinal tract is typically the entry point, but in some patients, particularly those with inserted central venous catheters, the skin is most likely to be the source of the infection. Although immunocompetent individuals may also develop candidemia, the presence of *Candida* species in blood is an important indicator of immunodeficiency. It has been reported that Candidemia affects 10%–20% of leukaemia patients (or myeloproliferative diseases) and up to 74% of AIDS patients. *Candida* spp. are among the most significant categories of opportunistic fungal pathogens, they represent 8%–10% of nosocomial bloodstream infections and are still significant pathogens in intensive care units (Bassetti *et al.*, 2006).

The encapsulated yeast *Cryptococcus neoformans* is another opportunistic fungal pathogen that affects the immunocompromised host almost everywhere in the world. *C. neoformans* is a neurotropic pathogen and a majority of cryptococcal meningitis patients experience deficiency in cellular immunity (Khawcharoenporn *et al.*, 2007). Moreover, during the last few decades, the frequency of infection has increased considerably. *Cryptococcus laurentii* and *Cryptococcus albidus* collectively account for 80% of instances that have been documented (Halliday *et al.*, 1999).

Aspergillus spp. are opportunistic filamentous fungi that cause infection as well as allergic reactions in the host. There are more than 175 *Aspergillus* species but only a small number has been associated with illnesses in humans. *Aspergillus fumigatus* is the primary cause of infections, followed by *Aspergillus flavus*, *Aspergillus terreus*, *Aspergillus niger*, and *Aspergillus nidulans*. *Aspergillus* species are found in varied habitats globally like soil, air, and decomposing waste matter (Denning, 1991). One of the major causes of morbidity and mortality in immunocompromised people is invasive aspergillosis. The most common clinical stains isolated from these patients are *A. fumigatus*, followed by *A. flavus*, *A. niger*, and *A. terreus* (Denning, 1998).

Mucormycoses are infections caused by saprophytic aerobic fungal pathogens called Mucorales. The disease is acute, fast spreading and often becomes fatal in immunocompromised individuals, and is common in people suffering from neutropenia. The fungus makes way into the host through inhalation of spores or due to the progression of previously localized skin sores (Roden *et al.*, 2005; Chayakulkeeree *et al.*, 2006). In addition, individuals with diabetes, those undergoing deferoxamine therapy (given to anaemic or thalassemia patients), injecting drug users, and those without evident immunological deficiency may develop fatal infections due to zygomycetes. In terms of clinical presentation, invasive zygomycosis resembles aspergillosis; fungi typically impact the skin (19%), lungs (24%), and paranasal sinuses (39%). Infection disseminates to other parts of the body in 23% of the cases, and 96% of cases result in death (Almyroudis *et al.*, 2006).

Fusarium spp. are important filamentous fungi, second to *Aspergillus* spp. in terms of infection-causing capacity. Although *Fusarium* spp. are mainly recognized as plant fungal pathogens, they also infect humans and other animals, causing both superficial and invasive diseases that become widely distributed (Nucci and Anaissie, 2007). One of the most significant risk factors for developing fusariosis is neutropenia. Other factors like organ transplantation, haematological malignancies, and acquired immune deficiency syndrome (AIDS) are also responsible (Muhammed *et al.*, 2011). The clinically significant isolates include the *Gibberella fujikuroi* species complex, the *Fusarium oxysporum* species complex, and the *Fusarium solani* species complex. The majority of *Fusarium* infections are caused by *Fusarium solani* species complex, followed by *F. oxysporum*, *G.fujikori*, *F. verticillioides*, and *F. proliferatum* (Nucci and Anaissie, 2007).

4.2 PATHOGENESIS OF OPPORTUNISTIC FUNGAL INFECTIONS

The rising incidence of fungal infections in recent times is a matter of grave concern and requires immediate action. These infections often go unnoticed during their initial stages, leading to increased severity and treatment complications. Fungal pathogens employ various strategies to elude the host immune system and develop resistance through a combination of physiological adaptations and genetic mutations aggravating the infection. For the treatment of both superficial and systemic fungal infections, several classes of antifungal drugs are available and administered accordingly (Reddy *et al.*, 2022). However, fungal resistance to most of these drugs and toxicity pose significant challenges to disease management and antifungal therapy. The limited availability of effective antifungal options, coupled with challenging biofilm-related resistance, emphasizes the pressing need for the development of more efficacious drugs and alternative approaches to achieve better outcomes in the treatment of fungal infections (mycoses) (Kangabam and Nethravathy, 2023). Addressing this issue requires a concerted effort in research, diagnostics, and the development of innovative therapies to combat these increasingly problematic infections.

The phyla Ascomycota and Basidiomycota are major contributors of human fungal pathogens. The organisms within the phylum Ascomycota are particularly notable for their involvement in a wide array of infections affecting various parts of the body. These fungi are known to cause infections in several areas, including the pharynx, larynx, skin, eyes, nerves, genitourinary tract, heart, and lungs, and can even lead to systemic infections. The phylum Basidiomycota includes a diverse range of fungal pathogens like *Cryptococcus* and *Malassezia* that are well-recognized for their role in causing specific infections in humans. *Cryptococcus* is notably associated with invasive meningitis, a life-threatening condition that affects the central nervous system. On the other hand, *Malassezia* is commonly known for causing superficial skin infections, which range from mild irritations to more severe dermatological conditions (Wan Mohtar *et al.*, 2022).

4.2.1 HOST IMMUNITY AND SUSCEPTIBILITY

Fungal pathogens primarily rely on two main routes for transmission: Direct contact and inhalation. Dermatophytic fungi, which include species of the genus *Microsporum, Epidermophyton*, and *Trichophyton*, as well as *Sporothrix* and *Malassezia* spp., cause infections through direct contact. These fungi secrete proteolytic enzymes that degrade keratinized tissues causing superficial mycoses (Jazdarehee *et al.*, 2022). Spore inhalation, on the other hand, generally leads to pulmonary infections which can then spread to other body organs. Fungi that are mainly spread through inhalation include *Blastomyces dermatitidis* (Blastomycosis), *Paracoccidioides brasiliensis* and *P. lutzii* (Paracoccidioidomycosis), *Histoplasma capsulatum* (Histoplasmosis), *Pneumocystis jirovecii* (Pneumocystis pneumonia), *A. fumigatus* and *A. flavus* (Aspergillosis), *Coccidioides immitis* and *C. posadasii* (Coccidioidomycosis), and *C. neoformans* and *C. gattii* (Cryptococcosis). The fungus *Talaromyces marneffei* (Talaromycosis) employs both direct contact and inhalation routes for transmission, setting it apart from other fungal pathogens (Langfeldt *et al.*, 2022).

The host immune system is equipped with pattern recognition receptors (PRRs) like dectins and Toll-like receptors (TLRs) that can identify pathogen-associated molecular pattern (PAMP) components found in the fungal cell wall, such as chitin and β-1,3-glucans. When these PAMPs are recognized by pattern recognition receptors, they trigger signalling pathways that initiate immune responses such as phagocytosis, release of reactive oxygen species (ROS) and reactive nitrogen species (RNS), and the production of cytokines and chemokines. These immune reactions are essential for neutralizing fungal pathogens (Hernández-Chávez *et al.*, 2017). The monocytes, macrophages, dendritic cells, and neutrophils play a crucial role as the first line of defence (innate immune system). They can eliminate fungal cells in phagolysosomes and contain them within granulomas, surrounded by macrophages. Monocytes also secrete chemokines and cytokines and serve as antigen presenting cells to T lymphocytes, which initiate an adaptive immune response that helps in clearing fungal cells (Marcos *et al.*, 2016).

Fungal pathogens, however, have developed various mechanisms to evade and outsmart the host immune system. They can modify their outer surface to shield the conserved PAMPs, preventing recognition by host PRRs. For instance, *Pneumocystis jirovecii* possesses a major surface glycoprotein (Msg or gpA) that masks the surface β 1,3-glucans as it lacks chitin exceptionally. Additionally, layers of fungal components like mannan, melanin, α-1,3-glucan, and hydrophobin (e.g. Rod A of *A. fumigatus*) acts as a shield to protect the surface polysaccharides, which act as PAMPs, allowing fungi to evade the host immune system (Ma *et al.*, 2016; Hatinguais *et al.*, 2020). In some cases, fungal pathogens alter their morphological structures to avoid immune recognition. For example, *C. neoformans* and *Sporothrix schenckii* form titan cells and asteroid bodies, respectively. Some fungal pathogens switch between different growth forms for immune evasion and increased virulence. *Candida* species are dimorphic and can switch from unicellular yeast form to filamentous hyphal form, enhancing tissue invasion and reducing phagocytosis. *C. immitis* and *C. posadasii* switch from chains of arthroconidia to conidia packed into spherules. These thick-walled spherules are associated with resistance to phagocytosis, production of RNS and enzymes like arginases and ureases (Li and Nielsen, 2017).

Macrophages are crucial for immunity against fungal infections. But certain fungi can target them and escape phagocytosis. *C. neoformans*, when infecting the lungs, depletes it of the alveolar macrophages. It parasitizes the macrophages, ejecting out of it unharmed (vomocytosis). The intact fungal cells, when released cross the blood–brain barrier, and affect the brain, cerebrospinal fluid, and central nervous system, resulting in cryptococcal meningitis. The spores of *B. dermatitidis* can similarly evade the attack by immune cells and circulates in the lymph and blood. The *Blastomyces* adhesin-1 (BAD1) protein can inactivate the complement cascade and increase the probability of systemic infections. In conclusion, these immune evasion strategies prove beneficial to the pathogen in several ways, including anti-phagocytic effects, suppression of T cell proliferation, reduced production of proinflammatory cytokines, and resistance to oxidative and nitrosative stress (Gilbert *et al.*, 2014). In addition to immune evasion, biofilm formation in certain fungal pathogens like *Candida*, *Cryptococcus*, *Histoplasma*, *Paracoccidioides*, and *Aspergillus* offers a significant advantage. Biofilms enable these fungi to thrive in hostile environments and render antifungal treatments ineffective (Desai *et al.*, 2014).

Fungal infections result due to the ability of these eukaryotic microbes to adapt and thrive in the challenging host environment. Only a few fungi are inherently pathogenic and can cause disease in healthy individuals (e.g. *C. neoformans*). On inhalation or exposure to large quantities of spores (conidia), the fungus may transition into the pathogenic forms, e.g. *C. immitis*, *H. capsulatum*, *B. dermatitidis*, and *P. brasiliensis*. In general, immunocompromised individuals are vulnerable to fungal infections. Opportunistic fungal pathogens include commensals like *C. albicans* and saprophytes like *A. fumigatus* and *C. neoformans* (Fernandes and Carter, 2020).

4.2.2 Fungal Virulence Factors

The pathogenicity of fungi is associated with several virulence factors like adhesion, biofilm formation, and secretion of hydrolytic enzymes. Virulence also depends on the ability of the fungus to switch to a more pathogenic form. There are several genes, the expression of which lead to increased virulence. Fungal cell walls, capsules, and plasmids are some other pathogenicity contributors.

4.2.2.1 Adhesion

The very first stage in fungal colonization is host-pathogen interaction and adhesion. Before invading the host tissues, the pathogen must adhere to surfaces to avoid being removed by mucociliary clearance, which is a defence mechanism of the lungs. *C. albicans* has several adhesion molecules which play a vital role in the formation of biofilms on surfaces, both biotic and abiotic (medical devices, catheters, etc.). Some such proteins include the Als family, Hwp1p, Eap1p, and Csh1p. The ALS (Agglutinin-Like Sequence) family of cell-surface glycoproteins are encoded by eight genes in *C. albicans*. The Als proteins facilitate adhesion by binding to host molecules like collagen, laminin, endothelial cells, epithelial cells, and by cell-to-cell aggregation. Hwp1p (Hyphal Wall Protein 1) is a hyphae-specific cell wall protein required for hyphal growth and *Candida* virulence while Int1p is required for morphogenesis and mediates adhesion to platelets. It is to be noted that the development of vaccines against invasive candidiasis is based on targeting Als3p (Angiolella, 2022). Other pathogenic fungi also employ specific adhesion molecules for pathogen host interactions. The conidia of *A. fumigatus* are covered with hydrophobic proteins called rodlets, encoded by RodA and RodB genes. These stable proteins facilitate adhesion to albumin and collagen. The hyphal surface has receptors (galactomannan and chitin) that mediate adhesion to complement proteins, fibrinogen, immunoglobulins, and surfactants A and D. The adherence of *Blastomyces* to host tissues is through BAD1, a high-capacity calcium binding protein and an essential virulence factor for this fungus. It binds to complement receptors CR3 and CD14 on phagocytes, and modulates host immune responses. *H. capsulatum* uses heat shock protein Hsp60, while *P. brasiliensis* employs glyceraldehyde-3-phosphate dehydrogenase (GAPDH) and polypeptides p19, p30 and p32 for adhesion. *Coccidioides* utilizes the outer wall of spherules for adhesion to host tissues. Adhesins and the mechanism of adhesion are essential for the successful colonization and survival of pathogenic fungi within the host, hence contributing to pathogenicity and the development of infections (Kumari *et al.*, 2021).

4.2.2.2 Dimorphism and Thermotolerance

Morphological transition plays a significant role in the virulence of pathogenic fungi. Some fungal pathogens exhibit dimorphism, switching between two different forms, one being more invasive. In the environment or as commensals inside the host, these fungi normally exist in the non-pathogenic yeast form. It switches to the elongated more invasive hyphal form in search for nutrition and penetration. After invasion of tissues (angioinvasion) and colonization, the dimorphic fungi prefer to live in the yeast form so that it disseminates to other body organs via blood. Dimorphic fungal species include *B. dermatitidis*, *C. immitis*, *H. capsulatum*, *P. brasiliensis*, and *C. albicans*. The large endosporulating spherules of *C. immitis* are part of this phenomenon. *C. albicans* exhibits an intermediate pseudohyphae in addition to yeast and hyphae. This adaptability allows it to thrive in various environmental conditions (Staniszewska, 2020).

The ability of pathogenic fungi to thrive at/around body temperature (37°C), is another important factor. The ability to survive and replicate at this and higher temperatures is known as thermotolerance and is a common property of *C. neoformans*, *H. capsulatum*, *S. schenckii*, and other pathogenic fungi. It has been observed that isolates of *C. neoformans*, that do not grow efficiently at 37°C, are unable to induce fatal infection in mice, while isolates that can grow at this temperature produce lethal infection. *H. capsulatum* strains of high virulence can withstand drastic temperature changes and undergo rapid yeast to hyphal morphological transition in comparison to strains of low virulence. Clinical isolates of *S. schenckii* from systemic lesions can grow at the temperature range 35–37°C, but isolates from fixed cutaneous lesions can grow only at 35°C (Gauthier, 2017). It has been observed that even small temperature differences in tolerance can influence the pathogenic potential of the microorganism, as well as the form of disease presented by the host. Resistance to temperature change is also related to the synthesis of heat-shock proteins. Production of these proteins seems to play an important role not only in thermo-adaptation but also in the yeast to mycelium phase transition in dimorphic fungi (Angiolella, 2022).

Systemic fungi are capable of growing at high temperatures, even in the febrile state (~38–42°C). *A. fumigatus* is an example of a particularly thermophilic fungus, capable of growing at temperatures as high as 55–77°C. Hsp70 has been observed to be necessary for fungal adaption to high temperatures (Bhabhra and Askew, 2005). In the case of dimorphic fungi, the transition from one form to another often occurs at different temperatures. While many fungi exist in the mould form at ambient temperatures, they change to the yeast form at higher temperatures typically encountered in mammals during infection. Studies have shown that this transition is a key factor in fungal pathogenicity. For instance, when the transition from mycelia to yeast is blocked in *H. capsulatum*, the organism can still grow at 37°C but loses its virulence. Both the morphological forms of *C. albicans* are pathogenic, and the fungus switches between yeast and hyphae in response to changes in the environment. It adopts the unicellular yeast form at lower temperatures and acidic pH, favouring its spread in the environment. The hyphal form is specifically useful during tissue invasion (Xiao *et al.*, 2022). The analysis of morphological transition and thermotolerance is crucial for understanding the pathogenicity related mechanisms and the development of strategies for controlling fungal infections.

4.2.2.3 Cell Walls and Capsules

Capsule formation is a common feature in bacteria but is rare in fungi. Fungal species of the genus *Cryptococcus* are covered by a protective capsule, typically composed of glucuronoxylomannan, which serves as a crucial virulence factor. The role of capsules is multifaceted, including in the resistance to phagocytosis, which allows the pathogen to persist in the host and evade immune responses (Angiolella, 2022). The thickness of the polysaccharide capsule varies among *C. neoformans*. Clinical isolates that causing infection usually have prominent, well-developed capsules, while the environmental strains have thinner and weaker capsules. Fungal strains that have loose capsules, are generally non-virulent as they can be easily phagocytosed by immune cells. Four genes have been identified for capsule formation and regulation in *C. neoformans*, of which CAP59 and CAP64 are particularly crucial as they play a critical role in determining capsule size and virulence. The capsule can deplete complement components, cause imbalance in cytokine production, and contribute to host immune dysregulation. This disrupts the ability of the host to generate effective defences against the pathogen (Zaragoza, 2019). The capsule may inhibit the migration of white blood cells (leukocytes) to the site of infection, further suppressing host immunity, as immune cells are crucial for eliminating pathogens. The capsule of *C. neoformans* is thus key to virulence and helps the fungus evade host defences, establish infection, and persist in the host.

Besides attributing virulence to the microbe, both the fungal cell wall and capsule also confer protection against host resistance. Hence, these structures are considered major targets for antifungal studies and pathogenicity. Cell wall polysaccharides have been constantly associated with increased virulence in several fungal strains. Reports suggest the involvement of α-1,3-glucans and ß-1,3-glucans in dimorphism and virulence of *P. brasiliensis*. Moreover, α-1,3-glucans protect the fungus against the harmful impact of host degradation enzymes, leukocytes and macrophages (Azevedo *et al.*, 2016). It was demonstrated that smooth variants of *H. capsulatum* which lacked α-1,3-glucans, were avirulent in comparison to the rough variants which possessed this polysaccharide. The rough variants were capable of destroying macrophages *in vitro*, suggesting a close association between α-1,3-glucans and virulence (Iyalla, 2017).

4.2.2.4 Secretion of Hydrolytic Enzymes

After adhesion, most pathogenic fungi secrete extracellular enzymes for host tissue invasion. They help in the establishment and dissemination of the disease deeper into the host. The hydrolytic enzymes cause tissue damage and impair host immune defences. *C. albicans* secrete phospholipases, lipases, and proteases. The phospholipases (A, B, C, and D) hydrolyse membrane phospholipids by breaking ester bonds. The pathogenic strains of this yeast release higher level of phospholipases in comparison to the commensal strains. Enzymes are also required for acquisition of nutrients and essential metals like iron (Fe). The secreted aspartyl proteases (SAP) of *C. albicans* hydrolyse various host proteins, including extracellular matrix proteins, coagulation factors (Hageman factor and factor X), host defence proteins (mucin, IgA, and lactoferrin), and complement components (Erum *et al.*, 2020). The proteases secreted by the mould *A. fumigatus* (serine proteases, aspartic proteases, and metalloproteases) can degrade elastin of lung tissue. Serine proteases are also involved in the degradation of collagen, fibrin, and fibrinogen (Namnuch *et al.*, 2020). *C. neoformans* also secretes proteases and phospholipases. The phospholipases include lysophospholipase and lysophospholipase-transacylase (LPTA), which can destroy lung surfactants and enhance adhesion (Almeida *et al.*, 2015). These degradative enzymes serve various functions, including in the breaking down of host tissues, disruption of the immune response, and promotion of the survival and dissemination of fungal cells in the host. Understanding the mechanism of action of these enzymes is critical in developing strategies to combat fungal infections.

4.2.2.5 Defence against Reactive Oxygen and Nitrogen Species

Neutrophils and macrophages are important components of the host immune defence system. They utilize oxidative mechanisms, involving ROS and NS), to damage fungi by causing lipid peroxidation and nucleic acid breaks. However, pathogenic fungi have developed protective mechanisms to defend against these oxidative attacks. *C. albicans* employ superoxide dismutase (SOD) and Hsp as part of their defence against ROS. The enzyme SOD helps in the breakdown of superoxide radicals, the primary ROS. Hsps play a significant role in protecting fungal cells from the damaging effects of oxidative stress. *C. neoformans* use the production of copper and zinc, as well as peroxidases, to resist oxidative damage. Copper and zinc are essential co-factors for enzymes that can help neutralize ROS. Peroxidases are enzymes that catalyse the breakdown of hydrogen peroxide, a non-radical ROS, into water and oxygen, thus reducing oxidative stress. *A. fumigatus* produces multiple catalases. Cat-A is associated with conidia, and Cat 1p and Cat 2p are associated with hyphae. Catalases are enzymes that break down hydrogen peroxide into water and oxygen, protecting the fungus from oxidative damage. *A. fumigatus* also produces SOD, which contain manganese (Mn), copper (Cu), and zinc (Zn). These enzymes help convert superoxide radicals into less harmful molecules (Warris and Ballou, 2019). The host defence mechanisms help pathogenic fungi counteract the oxidative damage caused by the host immune response. By neutralizing ROS and RNS, these fungi can evade the host immune defences and establish infection in the host.

4.2.3 FACTORS INFLUENCING COLONIZATION AND INVASION

Fungal infections are a matter of significant public health concern, with life-threatening implications. The occurrence of fungal infections, particularly in patients with pre-existing conditions like diabetes, cancer and more recently COVID-19, increases morbidity and the risk of mortality. These infections encompass a wide spectrum, ranging from superficial and cutaneous to subcutaneous, mucosal, and systemic infections, varying in severity. *Candida* species are commensals normally found as part of the human microbiota. These opportunistic organisms become pathogenic in immunocompromised patients such as those undergoing treatment for chronic diseases like AIIDS and cancer, and patients taking immunosuppressive drugs during organ transplant surgeries. Fungal pathogens like *Candida*, *Aspergillus*, *Fusarium*, *Mucorales*, and other moulds can give rise to healthcare-associated infections in patients with underlying medical conditions (Jacobsen, 2023). Furthermore, certain endemic fungi like *Blastomycosis*, *Coccidioidomycosis*, *Histoplasmosis*, *Talaromycosis*, *Paracoccidioidomycosis*, and *Sporotrichosis* are prevalent in specific geographic regions posing major health challenges. Systemic fungal infections are often diagnosed late increasing patient mortality. The Centers for Disease Control and Prevention (CDC) dedicates a fungal disease awareness week every year in September to educate and highlight the importance of early diagnosis of fungal infections to alleviate debilitating effects (CDC website) (Lionakis *et al.*, 2023). September 18–22, 2023 was dedicated to the theme 'The Growing Need to Think Fungus.'

Healthy, immunocompetent individuals do not usually develop mycoses, although constantly exposed to infectious propagules. It is only when fungi accidentally penetrate barriers such as the skin and mucous membranes, or when immunologic defects or other debilitating conditions exist in the host, that conditions become favourable for fungal colonization (Mahalingam *et al.*, 2022). Many fungi develop mechanisms that facilitate multiplication within the specific host tissues. Dermatophytes secrete enzymes such as keratinase, protease, phospholipase, lipase and elastase that help colonization. Keratinase specifically digests keratin of skin, hair, and nails. Disseminated fungal diseases usually indicate a breach in host defences which may be caused by endocrinopathies, immune disorders, or induced iatrogenically (Mukjang, 2022).

Effective management of the fungal infection requires a concerted effort to uncover and correct the underlying defects. When protective barriers of human defence are breached only then can fungi gain access to, colonize, and multiply in host tissues. Fungi gain access to host tissues by traumatic implantation or inhalation. The severity of disease caused by these organisms depends upon the size of the inoculum, the magnitude of tissue destruction, the ability of fungi to multiply in tissues, and the immunologic status of the host. Superficial fungal infections involve only the outermost layers of the stratum corneum of the skin (*Phaeoannellomyces werneckii* and *M. furfur*) or the cuticle of the hair shaft (*Trichosporon beigelii* and *Piedraia hortae*). These infections usually constitute cosmetic problems and rarely elicit an immune response from the host (except occasionally *M. furfur* infections). The dermatophytes possess greater invasive properties than those causing superficial infections, but they are limited to the keratinized tissues. They cause a wide spectrum of diseases that range from a mild scaling disorder to one that is generalized and highly inflammatory. Trauma plays an important role in infection (Patil *et al.*, 2023). Fungi implicated in subcutaneous mycoses are abundant and have a low degree of infectivity. These organisms gain access to the subcutaneous tissues through some kind of trauma. Histopathologic evidence indicates that these organisms survive in the subcutaneous tissue by producing proteolytic enzymes and maintaining a facultative microaerophilic existence because of the lowered redox potential of the damaged tissue. In specific groups like denture wearers, patients undergoing radiotherapy for head or neck cancer, and among HIV-positive individuals, fungal infections are prevalent. The relationship between cancer and microbial infections has attracted enormous attention. Transient immunosuppression is commonly seen in cancer patients undergoing chemotherapy, which consequently has been linked to bacterial and fungal infections (Cong *et al.*, 2023). Gastrointestinal diseases, predominantly IBD, have emerged as an important public health challenge by globally affecting over 1.5 million annually in North America alone. Several gastrointestinal diseases are interconnected with fungal dysbiosis, which has been shown to contribute to disease sensitivity and severity. Antibiotic exposure, genetic diversity, and diet are the underlying factors that promote fungal dysbiosis. Fungal genera, mostly belonging to the *Candida* and *Saccharomyces* species have been shown to correlate with chronic idiopathic disorders such as ulcerative colitis (UC) and Crohn's disease (CD), which together are known as IBD. *Candida* species overgrowth in the gut due to antibacterial exposure has been known for a long time. COVID-19 patients have shown changes in microbiota composition, due to secondary infections, including those caused by bacteria and fungi. Among the opportunistic infections in COVID-19 patients, a higher incidence of fungal co-infections was observed, primarily involving species of *Mucor*, *Aspergillus*, and *Candida* (Jacobsen, 2023).

4.3   RISK FACTORS FOR OPPORTUNISTIC FUNGAL INFECTION

Susceptibility to opportunistic fungal infections is based on several factors like disparities in host immune response, host-specific genetic variations, and epigenetic factors that may originate from societal health determinants and socioeconomic conditions. These factors play a role in increasing susceptibility to fungal infections particularly in certain ethnic groups that may be predisposed to conditions that elevate the risk. Disparities in healthcare access and health insurance can further aggravate the situation. Differences in environmental conditions and underlying societal health determinants are major risk factors for Aspergillosis and other mould infections (van Rhijn and Bromley, 2021). Other contributors include immunological irregularities and genetic variations such as defects in the CARD9 pathway and polymorphism in genes namely IL-1, IL-10, IL-15, IL-23, TNF-α, and INF-γ. On the contrary, there appears to be a higher risk of candidiasis and cryptococcosis in African Americans in comparison to white Americans. Although the exact reason for this discrepancy remains unclear, it may be primarily related to societal health determinants, disparities in healthcare access, and other socioeconomic factors, rather than solely genetics, since a substantial amount of genetic diversity occurs within these populations (Naik *et al.*, 2021). Hence, for most fungal diseases, differences in epidemiology and disease severity among various ethnic groups can often be attributed to other factors, such as environmental exposure and variation in societal health determinants rather than genetic factors (Jenks *et al.*, 2023).

4.3.1 IMMUNOCOMPROMISED CONDITIONS

Contemporary researchers have shown great interest in mycoses, primarily because they have emerged as prominent secondary infections affecting individuals with immunocompromised conditions like AIDS. Surprisingly, only a few studies focus on the direct impact of HIV on host-fungal interactions. The increased vulnerability of HIV-infected individuals to mycoses may be due to progressive depletion and qualitative dysfunction of CD4 cells, shift in pattern of lymphokine secretion from Th1 to Th2, dysfunctions of mononuclear and polymorphonuclear phagocytes, and uncontrolled secretion of cytokines. Moreover, research shows that fungi and fungal products can stimulate HIV replication in latently infected macrophages and lymphocytes, potentially exacerbating the progression of HIV infection. A more comprehensive understanding of the specific deficiencies that predispose HIV-infected individuals to mycoses should ultimately lead to improved treatment and prophylactic strategies (Fidel, 2011).

4.3.2 IMMUNOMODULATORY THERAPIES AND OTHER PREDISPOSING FACTORS

Treatment strategies for cancer (chemotherapy and radiation therapy), lower white blood cell count (neutropenia) and weaken patient immunity. This in turn increases the risk of secondary fungal infections. Although these therapies are effective against cancer, they are non-specific and hence produce potentially life-threatening side effects. Post-therapy, the patient becomes increasingly susceptible to opportunistic microbial infections. Candidemia (*Candida* in bloodstream) is highly prevalent in cancer patients and is associated with morbidity, mortality, and high healthcare costs (Teoh and Pavelka, 2016). Chemotherapy has detrimental effects on the epithelial cells and progenitors as it can non-specifically damage the rapidly dividing cells. Moreover, chemotherapy-induced changes in the tight junctions of gut epithelium, can also lead to increased gut permeability destroying barriers and rendering patients more susceptible to systemic infections. Furthermore, chemotherapy impacts the commensal host microbiota drastically. All these collectively increase the risk of opportunistic fungal infections in cancer patients (Lionakis *et al.*, 2023).

Acute leukemia patients, haematopoietic stem cell transplant recipients, and solid-organ transplant recipients are the three most common groups susceptible to invasive fungal infections. When dealing with such immunocompromised individuals, it is crucial to understand the specific etiologic agents, the duration and severity of underlying immunological deficiencies. External factors that influence the type of the pathogen involved include radiation, prescribed medications, surgical procedures that disrupt the mucocutaneous defences, the extent of immunosuppression (dosage, duration, and time of treatment), and environmental exposures (both in the community and healthcare setting). The overall risk of infection depends on overall immunosuppression and the epidemiological factors encountered (Lien *et al.*, 2018). In patients with acute leukemia, the risk of invasive fungal infection is closely linked to the type and status of the disease, the duration and severity of neutropenia (reduced neutrophil count), and the specific chemotherapeutic agents used for treatment. Neutrophils are especially effective in destroying fungal hyphae by the releasing the toxic oxygen radicals. Prolonged neutropenia, combined with disruptions in the skin and gut barriers, can make patients more vulnerable to candidemia and invasive aspergillosis.

In case of haematopoietic stem cell transplant recipients, the risk of invasive fungal infection varies over time due to factors like the health status of the recipient, the kind of graft they receive, and further complications arising during and after the transplant procedures, for instance using central venous catheters (Lionakis *et al.*, 2023). Host-related factors (old age, health, etc.) and transplant-related factors (leukocyte antigen mismatch, etc.) tend to influence the risk of invasive fungal infections early post-transplant. The use of immunosuppressive medications in the pre-engraftment phase, to prevent graft-versus-host disease, can lead to significant neutrophil depletion, increasing the chances of invasive candidiasis and aspergillosis (Gong *et al.*, 2020). Certain biological factors such as malnutrition, iron overload, diabetes mellitus, and blood cell deficiencies have moderate impact throughout the post-transplant period. Thus, the risk of acquiring fungal infections in patients undergoing stem cell transplant and those with haematologic cancers such as leukemia, lymphoma, or myeloma, depends on specific clinical characteristics and treatment therapies (Maertens *et al.*, 2001). Diabetes mellitus is another significant predisposing factor for fungal infections engendering challenging issues for diagnosis and treatment. They can result in severe health complications in diabetic patients on haemodialysis and kidney transplant procedures. Patients who receive kidney graft during pancreas transplantation, are at an additional risk for fungal infections in diabetics (Lao *et al.*, 2020).

4.4 CLINICAL MANIFESTATIONS

4.4.1 Candidiasis

The clinical manifestations of *Candida* vary from superficial mucocutaneous infections to deep seated invasive infections affecting multiple organs. Also known by the synonyms, moniliasis and thrush, this common commensal is the inhabitant of oral cavity, mucosa, skin vulvovaginal areas and other sensitive parts of the body. Poor oral hygiene, use of mobile prosthetic replacements, orthodontic appliances, dry mouth, smoking, using steroid inhalers, a diet high in carbohydrates, and disorders of the oral mucosa are the most prevalent local risk factors for candidiasis. Age (elders and neonates are at higher risk), pregnancy, systemic corticosteroid therapy, antibiotic therapy, chronic diseases and their treatments, digestive system disorders, nutritional deficiencies (iron, folic acid, and vitamins), endocrinopathy (diabetes, hypothyroidism, hypoparathyroidism, autoimmune diseases), HIV, and primary immunodeficiencies are the systemic predisposing factors for candidiasis (Talapko *et al.*, 2021). Table 4.1 compiles the common clinical manifestations associated with fungal diseases.

4.4.1.1 Oral Candidiasis

The oral microflora of all human contains *Candida* species as common constituents, fungal load varying between individuals. This has been confirmed by molecular identification techniques. These species vary depending on the health status of the individual and other environmental factors. The primary etiological agents include *C. albicans*, *C. glabrata*, *C. tropicalis*, *C. parapsilosis*, *C. krusei*, *C. dubliniensis*, and *C. guilliermondii*; the most prevalent species being *C. albicans* accounting for more than 80% of all oral fungal isolates (Sav *et al.*, 2020). Based on the range of clinical manifestations, oral candidiasis is called primary when it affects only the oral cavity and the perioral area; it is called secondary when it becomes systemic. Bad breath, abnormal metallic taste, difficulty in swallowing, erythma or redness, and dryness are some of the clinical manifestations of oral candidiasis.

4.4.1.2 Mucosal Candidiasis

The gastrointestinal tract is one of the primary reservoirs for *C. albicans*, from where it finds passage into the bloodstream leading to disseminated candidiasis. This mostly occur after the gastrointestinal tract is invaded which, of course, varies from person to person depending upon food habits, overall health, and hygiene (Tong and Tang, 2017). A broad range of clinical manifestations can be seen depending upon patient age and body site affected. The main predisposing factors that facilitate transformation of *C. albicans* from a commensal to a pathogen are immune dysfunction, mucosal barrier damage, and dysbiosis of the home microbiota (Allert *et al.*, 2018).

TABLE 4.1

Common Clinical Manifestations Associated with Fungal Diseases

S. No.	Disease	Clinical Manifestations	Main Etiological Agent
1.	Oral candidiasis	Bad breath, abnormal cottony metallic taste, pain during swallowing, erythema, and dry mouth	*C. albicans* *C. glabrata*
2.	Mucosal candidiasis	Immune system dysfunction, mucosal barrier damage, and dysbiosis	*C. tropicalis* *C. parapsilosis*
3.	Vulvovaginal candidiasis	Vulvodynia, soreness, vulvovaginal irritation and dysuria	*C. krusei*
4.	Invasive candidiasis	Dysphagia, skin, mucosal lesions, blindness, fever, shock, oliguria, renal shutdown, and disseminated intravascular coagulation	*C. dubliniensis* *C. guilliermondii*
5.	Cutaneous aspergillosis	Macules, papules, nodules, or plaques. Pustules or lesions with purulent discharge among neonates	*A. fumigatus* *A. terreus*
6.	Pulmonary aspergillosis	Worsening asthma, blood eosinophilia, sputum with brown mucus and plugs, low-grade chronic fevers, coughing, chest pain, and episodic wheezing	*A. niger* *A. flavus*
7.	Rhinosinusitis	Nasal obstruction, loss of smell, nasal discharge, nasal crust, pressure sensation on the face sinuses, double vision or visual loss (rare), facial asymmetry, and proptosis	
8.	Aspergilloma	Fever, chest pain, cough, coughing up blood, and shortness of breath	
9.	Pneumocystis pneumonia	Fever, cough, and dyspnea	*Pneumocystis jirovecii*
10.	Pulmonary mucormycosis	Dyspnea, coughing and chest pain	*Rhizopus* and *Mucor*
11.	Rhinocerebral mucormycosis	Eye and/or facial pain, facial numbness, and blurred vision	
12.	Cutaneous mucormycosis	Discomfort, warmth, intense redness, or edema surrounding an injury	
13.	Gastrointestinal mucormycosis	Bowel perforation, peritonitis, sepsis, and massive gastrointestinal haemorrhage	
14.	Disseminated mucormycosis	Nasal obstruction or congestion with noisy breathing, headache, odontalgia, maxillary pain, and hyposmia or anosmia	
15.	Fusariosis	Fever, skin lesions, superficial skin lesions with subsequent lymphangitis, and pneumonia	*Fusarium*

4.4.1.3 Vulvovaginal Candidiasis

C. albicans, followed by *C. glabrata*, *C. africana*, *C. dubliniensis*, *C. parapsilosis*, *C. tropicalis*, *C. krusei*, *C. guilliermondii*, and *C. lusitaniae* are the most frequently isolated species from patients of vulvovaginal candidiasis or vaginal thrush. Approximately 15%–30% of asymptomatic women are carriers, the percentage increasing further during pregnancy. Practically all women experience vulvovaginal candidiasis at least once in their lifetime. Males also experience infection in the genital areas (balanitis and balanoposthitis). Common clinical manifestations include vulvar burning, soreness, and irritation accompanied with dysuria (painful urination) and/or dyspareunia (painful intercourse) (Talapko *et al.*, 2021).

4.4.1.4 Invasive Candidiasis

C. albicans is the most prevalent species found in clinical setting followed by *C. glabrata*, *C. tropicalis*, *C. parapsilosis* and *C. krusei*. As several organs are affected the disease displays numerous clinical manifestations like pericarditis, pneumonia, empyema, mediastinitis, endophthalmitis, meningitis, osteoarticular infections, endocarditis, peritonitis, and infections of the urinary bladder, kidney and abdomen. Candidemia is common, especially in patients that are immunosuppressed and require intensive care. A localized or extensive lack of host defence combined with increased colonization is typically the cause of invasive candidiasis (Pappas *et al.*, 2018). The clinical presentation of invasive candidiasis depends on the site of infection and includes dysphagia, skin and mucosal lesions, blindness, fever, shock, oliguria, renal shutdown, and disseminated intravascular coagulation.

4.4.2 ASPERGILLOSIS

Aspergillosis is an infection caused by the mould *Aspergillus*, the spores of which are ubiquitous in the environment. The diseased condition may vary from non-invasive allergic responses to chronic and invasive lung infections. After inhalation, the infection may propagate locally or disperse to different organs, based on the host immune status (Thompson and Young, 2021). More invasive conditions include invasive and chronic pulmonary aspergillosis, while non-invasive forms include allergic bronchopulmonary aspergillosis and allergic fungal rhinosinusitis. Although, the disease can affect almost any organ of the human body, the most common sites of infection are the lungs and lower respiratory tract. Later, the skin, sinuses, and nasal cavities are the most susceptible. The central nervous system and the cardiovascular system may be affected either through direct infection or through haematogenous dissemination. The immune status of infected host and presence of a pre-existing pulmonary illness influences the disease spectrum. Other environmental factors that can increase fungal spores in the ambient air (renovation, demolition, and reconstruction), especially in the healthcare setting, can cause infections to become endemic (Rudramurthy *et al.*, 2016). Haematological malignancies, haematopoietic stem cell transplantation, solid-organ transplantation, patients receiving prolonged chemotherapy or steroids, and previous cases of pneumonia (COVID-19 or influenza) are the main risk factors (Latgé and Chamilos, 2019).

4.4.2.1 Superficial or Cutaneous Aspergillosis

The main etiological agent of cutaneous aspergillosis is *A. fumigatus*. It rarely penetrates deeper tissue and most often affects the outer layers of the skin, nails, cornea, or ear canal. These infections, which include fungal otomycosis, keratitis, onychomycosis, and cutaneous aspergillosis, are mainly contracted by direct trauma (injections, thorns, etc.). Immunocompetent patients frequently experience keratitis, otomycosis, and onychomycosis, but immunocompromised patients are more likely to experience cutaneous aspergillosis (Merad *et al.*, 2021). The initial lesions of cutaneous aspergillosis may appear as macules, papules, nodules, or plaques. Pustules or lesions with purulent discharge have generally occurred among neonates.

4.4.2.2 Pulmonary Aspergillosis

A. fumigatus, *A. flavus*, *A. terreus*, and *A. niger* are some common species causing pulmonary aspergillosis. *Aspergillus* species typically persist in pre-existing lung cavities that are caused by tuberculosis, bronchiectasis, sarcoidosis, or cavitary neoplasia. In immunocompetent individuals, these kinds of infections can result in conditions such as chronic pulmonary aspergillosis. Hypersensitive people frequently experience allergic rhinitis and severe asthma with fungal sensitization. In contrast, patients with underlying asthma or cystic fibrosis are often affected by allergic bronchopulmonary aspergillosis. As part of systemic involvement, invasive pulmonary aspergillosis can be observed in immunocompromised and neutropenic individuals, as well as in non-neutropenic patients that are undergoing mechanical ventilation, prolonged intensive care unit stays, chronic obstructive pulmonary disease (COPD), and more recently coronavirus disease (COVID-19) (Kanaujia *et al.*, 2023). Clinical manifestations include blood eosinophilia, worsened asthma, sputum with brown mucus and plugs, low-grade chronic fevers, coughing, chest pain, and episodic wheezing.

4.4.2.3 Rhinosinusitis

A non-invasive form of the disease known as allergic fungal rhinosinusitis is usually observed in younger, immunocompetent, atopic individuals. The frequently involved species include *A. fumigatus* and *A. flavus*. The disease is characterized by immune complex deposition, inflammation, and a hypersensitivity reaction to fungal antigens. Acute invasive fulminant rhinosinusitis, chronic invasive rhinosinusitis, and chronic granulomatous invasive fungal rhinosinusitis are some forms of invasive rhinosinusitis. Fungal ball and saprophytic fungal infestation are two forms that occur in non-invasive fungal rhinosinusitis. These conditions may be linked to prior mucosal trauma or surgery, especially dental procedures (Denning *et al.*, 2016). Nasal obstruction, loss of smell, nasal discharge, nasal crust, pressure sensation over face sinuses, double vision or loss of vision (rare), facial asymmetry, and proptosis are the common clinical presentations in these patients.

4.4.2.4 Aspergilloma

Aspergilloma involves colonization and proliferation of the fungus in a pre-existing pulmonary cavity (fungus ball), often formed due to a previous infection. In developing countries, over 90% cases of tuberculosis patients are susceptible to formation of aspergillomas or fungal balls. Its occurrence has been reported (20% patients) within three years after recovery from cavitary tuberculosis. Furthermore, chronic cavitary lung diseases, including sarcoidosis, and other fungal infections, become worse due to aspergilloma. Aspergilloma usually appears as a self-limiting condition with mild haemoptysis in 50%–90% patients. However, haemoptysis can occasionally become fatal when the patient starts coughing up large amounts of blood (Latgé and Chamilos, 2019). Some clinical features associated with aspergilloma are fever, chest pain, coughing up blood, and shortness of breath.

4.4.3 PNEUMOCYSTIS PNEUMONIA (PCP)

Pneumocystis pneumonia (PCP) is a potentially life-threatening disease caused by *Pneumocystis jirovecii*, a typical opportunistic fungus. Patients who are immunosuppressed or immunomodulated due to underlying illnesses and their treatment can develop PCP. Besides AIDS, cellular immunodeficiencies caused by haematological malignancies, T cell deficiency, and other serious illnesses that require corticosteroid treatment and premature or malnourished infants have reported PCP. It was reported to be the most prevalent and disease defining opportunistic infection in HIV-positive individuals in the 1980s, with high rates of morbidity (Xue *et al.*, 2023). Haematological malignancies, solid tumours, organ or haematopoietic stem cell transplantation, and connective tissue diseases treated with immunosuppressive drugs are some underlying conditions associated with PCP in non-HIV patients. The most important immunosuppressive drugs associated with PCP in non-HIV patients are corticosteroids, which are commonly used during organ and stem cell transplants, autoimmune and inflammatory diseases, and cancers. Corticosteroids are also administered to PCP patients who were not HIV positive, typically in combination with other immunosuppressive medications (Xue *et al.*, 2023). Fever, cough, and dyspnea are the typical clinical manifestations of PCP. The clinical characteristics of PCP differ significantly in HIV-positive and negative patients. Respiratory insufficiency, that develops suddenly, is a characteristic of PCP in non-HIV patients. In immunocompromised individuals without AIDS, PCP typically develops more quickly and impairs oxygenation more severely than in HIV-positive patients. In patients who are not HIV positive, the progression of disease occurs gradually over a week from fever and dry cough to respiratory failure, while in patients who are HIV positive, the illness takes 2 weeks to 2 months to manifest. In non-HIV patients, PCP mortality rates range from 30%–60%, whereas in HIV-positive patients, they are 10%–20%. The mortality rate in non-HIV patients varies depending on the population at risk; patients with cancer have a higher fatality rate than patients receiving transplants or those suffering from connective tissue diseases (Tasaka, 2020). About 30%–80% of PCP cases among non-HIV patients are caused by haematological malignancies, which include multiple myeloma, leukaemia, and lymphomas. These diseases continue to be the most common underlying causes of PCP predisposing patients to PCP (Liu *et al.*, 2017).

4.4.4 Zygomycosis (Mucormycosis)

The incidence of Mucormycosis is much less common in comparison to other opportunistic fungal infections like Candidiasis and Aspergillosis. Mucorales, the causative fungal pathogens, are also frequently encountered in patients receiving solid organ transplants, haematopoietic stem cell transplants, and haematological malignancies. Additionally, these infections are increasingly recognized in individuals with diabetes mellitus, after trauma or iatrogenic injury (Jeong *et al*., 2019). *Rhizopus oryzae* is the most frequently isolated organism from mucormycosis patients, accounting for ~70% of all cases (Ibrahim *et al*., 2012). In patients undergoing organ transplantation, haemodialysis or cancer therapy, it rapidly develops as a nosocomial infection. Patients with diabetes have an overall mortality rate of 44%, those without underlying medical conditions have a mortality rate of 35%, and patients with cancer have a death rate of 66%. The rate of mortality varied depending on the site of infection and host. The reported death rates of patients with disseminated infections, gastrointestinal infections and pulmonary infections were 96%, 85%, and 76%, respectively. The skin and gut in children are affected more frequently than in adults (Petrikkos *et al*., 2012). Mucormycosis can be classified into six categories based on clinical presentation and involvement of specific anatomical sites: (1) rhinocerebral; (2) pulmonary;,(3) cutaneous; (4) gastrointestinal; (5) disseminated; and (6) uncommon rare forms, such as endocarditis, osteomyelitis, peritonitis, and renal infection.

4.4.4.1 Pulmonary Mucormycosis

Mucormycosis is most frequently observed in the lungs of leukemia patients receiving chemotherapy, patients receiving treatment with haematopoietic stem cells, and neutropenic patients. This infection spread quickly involving other areas contiguously like cardiac muscles, bronchi and mediastinum (mediastinitis). Patients with pulmonary mucormycosis have an overall mortality rate of 48%–87%; the rate being higher in patients with severe immunosuppression. (Steinbrink and Miceli, 2021). The clinical features include dyspnea, cough, frequent fever and chest pain. Fungal invasion of blood vessels causes thrombosis, followed by cavitation, tissue necrosis, and ultimately fatal haemoptysis. Chest radiography may reveal wedge-shaped infarcts, isolated masses, nodular disease, cavities, or lobar consolidation.

4.4.4.2 Rhinocerebral Mucormycosis

This is most prevalent in diabetic patients and the causative agent is a saprophytic fungus like *Mucor, Rhizopus, Absidia, Cunninghamella* genera, and *Apophysomyces elegans* (Bhandari *et al*., 2020). Patients with suppressed immune system due underlying diseases are susceptible. Fungal sporangiospores, inhaled into the paranasal sinuses, germinate causing infection and then move laterally into the cavernous sinus affecting the orbits. The palate, sphenoid sinus, and the brain can also be affected. (Petrikkos *et al*., 2012). The mortality rate is ~25% less than in other types of mucormycosis. The initial clinical manifestations include eye and/or facial pain, facial numbness, and blurred vision, which are similar to those of sinusitis and periorbital cellulitis.

4.4.4.3 Cutaneous Mucormycosis

The spores of *Rhizopus oryzae, Lichtheimia corymbifera*, and *Apophysomyces elegans* directly inoculate the skin causing cutaneous mucormycosis, which eventually disseminates to other parts of the body (Castrejón-Pérez *et al*., 2017). The infection is localized when only skin and subcutaneous tissues are affected; deep seated when muscles, tendons, and bones are affected; and disseminated when other body organs are affected (Paes De Oliveira-Neto *et al*., 2006). The clinical manifestations of cutaneous mucormycosis varies, depending on the site of infection and severity of the disease. The disease may develop and progress slowly, or it may spread fast haematogenously, resulting in gangrene. It usually manifests as a necrotic eschar, with surrounding erythema and induration. Other less common symptoms of cutaneous mucormycosis include targetoid plaques with outer erythematous edges and ecchymotic or blackened necrotic centres, superficial lesions with only slightly elevated circinate and squamous borders resembling tinea corporis, and lesions that resemble bread mould and have a cotton-like texture (Petrikkos *et al.*, 2012). The affected area of cutaneous mucormycosis may turn black and resemble blisters or ulcers. Additionally, patient also experiences discomfort, warmth, intense redness, or oedema surrounding an infection site.

4.4.4.4 Gastrointestinal Mucormycosis

This infection is rare and seldom diagnosed. Mortality rate is very high (85%) as diagnosis in patients is often delayed (Roden *et al.*, 2005). Most of the cases have been documented in under-nourished children, premature neonates, people with diabetes mellitus, and a history of corticosteroid therapy. It is contracted by consuming infected food like dried bread products and fermented milk. The most frequently affected components of the digestive system are the stomach, colon, and ileum. The infection typically manifests as a gastric perforation or mass in the appendix, cecum, or ileum, which may then cause severe bleeding in the upper gastrointestinal tract also affecting the pancreas, spleen, and liver. The fungus can invade bowel walls and blood vessels, resulting in bowel perforation, peritonitis, sepsis, massive gastrointestinal haemorrhage, and distention associated with nausea and vomiting.

4.4.4.5 Disseminated Mucormycosis

Any primary site of infection can be the source of disseminated mucormycosis. Patients undergoing liver transplants are more likely to experience dissemination for unknown reasons. Metastatic lesions can also be found in the liver, spleen, heart, and other organs, even though the brain is a common site of spread. When brain dissemination occurs, the death rate is very close to 100% (Steinbrink and Miceli, 2021). Disseminated mucormycosis is commonly associated with patients who have iron overload (particularly those on deferoxamine), profound immunosuppression (such as graft versus host disease treated with corticosteroids in recipients of allogeneic stem cell transplants), profound neutropenia, or acute leukaemia. Nasal obstruction or congestion with noisy breathing, headache, odontalgia, maxillary pain, and hyposmia or anosmia are common symptoms to deal with.

4.4.5 Fusariosis

Fusariosis is caused by species of the genus *Fusarium*, also called hyphomycetes. This ubiquitous filamentous fungus is basically a plant pathogen and rarely becomes invasive in animals. However, in many countries, fusariosis ranks as the third most prevalent infection after aspergillosis and mucormycosis (Hoenigl *et al.*, 2023). Their wide range of infections in humans includes superficial, locally invasive, and disseminated infections; onychomycosis, skin infections, pneumonia, and keratitis are the most common infections caused by them. The primary entry point for the fungus is the airways, followed by the skin near the site of onychomycosis, tissue lesions, contact lenses, and probably mucous membranes (Batista *et al.*, 2020). Fever and metastatic skin lesions, pneumonia, cellulitis, or lymphangitis at sites of skin breakdown are among the common clinical presentations. Patients may present with catheter-related fungemia (less frequently), arthritis and sinusitis. The central nervous system is rarely affected reported (Nucci *et al.*, 2021). In immunocompromised people, the mortality rate in disseminated fusariosis is between 50–70%. Haematologic malignancy or allogeneic stem cell transplants, decreased cellular immunity, solid organ transplantation, and profound or prolonged neutropenia following cytotoxic chemotherapy are risk factors for the disease. The clinical manifestation (fever, skin lesions, superficial skin lesions with subsequent lymphangitis, and pneumonia) rely on patient immunity.

4.4.5.1 Skin Infections

This is the result of fungal dissemination, primarily in immunocompromised patients. The most typical disease pattern consists of several painful erythematous papules or nodules, frequently accompanied by central necrosis. These events propagate throughout the body and persistently discharge fungal cells, leading to positive blood culture and frequent pulmonary involvement, with or without the involvement of other sites (Batista *et al.*, 2020).

4.4.5.2 Keratitis

One of the most prevalent infections brought on by *Fusarium* species is keratitis, which is primarily caused by eye trauma, wear and tear by contact lenses, and corticosteroid therapy. Trauma is the key predisposing factor and occurs in 40%–60% of the patients.

4.4.5.3 Pneumonia

Invasive fusariosis can manifest as pneumonia, involving a single organ or as a component of a disseminated illness. The pathophysiology of invasive fusariosis is very similar to that of invasive aspergillosis, which involves inhalation of airborne conidia, alveoli colonization, and subsequent hyphae formation, bronchoalveolar dissemination, and angioinvasion with lung infarction. However, *Fusarium* species may also have a cutaneous portal of entry, with subsequent haematogenous dissemination to the lungs, in contrast to aspergillosis, in which the airways are the portal of entry in almost all cases (Nucci *et al.*, 2021).

4.4.5.4 Sinusitis

In patients with haematologic disorders, sinusitis is a common sign of invasive fusariosis. Clinical manifestations include nasal discharge, obstruction, and necrotic lesions. Periorbital and paranasal cellulitis can develop in the advanced stages of the illness.

4.4.5.5 Disseminated Infections

Disseminated infection is the most frequent clinical presentation of invasive fusariosis. Patients typically present with pneumonia, positive blood cultures, and metastatic skin lesions. A type of endophthalmitis or blindness due to retinal vascular thrombosis can affect the eye. A high fungal burden and severe immunosuppression are the hallmarks which have adverse outcomes (Nucci *et al.*, 2021).

4.5 DIAGNOSIS OF OPPORTUNISTIC FUNGAL INFECTIONS

Fungal infections do not always present with distinct clinical symptoms. The gold standard procedures for identification of pathogenic fungal species are still fungal culture, microscopy, and histopathology, used to provide a conclusive diagnosis. However, new techniques of higher sensitivity are required for identifying fungal pathogens. Several molecular and serological tests have been developed and many of them are under clinical evaluation such as the Galactomannan Antigen Test for Aspergillosis. PCR and allied molecular approaches, such as matrix-assisted laser desorption ionization (MALDI) and fluorescence *in situ* hybridization (FISH), have demonstrated reliable results in clinical trials but must be standardized prior to being introduced for clinical applications. Table 4.2 deliberates the advantages and disadvantages of some common diagnostic methods for fungal infections.

4.5.1 FUNGAL CULTURES

Growing cultures of clinical specimens (tissue, sputum, urine, wound, blood, and cerebrospinal fluid) is still the best diagnostic method, but only if the samples are from sterile sites. Although this method is the gold standard, it has low sensitivity. Blood cultures have an overall sensitivity of 50%–95% for yeasts, but the situation is complex for moulds that give only 1%–5% sensitivity (Arvanitis *et al.*, 2014).

TABLE 4.2

Overview of Advantages and Disadvantages of the Most Commonly Used Diagnostic Techniques for Fungal Infections

S. No.	Diagnostic Techniques	Advantages	Disadvantages	References
1.	Fungal cultures	Identification of pathogen at species level and detection of antifungal resistance	Time-consuming (in case of yeasts up to 5 days and mould up to 4 weeks) and prone to contamination	Arvanitis *et al.*, 2014
2.	Microscopy	Visualization of fungal shape, structure and biofilm formation	Due to similarities between structures of several fungus, does not allow genus or species identification	Jenks *et al.*, 2020
3.	Chromogenic media	Identification of species in polymicrobial samples; fast and cost-efficient	Ability to distinguish between related species at the phenotypic level may be hampered	Borman *et al.*, 2021
4.	MALDI-TOF MS	Identification of the pathogen at the genus, species and strain level; Quick and accurate species identification of *Aspergillus* and *Candida;* Differentiation between closely related species	Databases must be updated often to include the rarest and most emerging fungal species; High cost of instrumentation	Franco-Duarte *et al.*, 2019
5.	FISH	High specificity and sensitivity; time saving; accurate identification of *Candida* species	Reduced number of peptide nucleic acid (PNA) probes commercially available.	Franco-Duarte *et al.*, 2019
6.	Radiography	Monitors the progression of infection as well as patient response to antifungal therapy	Use of each radiolabelled technique is limited and specific	Alexander *et al.*, 2021
7.	Serological methods	Detection of relevant fungal pathogens; gives fast results; excellent for invasive fungal infections	Lower sensitivity and limited specificity; high number of false-positive results	Arvanitis *et al.*, 2014
8.	PCR method	High sensitivity and specificity, allows quantification; allows species identification and intraspecies differentiation	Contamination; lack of standardization of the fungal DNA isolation techniques; traditional PCR does not allow quantification	Arvanitis *et al.*, 2014

4.5.2 MICROSCOPY

Microscopic examination of sample smears using potassium hydroxide is the most efficient, economical, and sensitive way to identify the concerned pathogen based on its morphological features and fungal elements. Several dyes and fluorescent probes like India ink, Calcofluor white, periodic acid-Schiff, and Gomori's methenamine silver can be used for visualization with the help of appropriate microscopy techniques.

4.5.3 CHROMOGENIC MEDIA

The differential medium uses selective chromogenic substrates to specifically identify fungal species based on enzyme activity from a heterogeneous culture of microbes. Since such media are selective in promoting growth of only particular species while preventing the growth of other microorganisms, they are appropriate for non-sterile samples. Chromogenic Candida Agar and CHROMagar Candida are some ready to use chromogenic media, commercially available for the identification of *Candida* species giving different colour-specific colonies. Chromogenic media has several advantages including enhanced accuracy, easy detection, cost-effective, gives faster results, and requires only basic microbiology skills. It is considered as an alternative method since it may give false-positive results (Borman *et al.*, 2021).

MALDI-TOF MS (Matrix-Assisted Laser Desorption/Ionization): Microbiologists have recently started using mass spectrometry to identify microbes as it has proved to be a faster more sensitive technique to identify a wide range of microorganisms. The most popular mass spectrometry variant for the identification of fungal species is MALDI-TOF MS, which is based on the recognition of extracted protein fingerprints, primarily those of ribosomal and membrane proteins. Identification at the species and genus levels is made possible by comparing protein profile of each isolate with the universal profile databases (Franco-Duarte *et al.*, 2019). This technique is widely employed for identification of common yeast, but it is still a challenge to identify pathogenic filamentous fungi.

FISH (Fluorescence *in situ* hybridization): This is probably the most conclusive technique for diagnosis by locating specific DNA sequences. The DNA (deoxyribonucleic acid)-based FISH probes are comparatively cheaper than the PNA (peptide nucleic acid)-based FISH probes, which are commercially more easily available and have higher affinity, specificity, and resistance to degradation. Though not very common at present, this method gives valuable information in some cases (Da Silva *et al.*, 2015).

4.5.4 RADIOGRAPHY

X-rays, high-resolution computed tomography, magnetic resonance imaging (MRI), and positron emission tomography (PET) are some other techniques that can help in diagnosis of fungal infections. Pulmonary fungal infections such as aspergillosis, fusariosis, scedosporiosis, or zygomycosis show cavitary lung lesions, infiltration, pulmonary nodules, and air crescents as characteristic features. However, radiation exposure to children and pregnant women needs to be regulated and thus these techniques would have limitations for this tomography (Brenner and Hall, 2007).

4.5.5 SEROLOGICAL METHODS

Exposure to fungal infections can be easily diagnosed by detecting the presence of antibodies in blood samples. The serological methods give much faster and explicit results than culture methods. Serum IgE levels and skin tests are useful diagnostic tools. For rapid diagnosis in critically ill patients, antigens can be quickly detected in blood, bronchoalveolar lavage, or urine using enzyme immunoassay and latex agglutination test. Galactomannans, β-1,3-D-glucans, and mannans are some fungal cell wall markers that have been found in serum.

Galactomannan can be detected in specimens like urine, bronchoalveolar lavage, cerebrospinal fluid using enzyme immunoassay. It is comparatively specific to the *Aspergillus* species with varying sensitivity rates of 30%–100% (Viscoli *et al.*, 2002; Mennink-Kersten and Verweij, 2006). β-1,3 D-glucans, present in the cell wall of most pathogenic fungi, are not species-specific but have high sensitivity (Ostrosky-Zeichner *et al.*, 2005; Badiee and Hashemizadeh, 2014). In patients with neutropenia post myeloablative chemotherapy, presence of circulating *Candida* mannans and anti-mannan antibodies has been observed and utilized as a diagnostic marker for candidemia (Ellis *et al.*, 2009). The overall sensitivity of mannan detection in patients with candidemia has been reported between 69%–91%, and specificity between 46%–89% compared to the culture method (Kurita *et al.*, 2009). The serologic tests show limited specificity when diagnosing Candidemia due to the presence of *Candida* in host microflora. They also show limited sensitivity in high-risk immunocompromised patients, and different sensitivity for different species.

4.5.6 PCR METHODS

Polymerase chain reaction or PCR-based techniques are frequently used for direct detection of fungal DNA using samples like bronchoalveolar lavage, blood and cerebrospinal fluid. Novel and increasingly useful variations of conventional PCRs have been developed over time like real-time PCR, reverse-transcriptase PCR, and nested PCR. The most popular methods for detecting fungal pathogens are real-time PCR and conventional PCR, which easy to handle, have high sensitivity, and give results in less time (Badiee and Hashemizadeh, 2014).

4.6 TREATMENT STRATEGIES

4.6.1 CONVENTIONAL ANTIFUNGAL DRUGS

For the treatment of invasive fungal infections, there are five conventional antifungal drugs available that have been approved for use in humans. Antifungal medications are often developed against cell wall components (mannans, glucans, and chitins) and other molecules that are unique to fungi like ergosterol and enzymes crucial to their biosynthesis (Georgiev, 2000). The most used antifungal drugs are classified as azoles, polyenes, echinocandins, pyrimidine analogues, allylamines, thiocarbamates, and morpholines (Castelli *et al.*, 2014). Figure 4.1 shows the different classes of antifungal drugs and their general mechanisms of action.

4.6.1.1 Azoles

The first azoles were developed based on imidazoles. Currently, the most commonly prescribed azoles are marketed as miconazole, clotrimazole, econazole, ketoconazole, tioconazole, and

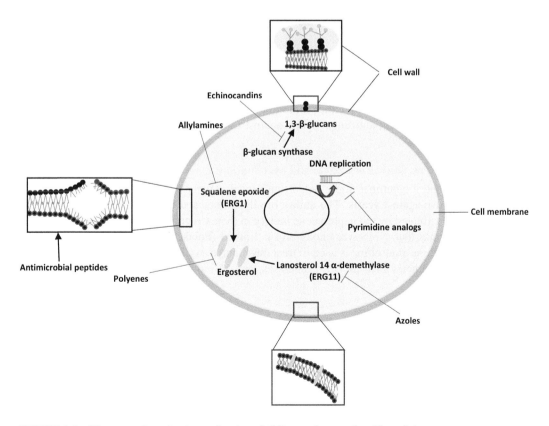

FIGURE 4.1   The general mechanisms of action of different classes of antifungal drugs.

sulconazole. Luliconazole, the most recent imidazole to be licensed in 2013, is available as a topical ointment for the treatment of dermatophytes. Triazole-based medications were developed later having a wider activity range, such as fluconazole and itraconazole. Imidazoles are primarily used to treat mucosal fungal infections, while triazoles are given for both systemic and mucosal infections (Sanglard and Coste, 2016). In the fungal cell membranes, lanosterol is converted to ergosterol by the enzyme lanosterol-14 α-demethylase, encoded by the *ERG11* gene. This enzyme contains an iron protoporphyrin ring in its active site. Azoles competitively bind at the active site blocking ergosterol biosynthesis. This results in the accumulation of 14-methyl sterol that alters membrane stability, permeability, and function (Cowen *et al.*, 2015).

4.6.1.2 Polyenes

These are polyunsaturated organic compounds that bind to ergosterol making the membrane porous. The polyenes used as antifungals include Amphotericin B, Nystatin, Candicidin, Natamycin, Hachimycin majorly derived from *Streptomyces*. Amphotericin B is an amphipathic macrolide, administered intravenously or topically due to poor absorption on oral consumption (Lemke *et al.*, 2005; Sanglard *et al.*, 2009; Vandeputte *et al.*, 2012).

4.6.1.3 Echinocandins

The members of this relatively recent class of antifungals are lipopeptides and function as noncompetitive inhibitors of β-1,3-D-glucan synthase, the enzyme required for β-glucan synthesis in fungal cell walls (Shapiro *et al.*, 2011). Impaired cell wall synthesis affects the integrity of fungal cells, resulting in extreme cell wall stress. Thus, the echinocandin treatment gives thicker cell walls, lower sterol levels, pseudohyphae development, and increased osmotic sensitivity. Echinocandins are fungicidal against many fungi, including *Candida,* and are often non-toxic to mammalian cells as the target is unique to fungi (Ghannoum and Rice, 1999; Perlin, 2011; Szymański *et al.*, 2022).

4.6.1.4 Pyrimidine Analogues

The pyrimidine analogues are synthetic counterparts of the nucleotide cytosine, such as 5-fluorocytosine (5-FC) which is converted to another analogue 5-fluorouracil (5-FU) by cytidine deaminase. 5-FU subsequently becomes incorporated into nucleotides (DNA and RNA) during their synthesis and suppresses cellular activity by either inhibiting DNA replication or protein synthesis. These analogues have shown potent antifungal action against *Candida* and *Cryptococcus* (Lemke *et al.*, 2005; Sanglard *et al.*, 2009).

4.6.1.5 Allylamines, Thiocarbamates, and Morpholines

Thiocarbamates and allylamines show potent antifungal activity against dermatophytes but are moderately effective against yeasts. The ability of allylamines and thiocarbamates to bind to the enzyme may be facilitated by the naphthalene moiety in their structures. Their interaction with the mammalian enzymes that synthesize cholesterol is limited (Polak, 1990). The topical antifungal medication morpholine amorolfine is used to treat onychomycosis. It inhibits the enzymes required for ergosterol biosynthesis such as 7, 8-isomerases, and 14-reductases (Mercer, 1991; Haria and Bryson, 1995).

4.6.2 Combinational Therapy

Due to the limited number of antifungal drugs and increasing cases of drug resistance, single drug-based monotherapy does not give the required results. Synergistic combination therapy improves therapeutic efficacy of conventional drugs and reduces the development of drug resistance considerably. Several *in vitro* and *in vivo* studies have shown that combinatorial approaches are much more effective in treating fungal infections and enhance bioavailability (Spitzer *et al.*, 2017). During the treatment of cryptococcal meningitis, it was observed that amphotericin B monotherapy had a higher relapse rate and a higher incidence of recurrence compared to flucytosine given in combination with amphotericin B. Fluconazole can also be given in combination with other non-antifungal drugs like calcineurin inhibitors, heat shock protein 90 (Hsp90) inhibitors, calcium homeostasis modulators, and conventional therapy. Drugs administered in combination showed much greater antifungal activity in terms of reducing growth and virulence of the pathogenic fungus (Garbati *et al.*, 2012). To evaluate drug-drug interactions, the checkerboard microdilution method is frequently employed (Fatsis-Kavalopoulos *et al.*, 2020). Azoles in combination with Harmine hydrochloride have a synergistically higher antifungal effect during the early formation phase of resistant *Candida* biofilms (Li *et al.*, 2019). Combining amorolfine with other conventional medications, such as ketoconazole, terbinafine, itraconazole, and griseofulvin, has synergistic antifungal efficacy but has shown increased fungistatic activity against dermatophytes (Brescini *et al.*, 2021). Interestingly, the antifungal activity of fluconazole showed increased efficacy against resistant clinical strains, when given in combination with *Ocimum sanctum* essential oil, linalool, and geraniol observed in terms of lower MIC and MFC values (Cardoso *et al.*, 2016). The fractional inhibitory concentration index (FICI) of thymol eugenol and methanol in combination with fluconazole was reported to be less than 0.5. Similarly, the effect on biofilms was also reduced synergistically (Pemmaraju *et al.*, 2013). *A. cryptantha* essential oil and fluconazole together have a synergistic effect against *C. albicans* (FICI=0.31) (Lima *et al.*, 2022). A different study used the checkerboard assay to demonstrate the synergistic impact of perillyl alcohol with the antifungal miconazole against *C. glabrata* (Gupta and Poluri, 2021).

4.7 CONCLUSION

Opportunistic infections are clinically significant in immunocompromised individuals. Aspergillosis and candidiasis are the most prevalent invasive infections that adversely impact critically ill patients. The number of effective antifungal drugs is limited, despite the unprecedented rise in the incidence and severity of infections caused by newly emerging fungal diseases. A wide range of fungal species demonstrate resistance to conventional antifungal drugs like azoles and more recently developed echinocandins. Successful and timely treatment of these infections mainly depends on early diagnosis and identification of fungal species. Improved techniques for early diagnosis and unconventional antifungal therapies, along with the reconstitution of host immunity can together improve treatment outcomes tremendously.

REFERENCES

Alexander, B. D., Lamoth, F., Heussel, C. P., Prokop, C. S., Desai, S. R., Morrissey, C. O., & Baddley, J. W. (2021). Guidance on imaging for invasive pulmonary aspergillosis and mucormycosis: From the imaging working group for the revision and update of the consensus definitions of fungal disease from the EORTC/MSGERC. *Clinical Infectious Diseases*, 72(suppl_2), S79–S88.

Allert, S., Förster, T. M., Svensson, C. M., Richardson, J. P., Pawlik, T., Hebecker, B., & Hube, B. (2018). *Candida albicans*-induced epithelial damage mediates translocation through intestinal barriers. *MBio*, 9(3), 10–1128.

Almeida, F., Wolf, J. M., & Casadevall, A. (2015). Virulence-associated enzymes of *Cryptococcus neoformans*. *Eukaryotic Cell*, 14(12), 1173–1185.

Almyroudis, N. G., Sutton, D. A., Linden, P., Rinaldi, M. G., Fung, J., & Kusne, S. (2006). Zygomycosis in solid organ transplant recipients in a tertiary transplant center and review of the literature. *American Journal of Transplantation*, 6(10), 2365–2374.

Angiolella, L. (2022). Virulence regulation and drug-resistance mechanism of fungal infection. *Microorganisms*, 10(2), 409.

Arvanitis, M., Anagnostou, T., Fuchs, B. B., Caliendo, A. M., & Mylonakis, E. (2014). Molecular and non-molecular diagnostic methods for invasive fungal infections. *Clinical Microbiology Reviews*, 27(3), 490–526.

Azevedo, R. V. D. M., Rizzo, J., & Rodrigues, M. L. (2016). Virulence factors as targets for anticryptococcal therapy. *Journal of Fungi*, *2*(4), 29.

Badiee, P., Alborzi, A., Moeini, M., Haddadi, P., Farshad, S., Japoni, A., & Ziyaeyan, M. (2012). Antifungal susceptibility of the *Aspergillus* species by Etest and CLSI reference methods. *Archives of Iranian Medicine*, *15*(7), 0–0.

Badiee, P., & Hashemizadeh, Z. (2014). Opportunistic invasive fungal infections: Diagnosis & clinical management. *Indian Journal of Medical Research*, *139*(2), 195.

Bassetti, M., Righi, E., Costa, A., Fasce, R., Molinari, M. P., Rosso, R., & Viscoli, C. (2006). Epidemiological trends in nosocomial candidemia in intensive care. *BMC Infectious Diseases*, *6*(1), 1–6.

Batista, B. G., Chaves, M. A. D., Reginatto, P., Saraiva, O. J., & Fuentefria, A. M. (2020). Human fusariosis: An emerging infection that is difficult to treat. *Revista da Sociedade Brasileira de Medicina Tropical*, *53*, e20200013.

Bhabhra, R., & Askew, D. S. (2005). Thermotolerance and virulence of Aspergillus fumigatus: Role of the fungal nucleolus. *Medical Mycology*, *43*(Suppl 1), S87–S93.

Bhandari, J., Thada, P. K., & Nagalli, S. (2023). Rhinocerebral mucormycosis. In *StatPearls*. Treasure Island, FL: StatPearls Publishing.

Bongomin, F., Gago, S., Oladele, R. O., & Denning, D. W. (2017). Global and multi-national prevalence of fungal diseases—Estimate precision. *Journal of Fungi*, *3*(4), 57.

Borman, A. M., Fraser, M., & Johnson, E. M. (2021). CHROMagarTM *Candida* Plus: A novel chromogenic agar that permits the rapid identification of *Candida auris*. *Medical Mycology*, *59*(3), 253–258.

Brenner, D. J., & Hall, E. J. (2007). Computed tomography—An increasing source of radiation exposure. *New England Journal of Medicine*, *357*(22), 2277–2284.

Brescini, L., Fioriti, S., Morroni, G., & Barchiesi, F. (2021). Antifungal combinations in dermatophytes. *Journal of Fungi*, *7*(9), 727.

Brown, G. D., Denning, D. W., & Levitz, S. M. (2012). Tackling human fungal infections. *Science*, *336*(6082), 647–647.

Cafarchia, C., Figueredo, L. A., & Otranto, D. (2013). Fungal diseases of horses. *Veterinary Microbiology*, *167*(1–2), 215–234.

Cardoso, N. N., Alviano, C. S., Blank, A. F., Romanos, M. T. V., Fonseca, B. B., Rozental, S., & Alviano, D. S. (2016). Synergism effect of the essential oil from *Ocimum basilicum* var. Maria Bonita and its major components with fluconazole and its influence on ergosterol biosynthesis. *Evidence-Based Complementary and Alternative Medicine*, *2016*, 5647182.

Castelli, M. V., Butassi, E., Monteiro, M. C., Svetaz, L. A., Vicente, F., & Zacchino, S. A. (2014). Novel antifungal agents: A patent review (2011–present). *Expert Opinion on Therapeutic Patents*, *24*(3), 323–338.

Castrejón-Pérez, A. D., Welsh, E. C., Miranda, I., Ocampo-Candiani, J., & Welsh, O. (2017). Cutaneous mucormycosis. *Anais brasileiros de dermatologia*, *92*, 304–311.

Chayakulkeeree, M., Ghannoum, M. A., & Perfect, J. R. (2006). Zygomycosis: The re-emerging fungal infection. *European Journal of Clinical Microbiology and Infectious Diseases*, *25*, 215–229.

Cong, L., Chen, C., Mao, S., Han, Z., Zhu, Z., & Li, Y. (2023). Intestinal bacteria—A powerful weapon for fungal infections treatment. *Frontiers in Cellular and Infection Microbiology*, *13*, 674.

Cowen, L. E., Sanglard, D., Howard, S. J., Rogers, P. D., & Perlin, D. S. (2015). Mechanisms of antifungal drug resistance. *Cold Spring Harbor Perspectives in Medicine*, *5*(7).

Da Silva, R. M., Da Silva Neto, J. R., Santos, C. S., Frickmann, H., Poppert, S., Cruz, K. S., & De Souza, J. V. B. (2015). Evaluation of fluorescence in situ hybridisation (FISH) for the detection of fungi directly from blood cultures and cerebrospinal fluid from patients with suspected invasive mycoses. *Annals of Clinical Microbiology and Antimicrobials*, *14*, 1–6.

Denning, D. W. (1991). Epidemiology and pathogenesis of systemic fungal infections in the immunocompromised host. *Journal of Antimicrobial Chemotherapy*, *28*(suppl_B), 1–16.

Denning, D. W. (1998). Invasive aspergillosis. *Clinical Infectious Diseases*, *26*, 781–803.

Denning, D. W., Cadranel, J., Beigelman-Aubry, C., Ader, F., Chakrabarti, A., Blot, S., & Lange, C. (2016). Chronic pulmonary aspergillosis: Rationale and clinical guidelines for diagnosis and management. *European Respiratory Journal*, *47*(1), 45–68.

Desai, J. V., Mitchell, A. P., & Andes, D. R. (2014). Fungal biofilms, drug resistance, and recurrent infection. *Cold Spring Harbor Perspectives in Medicine*, *4*(10), a019729.

Ellis, M., Al-Ramadi, B., Bernsen, R., Kristensen, J., Alizadeh, H., & Hedstrom, U. (2009). Prospective evaluation of mannan and anti-mannan antibodies for diagnosis of invasive *Candida* infections in patients with neutropenic fever. *Journal of Medical Microbiology*, *58*(5), 606–615.

Erum, R., Samad, F., Khan, A., & Kazmi, S. U. (2020). A comparative study on production of extracellular hydrolytic enzymes of Candida species isolated from patients with surgical site infection and from healthy individuals and their co-relation with antifungal drug resistance. *BMC Microbiology*, *20*(1), 368.

Fatsis-Kavalopoulos, N., Roemhild, R., Tang, P. C., Kreuger, J., & Andersson, D. I. (2020). CombiANT: Antibiotic interaction testing made easy. *PLOS Biology*, *18*(9), e3000856.

Fernandes, K. E., & Carter, D. A. (2020). Cellular plasticity of pathogenic fungi during infection. *PLOS Pathogens*, *16*(6), e1008571.

Fidel, P. L. Jr. (2011). Candida-host interactions in HIV disease: Implications for oropharyngeal candidiasis. *Advances in Dental Research*, *23*(1), 45–49.

Franco-Duarte, R., Černáková, L., Kadam, S., Kaushik, K. S., Salehi, B., Bevilacqua, A., & Rodrigues, C. F. (2019). Advances in chemical and biological methods to identify microorganisms—From past to present. *Microorganisms*, *7*(5), 130.

Fridkin, S. K., & Jarvis, W. R. (1996). Epidemiology of nosocomial fungal infections. *Clinical Microbiology Reviews*, *9*(4), 499–511.

Garbati, M. A., Alasmari, F. A., Al-Tannir, M. A., & Tleyjeh, I. M. (2012). The role of combination antifungal therapy in the treatment of invasive aspergillosis: A systematic review. *International Journal of Infectious Diseases*, *16*(2), e76–e81.

Gauthier, G. M. (2017). Fungal dimorphism and virulence: Molecular mechanisms for temperature adaptation, immune evasion, and in vivo survival. *Mediators of Inflammation*, 2017, 8491383.

Georgiev, V. S. (2000). Membrane transporters and antifungal drug resistance. *Current Drug Targets*, *1*(3), 261–284.

Ghannoum, M. A., & Rice, L. B. (1999). Antifungal agents: Mode of action, mechanisms of resistance, and correlation of these mechanisms with bacterial resistance. *Clinical Microbiology Reviews*, *12*(4), 501–517.

Gilbert, A. S., Wheeler, R. T., & May, R. C. (2014). Fungal pathogens: Survival and replication within macrophages. *Cold Spring Harbor Perspectives in Medicine*, *5*(7), a019661.

Gong, Y., Li, C., Wang, C., Li, J., Ding, M., Chen, D., & Lao, M. (2020). Epidemiology and mortality-associated factors of invasive fungal disease in elderly patients: A 20-year retrospective study from Southern China. *Infection and Drug Resistance*, *13*, 711–723.

Groll, A. H., Shah, P. M., Mentzel, C., Schneider, M., Just-Nuebling, G., & Huebner, K. (1996). Trends in the postmortem epidemiology of invasive fungal infections at a university hospital. *Journal of Infection*, *33*(1), 23–32.

Gupta, P., & Poluri, K. M. (2021). Elucidating the eradication mechanism of perillyl alcohol against *Candida glabrata* biofilms: Insights into the synergistic effect with azole drugs. *ACS Bio & Med Chem Au*, *2*(1), 60–72.

Hajjeh, R. A., Sofair, A. N., Harrison, L. H., Lyon, G. M., Arthington-Skaggs, B. A., Mirza, S. A., & Warnock, D. W. (2004). Incidence of bloodstream infections due to *Candida* species and in vitro susceptibilities of isolates collected from 1998 to 2000 in a population-based active surveillance program. *Journal of Clinical Microbiology*, *42*(4), 1519–1527.

Halliday, C. L., Bui, T., Krockenberger, M., Malik, R., Ellis, D. H., & Carter, D. A. (1999). Presence of α and a mating type in environmental and clinical collections of *Cryptococcus neoformans* var. Gattii strains from Australia. *Journal of Clinical Microbiology*, *37*(9), 2920–2926.

Haria, M., & Bryson, H. M. (1995). Amorolfine: A review of its pharmacological properties and therapeutic potential in the treatment of onychomycosis and other superficial fungal infections. *Drugs*, *49*, 103–120.

Hatinguais, R., Willment, J. A., & Brown, G. D. (2020). PAMPs of the fungal cell wall and mammalian PRRs. *Current Topics in Microbiology and Immunology*, *425*, 187–223.

Hernández-Chávez, M. J., Pérez-García, L. A., Niño-Vega, G. A., & Mora-Montes, H. M. (2017). Fungal strategies to evade the host immune recognition. *Journal of Fungi (Basel, Switzerland)*, *3*(4), 51.

Hoenigl, M., Jenks, J. D., Egger, M., Nucci, M., & Thompson, G. R. III (2023). Treatment of *fusarium* infection of the Central nervous system: A review of past cases to guide therapy for the ongoing 2023 outbreak in the United States and Mexico. *Mycopathologia*, 1–9.

Ibrahim, A. S., Spellberg, B., Walsh, T. J., & Kontoyiannis, D. P. (2012). Pathogenesis of mucormycosis. *Clinical Infectious Diseases*, *54*(suppl_1), S16–S22.

Iyalla, C. (2017). A review of the virulence factors of pathogenic fungi. *African Journal of Clinical and Experimental Microbiology*, *18*(1), 53–58.

Jacobsen, I. D. (2023). The role of host and fungal factors in the commensal-to-pathogen transition of *candida albicans*. *Current Clinical Microbiology Reports*, *10*, 55–65.

Jazdarehee, A., Malekafzali, L., Lee, J., Lewis, R., & Mukovozov, I. (2022). Transmission of onychomycosis and dermatophytosis between household members: A scoping review. *Journal of Fungi*, *8*(1), 60.

Jenks, J. D., Aneke, C. I., Al-Obaidi, M. M., Egger, M., Garcia, L., Gaines, T., Hoenigl, M., & Thompson, G. R. 3rd (2023). Race and ethnicity: Risk factors for fungal infections. *PLOS Pathogens*, *19*(1), e1011025.

Jenks, J. D., Gangneux, J. P., Schwartz, I. S., Alastruey-Izquierdo, A., Lagrou, K., Thompson, G. R. III, & European Confederation of Medical Mycology (ECMM) Council Investigators (2020). Diagnosis of breakthrough fungal infections in the clinical mycology laboratory: An ECMM consensus statement. *Journal of Fungi*, *6*(4), 216.

Jeong, W., Keighley, C., Wolfe, R., Lee, W. L., Slavin, M. A., Kong, D. C. M., & Chen, S. A. (2019). The epidemiology and clinical manifestations of mucormycosis: A systematic review and meta-analysis of case reports. *Clinical Microbiology and Infection*, *25*(1), 26–34.

Kanaujia, R., Singh, S., & Rudramurthy, S. M. (2023). Aspergillosis: An update on clinical spectrum, diagnostic schemes, and management. *Current Fungal Infection Reports*, 1–12.

Kangabam, N., & Nethravathy, V. (2023). An overview of opportunistic fungal infections associated with COVID-19. *3 Biotech*, *13*(7), 231.

Khawcharoenporn, T., Apisarnthanarak, A., & Mundy, L. M. (2007). Treatment of cryptococcosis in the setting of HIV coinfection. *Expert Review of Anti-Infective Therapy*, *5*(6), 1019–1030.

Kumari, A., Tripathi, A. H., Gautam, P., Gahtori, R., Pande, A., Singh, Y., Madan, T., & Upadhyay, S. K. (2021). Adhesins in the virulence of opportunistic fungal pathogens of human. *Mycology*, *12*(4), 296–324.

Kurita, H., Kamata, T., Zhao, C., Narikawa, J. N., Koike, T., & Kurashina, K. (2009). Usefulness of a commercial enzyme-linked immunosorbent assay kit for *Candida* mannan antigen for detecting *Candida* in oral rinse solutions. *Oral Surgery, Oral Medicine, Oral Pathology, Oral Radiology, and Endodontology*, *107*(4), 531–534.

Langfeldt, A., Gold, J. A., & Chiller, T. (2022). Emerging fungal infections: From the fields to the clinic, resistant *Aspergillus fumigatus* and dermatophyte species: A one health perspective on an urgent public health problem. *Current Clinical Microbiology Reports*, *9*(4), 46–51.

Lao, M., Li, C., Li, J., Chen, D., Ding, M., & Gong, Y. (2020). Opportunistic invasive fungal disease in patients with type 2 diabetes mellitus from Southern China: Clinical features and associated factors. *Journal of Diabetes Investigation*, *11*(3), 731–744.

Latgé, J. P., & Chamilos, G. (2019). *Aspergillus fumigatus* and Aspergillosis in 2019. *Clinical Microbiology Reviews*, *33*(1), 10–1128.

Lemke, A., Kiderlen, A. F., & Kayser, O. (2005). Amphotericin B. *Applied Microbiology and Biotechnology*, *68*, 151–162.

Li, Z., & Nielsen, K. (2017). Morphology changes in human fungal pathogens upon interaction with the host. *Journal of Fungi (Basel, Switzerland)*, *3*(4), 66.

Lien, M. Y., Chou, C. H., Lin, C. C., Bai, L. Y., Chiu, C. F., Yeh, S. P., & Ho, M. W. (2018). Epidemiology and risk factors for invasive fungal infections during induction chemotherapy for newly diagnosed acute myeloid leukemia: A retrospective cohort study. *PLOS One*, *13*(6), e0197851.

Lima, B., Sortino, M., Tapia, A., & Feresin, G. E. (2022). Synergistic antifungal effectiveness of essential oils from Andean plants combined with commercial drugs. *International Journal of Pharmaceutical Sciences and Developmental Research*, *8*(1), 023–031.

Li, X., Wu, X., Gao, Y., & Hao, L. (2019). Synergistic effects and mechanisms of combined treatment with harmine hydrochloride and azoles for resistant *Candida albicans*. *Frontiers in Microbiology*, *10*, 2295.

Lionakis, M. S., Drummond, R. A., & Hohl, T. M. (2023). Immune responses to human fungal pathogens and therapeutic prospects. *Nature Reviews. Immunology*, *23*(7), 433–452.

Liu, Y., Su, L., Jiang, S. J., & Qu, H. (2017). Risk factors for mortality from *Pneumocystis carinii pneumonia* (PCP) in non-HIV patients: A meta-analysis. *Oncotarget*, *8*(35), 59729.

Ma, L., Chen, Z., Huang, daW, Kutty, G., Ishihara, M., Wang, H., Abouelleil, A., Bishop, L., Davey, E., Deng, R., Deng, X., Fan, L., Fantoni, G., Fitzgerald, M., Gogineni, E., Goldberg, J. M., Handley, G., Hu, X., Huber, C., Jiao, X., Kovacs, J. A. (2016). Genome analysis of three *Pneumocystis* species reveals adaptation mechanisms to life exclusively in mammalian hosts. *Nature Communications* 7, 10740.

Maertens, J., Vrebos, M., & Boogaerts, M. (2001). Assessing risk factors for systemic fungal infections. *European Journal of Cancer Care*, *10*(1), 56–62.

Mahalingam, S. S., Jayaraman, S., & Pandiyan, P. (2022). Fungal colonization and infections—Interactions with other human diseases. *Pathogens*, *11*(2), 212.

Marcos, C. M., de Oliveira, H. C., de Melo, W. C., da Silva, J. F., Assato, P. A., Scorzoni, L., Rossi, S. A., de Paula E Silva, A. C., Mendes-Giannini, M. J., & Fusco-Almeida, A. M. (2016). Anti-immune strategies of pathogenic fungi. *Front Cell Infect Microbiol*, *6*, 142.

Mennink-Kersten, M. A., & Verweij, P. E. (2006). Non–culture-based diagnostics for opportunistic fungi. *Infectious Disease Clinics*, *20*(3), 711–727.

Merad, Y., Derrar, H., Belmokhtar, Z., & Belkacemi, M. (2021). *Aspergillus* genus and its various human superficial and cutaneous features. *Pathogens*, *10*(6), 643.

Mercer, E. I. (1991). Morpholine antifungals and their mode of action. *Biochemical Society Transactions*, *19*(3), 788–793.

Mora, C., Tittensor, D. P., Adl, S., Simpson, A. G., & Worm, B. (2011). How many species are there on Earth and in the ocean? *PLOS Biology*, *9*(8), e1001127.

Muhammed, M., Carneiro, H., Coleman, J., & Mylonakis, E. (2011). The challenge of managing fusariosis. *Virulence*, *2*(2), 91–96.

Mukjang, N. (2022). Assessing the factors affecting dispersal and colonisation of fungal plant pathogens across multiple ecological scales. (Doctoral dissertation, Imperial College London).

Naik, B., Ahmed, S. M. Q., Laha, S., & Das, S. P. (2021). Genetic susceptibility to fungal infections and links to human ancestry. *Frontiers in Genetics*, *12*, 709315. https://doi.org/10.3389/fgene.2021.709315

Namnuch, N., Thammasittirong, A., & Thammasittirong, S. N. (2020). Lignocellulose hydrolytic enzymes production by *Aspergillus flavus* KUB2 using submerged fermentation of sugarcane bagasse waste. *Mycology*, *12*(2), 119–127.

Nucci, M., & Anaissie, E. (2007). *Fusarium* infections in immunocompromised patients. *Clinical Microbiology Reviews*, *20*(4), 695–704.

Nucci, M., Barreiros, G., Akiti, T., Anaissie, E., & Nouér, S. A. (2021). Invasive fusariosis in patients with hematologic diseases. *Journal of Fungi*, *7*(10), 815.

Ostrosky-Zeichner, L., Alexander, B. D., Kett, D. H., Vazquez, J., Pappas, P. G., Saeki, F., … Rex, J. H. (2005). Multicenter clinical evaluation of the $(1\rightarrow 3)$ β-D-glucan assay as an aid to diagnosis of fungal infections in humans. *Clinical Infectious Diseases*, *41*(5), 654–659.

Paes De Oliveira-Neto, M., Da Silva, M., Cezar Fialho Monteiro, P., Lazera, M., De Almeida Paes, R., Beatriz Novellino, A., & Cuzzi, T. (2006). Cutaneous mucormycosis in a young, immunocompetent girl. *Medical Mycology*, *44*(6), 567–570.

Pappas, P. G., Lionakis, M. S., Arendrup, M. C., Ostrosky-Zeichner, L., & Kullberg, B. J. (2018). Invasive candidiasis. *Nature Reviews Disease Primers*, *4*(1), 1–20.

Patil, K., Bhale, G., & Kale, M. (2023). Dermatophytoses' immunopathogenesis and risk factors for persistent infections. *Journal of Coastal Life Medicine*, *11*, 61–65.

Pemmaraju, S. C., Pruthi, P. A., Prasad, R., & Pruthi, V. (2013). Candida albicans biofilm inhibition by synergistic action of terpenes and fluconazole. *Indian Journal of Experimental Biology*, *51*(11), 1032–1037.

Perlin, D. S. (2011). Current perspectives on echinocandin class drugs. *Future Microbiology*, *6*(4), 441–457.

Petrikkos, G., Skiada, A., Lortholary, O., Roilides, E., Walsh, T. J., & Kontoyiannis, D. P. (2012). Epidemiology and clinical manifestations of mucormycosis. *Clinical Infectious Diseases*, *54*(suppl_1), S23–S34.

Polak, A. (1990). Mode of action studies. In *Chemotherapy of fungal diseases* (pp. 153–182). Berlin, Heidelberg: Springer Berlin Heidelberg.

Queiroz-Telles, F., Fahal, A. H., Falci, D. R., Caceres, D. H., Chiller, T., & Pasqualotto, A. C. (2017). Neglected endemic mycoses. *The Lancet Infectious Diseases*, *17*(11), e367–e377.

Reddy, G. K. K., Padmavathi, A. R., & Nancharaiah, Y. V. (2022). Fungal infections: Pathogenesis, antifungals, and alternate treatment approaches. *Current Research in Microbial Sciences*, *3*, 100137.

Roden, M. M., Zaoutis, T. E., Buchanan, W. L., Knudsen, T. A., Sarkisova, T. A., Schaufele, R. L., & Walsh, T. J. (2005). Epidemiology and outcome of zygomycosis: A review of 929 reported cases. *Clinical Infectious Diseases*, *41*(5), 634–653.

Rudramurthy, S. M., Singh, G., Hallur, V., Verma, S., & Chakrabarti, A. (2016). High fungal spore burden with predominance of *Aspergillus* in hospital air of a tertiary care hospital in Chandigarh. *Indian Journal of Medical Microbiology*, *34*(4), 529–532.

Sanglard, D., & Coste, A. T. (2016). Activity of isavuconazole and other azoles against Candida clinical isolates and yeast model systems with known azole resistance mechanisms. *Antimicrobial Agents and Chemotherapy*, *60*(1), 229–238.

Sanglard, D., Coste, A., & Ferrari, S. (2009). Antifungal drug resistance mechanisms in fungal pathogens from the perspective of transcriptional gene regulation. *FEMS Yeast Research*, *9*(7), 1029–1050.

Sav, H., Altinbas, R., & Bestepe Dursun, Z. (2020). Fungal profile and antifungal susceptibility pattern in patients with oral candidiasis. *Le Infezioni in Medicina*, *28*, 392–396.

Shapiro, R. S., Robbins, N., & Cowen, L. E. (2011). Regulatory circuitry governing fungal development, drug resistance, and disease. *Microbiology and Molecular Biology Reviews*, *75*(2), 213–267.

Spitzer, M., Robbins, N., & Wright, G. D. (2017). Combinatorial strategies for combating invasive fungal infections. *Virulence*, *8*(2), 169–185.

Staniszewska, M. (2020). Virulence factors in *Candida* species. *Current Protein & Peptide Science*, *21*(3), 313–323.

Steinbrink, J. M., & Miceli, M. H. (2021). Mucormycosis. *Infectious Disease Clinics*, *35*(2), 435–452.

Szymański, M., Chmielewska, S., Czyżewska, U., Malinowska, M., & Tylicki, A. (2022). Echinocandins - structure, mechanism of action and use in antifungal therapy. *Journal of Enzyme Inhibition and Medicinal Chemistry*, *37*(1), 876–894.

Talapko, J., Juzbašić, M., Matijević, T., Pustijanac, E., Bekić, S., Kotris, I., & Škrlec, I. (2021). *Candida albicans*—The virulence factors and clinical manifestations of infection. *Journal of Fungi*, *7*(2), 79.

Tasaka, S. (2020). Recent advances in the diagnosis and management of *Pneumocystis pneumonia*. *Tuberculosis and Respiratory Diseases*, *83*(2), 132.

Teoh, F., & Pavelka, N. (2016). How chemotherapy increases the risk of systemic candidiasis in cancer patients: Current paradigm and future directions. *Pathogens*, *5*(1), 6.

Thompson, G. R. III, & Young, J. A. H. (2021). *Aspergillus* infections. *New England Journal of Medicine*, *385*(16), 1496–1509.

Tong, Y., & Tang, J. (2017). *Candida albicans* infection and intestinal immunity. *Microbiological Research*, *198*, 27–35.

van Rhijn, N., & Bromley, M. (2021). The consequences of our changing environment on life threatening and debilitating fungal diseases in humans. *Journal of Fungi*, *7*(5), 367.

Vandeputte, P., Ferrari, S., & Coste, A. T. (2012). Antifungal resistance and new strategies to control fungal infections. *International Journal of Microbiology*, 2012, 713687.

Viscoli, C., Machetti, M., Gazzola, P., De Maria, A., Paola, D., Van Lint, M. T., ... Bacigalupo, A. (2002). *Aspergillus* galactomannan antigen in the cerebrospinal fluid of bone marrow transplant recipients with probable cerebral aspergillosis. *Journal of Clinical Microbiology*, *40*(4), 1496–1499.

Wan Mohtar, W. H. M., Wan-Mohtar, W. A. A. Q. I., Zahuri, A. A., Ibrahim, M. F., Show, P. L., Ilham, Z., Jamaludin, A. A., Abdul Patah, M. F., Ahmad Usuldin, S. R., & Rowan, N. (2022). Role of ascomycete and basidiomycete fungi in meeting established and emerging sustainability opportunities: A review. *Bioengineered*, *13*(7–12), 14903–14935.

Warris, A., & Ballou, E. R. (2019). Oxidative responses and fungal infection biology. *Seminars in Cell & Developmental Biology*, *89*, 34–46.

Xiao, W., Zhang, J., Huang, J., Xin, C., Li, M. J., & Song, Z. (2022). Response and regulatory mechanisms of heat resistance in pathogenic fungi. *Applied Microbiology and Biotechnology*, *106*(17), 5415–5431.

Xue, T., Kong, X., & Ma, L. (2023). Trends in the epidemiology of *Pneumocystis Pneumonia* in immunocompromised patients without HIV infection. *Journal of Fungi*, *9*(8), 812.

Zaragoza, O. (2019). Basic principles of the virulence of *Cryptococcus*. *Virulence*, *10*(1), 490–501.

5 Allergic Fungal Infections

*Saiema Ahmedi and Nikhat Manzoor**

5.1 INTRODUCTION

Fungi are eukaryotic microorganisms found inhabiting different types of habitats varying from soil, water, and air to plants, animals, and humans. They are omnipresent and hence are encountered frequently. Humans use both innate and cell-mediated immunity as primary defense mechanisms against fungal allergies and associated infections (Lord and Vyas, 2019; Rizzetto et al., 2014). Most opportunistic fungal infections tend to affect individuals with compromised innate and/or adaptive cellular immunity (Wheeler et al., 2017). Fungi exhibit growth across a broad temperature range, typically between 25–50°C and sometimes even higher (Sussman and Airworth, 2013). Moisture levels represent one of the foremost physical factors that significantly influence fungal growth (Choidis et al., 2021). Although all fungi are not allergenic, some species produce proteins and substances that can trigger immune responses in susceptible individuals, leading to a range of allergic reactions (Platts-Mills and Woodfolk, 2011; Simon-Nobbe et al., 2007). Fungal allergens are specific components produced by fungi that elicit allergic response in humans. These allergens can be found in various fungal components, including spores, hyphae, and hyphal fragments (Crameri et al., 2014). Unlike other commonly allergenic sources like dust mites and pollen, fungi are universally abundant in the environment, and hence there is a constant threat of exposure to airborne spores. Another important difference is that fungi can form colonies in different parts of the human host, causing potential damage by releasing mycotoxins and hydrolytic enzymes (Nevalainen et al., 2015). The major fungal classes, including Ascomycetes, Basidiomycetes, and Zygomycetes are known to produce allergens and trigger an allergic response (Amado and Barnes, 2016). Mold and some yeast species have a substantially more pronounced effect on the immune system of individuals in contrast to pollen or other allergens (Rudert and Portnoy, 2017; Twaroch et al., 2015). Molds are commonly found both indoors and outdoors, their allergenic potential varying with different species. Common fungi associated with allergenicity include *Aspergillus, Penicillium, Alternaria, Cladosporium* and *Candida* species (Singh and Mathur, 2021; Twaroch et al., 2015). These allergens are predominantly proteins or glycoproteins, recognized by the immune system as foreign invaders (Portnoy et al., 2016). Fungal spores are increasingly acknowledged as significant contributors to respiratory allergies (Ahmad et al., 2022). Due to variable sizes, these spores are associated with both upper and lower respiratory symptoms. Allergic reactions typically manifest at specific locations where the allergens are deposited. In case of inhaled particles larger than 10 mm, the deposition in the nasopharynx is commonly linked to symptoms affecting the nose and eyes, often referred to as 'hay fever' or allergic rhinitis (Blanco and Garcia, 2008). On the other hand, smaller particles, especially those less than 5 mm, can penetrate deeper into the lower airways, where allergic reactions frequently present as asthma (Blanco and Garcia, 2008; Yamamoto et al., 2012). Fungal allergens can be grouped into different classes based on their source, structure, and clinical relevance (Table 5.1). This classification can help researchers and healthcare professionals to understand and manage fungal allergies better, as well as develop targeted diagnostic and treatment approaches that are better and more effective (Levetin et al., 2016). Fungal allergens form a diverse group, and ongoing research will continue to expand our knowledge regarding of their classification and clinical significance.

---

* Corresponding Author: nmanzoor@jmi.ac.in

DOI: 10.1201/9781032642864-5

TABLE 5.1
Classification of Fungal Allergens

S. No.		Fungal Allergens	References
1.	Source-based	• *Mold*: Allergens derived from various molds such as *Aspergillus*, Penicillium, Alternaria, and Cladosporium • *Yeast*: Allergenic proteins derived from *Candida*	Levetin et al., 2016; Vijay and Kurup, 2008
2.	Structural-based	• *Proteins*: Many fungal allergens are proteins categorized based on specific structural properties like glycoproteins • *Polysaccharides*: Some fungal allergens are carbohydrates like β-glucans.	Bowyer et al., 2006; Kurup et al., 2002
3.	Clinical relevance-based	• *Major Allergens*: Commonly associated with severe allergic reactions and are well-characterized, e.g. *Aspergillus fumigatus* allergen Asp f 1 • *Minor Allergens*: Less common and may trigger milder allergic responses in individuals	Crameri et al., 2006; Kurup et al., 2002
4.	Taxonomy-based	• *Ascomycetes*: Allergens derived from the phylum Ascomycota, e.g. *Aspergillus* species • *Basidiomycetes*: Allergens derived from the phylum Basidiomycota, e.g. Agaricus • *Zygomycetes*: Allergens from the phylum Zygomycota, e.g. *Rhizopus* and *Mucor* species	Levetin et al., 2016
5.	Geographic-based	• Some fungal allergens are more prevalent in specific geographic regions due to environmental factors and fungal species distribution	Kwong et al., 2023; Yamamoto et al., 2012

5.2   CLASSIFICATION OF MAJOR ALLERGENIC FUNGI

There is a range of fungal species that are associated with allergic reactions and hypersensitivity responses in individuals. This can trigger allergic responses primarily affecting the respiratory system and sometimes the skin surface (Crameri et al., 2006; Kurup et al., 2002; Kwong et al., 2023). Table 5.2 summarizes the major allergenic fungi and symptoms observed after the allergic response.

TABLE 5.2
Major Allergenic Fungi and Symptoms Observed after Allergic Response

S. No.	Species	Symptoms	References
1.	**Aspergillus species** *Aspergillus fumigatus*, the most significant fungal allergen, causes a spectrum of allergic diseases, e.g. ABPA, *Aspergillus* Sinusitis	Coughing, wheezing, lung inflammation, chronic sinusitis, nasal congestion, and sinus headaches	Hedayati et al., 2007

(Continued)

TABLE 5.2 *(Continued)*

Major Allergenic Fungi and Symptoms Observed after Allergic Response

S. No.	Species	Symptoms	References
2.	***Penicillium* species** Exposure to *Penicillium chrysogenum* allergens is associated with conditions like allergic rhinitis and asthma.	Coughing, nasal congestion, and lung inflammation	Reboux et al., 2019; Levetin et al., 2016
3.	***Alternaria alternata*** Most common outdoor mold that causes allergic rhinitis, asthma and hay fever (allergic rhinitis).	Wheezing, and lung inflammation	Gabriel et al., 2016; Abel-Fernández et al., 2023
4.	***Cladosporium* species** Allergic rhinitis and asthma are common allergic manifestations.	Coughing, wheezing	Ogórek et al., 2012; Levetin et al., 2016
5.	***Fusarium* species** Found in soil, plants, and indoor environments. Exposure to *Fusarium* allergens may lead to symptoms like allergic rhinitis and asthma.	Wheezing, and lung inflammation	Chou et al., 2014; Levetin et al., 2016
6.	***Candida* species** *Candida albicans* is majorly associated with allergic responses. contribute to conditions like oral thrush and fungal dermatitis.	Oral thrush, dermatitis and inflammation	Fukutomi and Taniguchi, 2015
7.	**Mucorales (Zygomycetes)** Known to cause allergic reactions and skin infections, e.g. *Rhizopus* and *Mucor*.	Dermatitis and inflammation	Levetin et al., 2016
8.	***Agaricus bisporus*** Exposure to allergenic proteins produced by this mushroom causes allergic rhinitis and skin allergies.	Coughing, nasal congestion, and dermatitis	Singh et al., 2023; Wang et al., 2020

5.3 SENSITIZATION AND ANTIBODY REACTIVITY TOWARD FUNGAL ALLERGENS

Sensitization is the initial step in the development of fungal allergies. It requires the immune system to recognize fungal antigens as potentially harmful and initiate an immune response for their elimination. Sensitization can occur through various routes of exposure, including inhalation of fungal spores (conidia) or interaction with fungal elements present in the environment (Fernández-Soto et al., 2018; Mari et al., 2003). Since fungal spores are airborne, they can be easily inhaled and the immune system can then detect them as foreign invaders, leading to sensitization (Lokhande, 2018). Individuals can also become sensitized by direct contact with fungal elements present in the environment during handling of contaminated materials like compost or water-damaged surfaces (Viegas et al., 2015). Several key immunological mechanisms are involved during fungal sensitization. The activation of antigen-presenting cells, primarily dendritic cells play a crucial role in the capture of fungal antigens and their presentation to the T cells. T cells (namely, CD4+ cells) become activated upon recognition of the fungal antigens, triggering an immune response subsequently (Pathakumari et al., 2020; Wüthrich et al., 2012). Sensitization results in the production of allergen-specific IgE antibodies, which recognize allergens and bind to these fungal proteins (Figure 5.1) (Bartemes and Kita, 2018; Caraballo et al., 2020; Soeria-Atmadja et al., 2010).

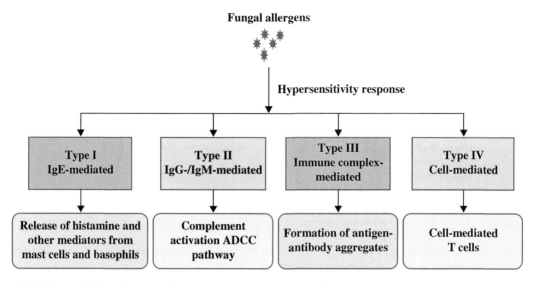

FIGURE 5.1   Various hypersensitivity responses against fungal allergens.

5.4   CLINICAL MANIFESTATION OF FUNGAL ALLERGIC DISEASES

5.4.1   Allergic Bronchopulmonary Aspergillosis (ABPA)

5.4.1.1   Introduction

Aspergillus species cause a broad spectrum of illnesses. The individuals most affected are those with underlying conditions such as asthma and cystic fibrosis, and experience a range of allergic respiratory symptoms (Agarwal, 2009; Greenberger et al., 2014). ABPA is a complex and potentially severe respiratory disorder that arises due to an exaggerated and inappropriate immune response against *Aspergillus fumigatus* (Agarwal et al., 2013). This condition is characterized by a range of symptoms and pulmonary abnormalities that make its diagnosis and effective management very challenging. *Aspergillus* spores are widely distributed throughout the environment, and hence it is nearly impossible to avoid their inhalation (Greenberger et al., 2014). *A. fumigatus* is an airborne filamentous mold that majorly thrives in the soil. In healthy individuals, the inhaled *Aspergillus* conidia are efficiently removed from the air passages by the host immune system without causing any serious health issues. However, in immunocompromised patients, *A. fumigatus* expresses several virulence factors and immune-evasion mechanisms that contribute to pathogenesis, both in the form of allergies and invasive infections (Patterson and Strek, 2010). As the spores of *A. fumigatus* measure only between 3–5 μm in size, they have the capacity to easily settle within the lower bronchial airways, particularly in individuals who are susceptible or immunocompromised. Patients of cystic fibrosis and asthma exhibit anomalies in their air passages and mucosal defense mechanisms (Muthu et al., 2020; Shah and Panjabi, 2016). This leads to impairment in mucociliary clearance and epithelial cell function (Moss, 2005). Exposure to increased number of *Aspergillus* conidia has been linked to instances of allergic bronchopulmonary aspergillosis (Patterson and Strek, 2010). A systematic review and meta-analysis, involving 21 studies with asthma patients, reported incidence of *Aspergillus* sensitization in 28% individuals with a prevalence of 12.9% ABPA (Muthu et al., 2020; Tracy et al., 2016). A similar meta-analysis was done with 64 individuals having cystic fibrosis. The prevalence of *Aspergillus* sensitization was 39.1%, while that of ABPA it was 8.9%. This study also observed that adults had a slightly higher prevalence of ABPA (10.1%) in comparison to children (8.9%) (Steels et al., 2023; Tracy et al., 2016).

5.4.1.2 Immune Responses Associated with ABPA

A. fumigatus and other *Aspergillus* species produce conidia, the cell wall of which has an outer layer made up of proteins and carbohydrates (Mouyna and Fontaine, 2008). The cell wall components are hydrophobic and typically do not trigger an inflammatory response in the host (Greenberger et al., 2014). However, in individuals who are susceptible, these conidia swell and germinate to give rise to elongated hyphae which then lead to a pronounced inflammatory host response (Agarwal et al., 2013; Greenberger et al., 2014). The innate immune cells recognize newly exposed pathogen-associated molecular patterns (PAMPs). The pattern recognition receptors (PRRs) found on both the epithelial cells and the specialized antigen-presenting cells, such as dendritic cells, play a crucial role in the recognition of PAMPs by innate immune cells. In case of invasive aspergillosis, PRRs include C-type lectin receptors (e.g. dectin-1), Toll-like receptors (especially TLR2 and TLR4), and nucleotide-binding oligomerization domain-like receptors (Agarwal et al., 2013; Muthu et al., 2020; Shah and Panjabi, 2016). When activated, these PRRs stimulate antigen-presenting cells, primarily dendritic cells, to release chemokines and cytokines, ultimately leading to adaptive immune T-helper cell responses (Agarwal et al., 2013; Patterson and Strek, 2010). Activation of T-helper 1 (Th1) immune response is linked to an effective pro-inflammatory reaction characterized by the phagocytosis and elimination of conidia by macrophages and neutrophils. Activated Th2 cells release cytokines, namely IL-4, IL-5, and IL-13, which induce the activation of eosinophils and the differentiation of B cells. This process results in the production of IgE antibodies after exposure to the allergens, initiating the degranulation of mast cells and basophils. This further results in the production of mucus in the air passages, hypersensitivity response, inflammation, and bronchiectasis (Patterson and Strek, 2010; Shah and Panjabi, 2016; Wark and Gibson, 2010).

5.4.1.3 Clinical Manifestations of ABPA

ABPA patients experience repeated wheezing, pulmonary infiltration, elevated eosinophil levels in their blood and sputum, and the presence of brown plugs or specks in coughed-up mucus (Agarwal et al., 2011; Zhang et al., 2020). Other major symptoms include frequent coughing, wheezing, breathlessness, fever, and loss of appetite (Agarwal et al., 2011). Immunologically, ABPA involves the presence of *Aspergillus*-specific antibodies, both IgE and IgG, can be observed in the local and systemic circulation. High-resolution CT chest scans of ABPA patients typically reveal central bronchiectasis, with a notable prevalence in the upper lobes and thickening of the bronchial walls which ultimately damages the lung airways (Figure 5.2) (Muthu et al., 2020; Shah and Panjabi, 2016; Zhang et al., 2020).

5.4.1.4 Treatment of ABPA

The primary objective of ABPA treatment is to alleviate the symptoms associated with asthma or cystic fibrosis, mitigating pulmonary inflammation, and effective management to prevent the advancement of end-stage pulmonary fibrosis (Muthu et al., 2020; Wark and Gibson, 2010). The use of intravenous pulse glucocorticosteroids, along with oral or inhaled antifungal agents and anti-IgE therapy, have been demonstrated as safe and effective supplementary approaches for the treatment for ABPA (Muthu et al., 2020; O'Reilly and Dunican, 2021). The assessment of treatment response in ABPA requires a comprehensive evaluation across various domains and different time periods. Studies involve clinical aspects, immunological factors (total IgE levels), physiological parameters (spirometry), and structural assessments (chest CT scans or radiographs) (Zhang et al., 2020). For current and developing therapies, well-designed and controlled trials are needed for the ABPA patient population (Table 5.3).

Aspergillus fumigatus spore

Hypersensitivity reactions

Immunopathogenesis of ABPA

- **Aspergillus-specific immunoglobulin (Ig)-E–mediated hypersensitivity**

- **IgG-mediated immune complex hypersensitivity**

- **Cell-mediated immune response and innate immunity**

Hypersensitivity responses

Bronchiectasis

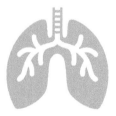

Lung damage

FIGURE 5.2   Overview of allergic bronchopulmonary aspergillosis (ABPA).

TABLE 5.3

Current Approaches for the Treatment of ABPA

S. No.	Treatment for ABPA	Mechanism	Side Effects	References
1.	Glucocorticosteroids (Oral or intravenous)	Lower hypersensitivity response and mucous deposition	Prolonged use of steroids causes hyperglycemia, Cushing's disease, weight gain, and osteoporosis	Patel et al., 2019; Wark and Gibson, 2010
2.	*Antifungal drugs*: Amphotericin B and azoles	Interfere with fungal cell membrane integrity	Fever, chills, anaphylaxis, cardiac arrhythmia, stomach upset, liver toxicity	Gilley et al., 2010; Patel et al., 2019
3.	*Immunotherapy*: Monoclonal antibody	Omalizumab binds to free serum IgE, interfering with IgE binding to its high-affinity receptor on mast cells and basophils	Headache, joint pain, nausea	O'Reilly and Dunican, 2021

5.4.2 Allergic Fungal Rhinosinusitis (AFRS)

5.4.2.1 Introduction

AFRS is a condition that involves a complex interplay between allergic responses and fungal sinus infections (Dykewicz et al., 2018). It is a distinct form of chronic rhinosinusitis characterized by unique clinical, immunological, and pathological features that occur in areas characterized by high humidity, where the number of mold colonies are elevated compared to drier areas (Glass and Amedee, 2011). AFRS is characterized by type 1 hypersensitivity response to fungal allergens. Significantly elevated serum IgE levels, accumulation of dense mucin containing eosinophils, non-invasive fungal hyphae in the paranasal sinuses, presence of nasal polyps, and structural alterations in the sinus bones are some other clinical features that characterize AFRS (Glass and Amedee, 2011; Hoyt et al., 2016). The condition was initially reported in a patient who already had ABPA, and was producing similar mucus within the paranasal sinuses (Tyler and Luong, 2018). Although AFRS and ABPA exhibit several clinical and pathological similarities, alterations in anatomical features of the paranasal sinuses and lungs can help in the application of different treatment strategies. AFRS initiates with the inhalation of fungal spores, which enter the nasal passages during normal respiration (Knutsen and Kariuki, 2019). These spores serve as the initial trigger for the immune response.

5.4.2.2 Immune Responses Associated with AFRS

The development of AFRS is an outcome of multiple factors, including local environmental, anatomical, and host immunological factors (Tyler and Luong, 2018). Fungi enter the nasal and sinus passages during regular respiration, inducing an inflammatory response. The overall fungal dose during exposure may explain the uneven geographical distribution of the disease. Although the potential inflammatory mechanisms in AFRS remain unclear, the conventional explanation suggests the involvement of the activation of Gell and Coombs types I and III hypersensitivity reactions by the fungal antigens (Glass and Amedee, 2011; Saravanan et al., 2006). This activation leads to the release of Th2-type cytokines and subsequent eosinophilic inflammation which then impairs the intrinsic host defense mechanisms like mucociliary clearance (Laury and Wise, 2013). Stagnant mucus and obstruction of the sinus tracts due to the formation of polyps, contribute to fungal proliferation and accumulation of eosinophils within the trapped mucus (Hutcheson et al., 2010). This entrapped mucin containing the fungal elements serves as a persistent stimulus for ongoing inflammation (Dutre et al., 2013; Houser and Corey, 2000). Numerous studies have reported elevated levels of antibodies, specifically IgE and IgG. This provides support for the hypothesis that Gell and Coombs types I and III hypersensitivity reactions play a significant role in the pathophysiology of the disease (Schubert, 2004). The precise immunological dysfunction accountable for AFRS is currently the subject of investigation. Recent research has shown a growing interest in dendritic cells due to their substantial role in allergic rhinitis. They serve as the principal antigen-presenting cells engaged in interaction with T cells, ultimately influencing the proliferation of T helper cell subtypes, either Th1 or Th2 (Pant et al., 2009; Tyler and Luong, 2020).

The T cells, specifically CD4+ cells, are activated and differentiate into Th1 or Th2 subtypes depending on the nature of the antigenic stimulus. Specific cytokines are released depending on the activated T cell subtypes (Bartemes and Kita, 2018; Plonk and Luong, 2014). Th1-driven cytokines (IFN-γ, IL-2, and TNF-α) are associated with pro-inflammatory and cell-mediated immune responses (Chakrabarti and Kaur, 2016). On the other hand, Th2-type responses lead to the production of cytokines like IL-4, IL-5, and IL-13 (Plonk and Luong, 2014). Th2-driven cytokines stimulate the recruitment and activation of eosinophils, which are hallmark cells in AFRS. Eosinophils contribute to the eosinophilic inflammation observed in the sinuses (Figure 5.3) (Bartemes and Kita, 2018; Plonk and Luong, 2014).

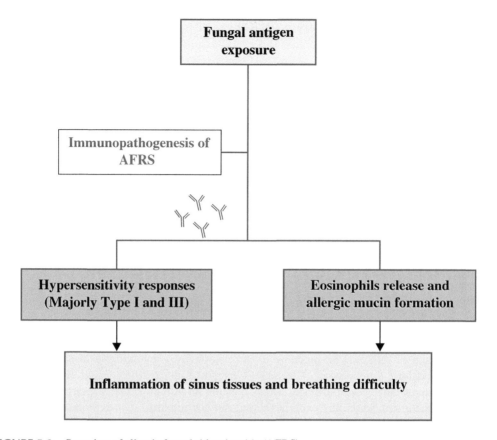

FIGURE 5.3   Overview of allergic fungal rhinosinusitis (AFRS).

5.4.2.3   Clinical Manifestations of AFRS

Eosinophilic inflammation and allergic responses lead to typical AFRS symptoms like nasal congestion, discharge, loss of smell, and cough. Asthma patients affected with AFRS simultaneously show aggravated symptoms (Glass and Amedee, 2011; Marple, 2006). An essential diagnostic criteria for AFRS is the presence of "eosinophilic mucin" characterized by thick, tenacious, and often dark yellow-brown or green-colored "allergic mucin," filling the affected sinuses (Oliveira et al., 2023; Salamah, 2020). The presence of this distinctive mucin in AFRS, reflects eosinophilic inflammation and characteristic fungal elements in upper respiratory tract (Oliveira et al., 2023; Plonk and Luong, 2014).

5.4.2.4   Treatment of AFRS

Understanding immune dysregulation in AFRS is crucial for the development of effective treatment strategies. Medical therapy for AFRS focuses on several key parameters, including suppression of inflammation, prevention of re-accumulation of allergic mucin, and maintenance of effective sinus drainage (Oliveira et al., 2023; Salamah, 2020). These therapeutic strategies are essential for managing the condition and alleviating symptoms in individuals with AFRS (Table 5.4).

   The pathogenesis of ABPA and AFRS exhibits several common features, such that AFRS can be regarded as the upper airway counterpart of ABPA (Barac et al., 2018; Plonk and Luong, 2014). The diagnosis of ABPA relies on a combination of clinical, radiological, and immunological findings. Table 5.5 shows some distinguishing clinical features of ABPA and AFRS.

TABLE 5.4
Current Approaches for the Treatment of AFRS

S. No.	Treatment Options	Mechanism	Side Effects	References
1.	Corticosteroid oral or nasal spray, e.g. Prednisone	Reduces nasal polyps and drains accumulated mucous	Excessive use causes vomiting, eye-related problems	Cameron and Luong, 2023; Marglani, 2014
2.	Antihistamines	Prevents allergic reactions and inflammation	Headache, nausea, watery eyes	Ada et al., 2023; Marglani, 2014
3.	Immunotherapy	Monoclonal antibodies	Joint pain, headache	DeYoung et al., 2014
4.	Antifungals like azoles, e.g. Itraconazole	Targets fungal cell membrane by inhibiting ergosterol biosynthesis pathway	Stomach upset and liver toxicity	Ryan and Clark, 2015
5.	Surgery	Removal of allergic mucin and sinus drainage	Pain and swelling around the incision site	Oliveira et al., 2023; Marglani, 2014

TABLE 5.5
Characteristic Distinguishing Clinical Features of ABPA and AFRS

S. No.	Clinical Features	ABPA	AFRS
1.	Asthma	Present	Present in some patients
2.	Cystic fibrosis	Present	Present in some patients
3.	Pulmonary infiltration	Present	Absent
4.	Antibodies to *A. fumigatus* antigen	Present	Present in some patients
5.	Increased IgE levels	Present	Present
6.	Inflammation of nasal and lung airways	Present	Present
7.	Brown mucus	Present	Present
8.	Paranasal sinuses	Absent	Present
9.	Eosinophilic mucin	Absent	Present
10.	Nasal polyps	Absent	Present

5.4.3 Hypersensitivity Pneumonitis (HP)

5.4.3.1 Introduction
Hypersensitivity pneumonitis, also known as extrinsic allergic alveolitis, is a type of interstitial lung disease (ILD) that results from the inhalation of allergens and environmental antigens (Miller et al., 2018; Watts and Grammer, 2019). These antigens trigger an exaggerated immune response in susceptible individuals, leading to inflammation and lung damage. HP is often categorized into acute, subacute, and chronic forms, each having varying degrees of severity and duration (Figure 5.4) (Soumagne and Dalphin, 2018; Watts and Grammer, 2019).

5.4.3.2 Immune Responses Associated with HP
Antigens with a diameter of less than 5 μm, when inhaled, have the potential to reach the lung parenchyma. From here they migrate to the lymphatic vessels and settle in the respiratory bronchioles (Soumagne and Dalphin, 2018). Pathogens associated with hypersensitivity pneumonitis (HP) elicit similar clinical features, primarily affecting the distal airways (Ojanguren et al., 2019). This results in the infiltration of inflammatory cells in the alveolar and interstitial areas, along with elevated levels of serum precipitating antibodies (precipitins) against the antigens responsible for alveolar inflammation (Miyazaki et al., 2016). Notably, these cases typically exhibit normal levels

Inhalation of allergens and environmental antigens

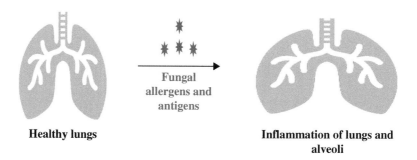

Fungal
allergens and
antigens

Healthy lungs   **Inflammation of lungs and
alveoli**

FIGURE 5.4   Hypersensitivity pneumonitis causes lung inflammation.

of IgE and eosinophils (Miyazaki et al., 2016; Ojanguren et al., 2019). Regardless of the causative agent, the pathogenesis of HP is quite similar, involving an inflammatory response within the
alveolar mucosa, characterized as a type 3 (immune complex-mediated) or type 4 (T lymphocyte-
mediated) hypersensitivity reaction (Miyazaki et al., 2016; Ojanguren et al., 2019). In type III
immune response, the immune complexes formed after antigen-antibody interaction deposit in the
lungs triggering inflammation (Ojanguren et al., 2019). Cytokines involved such cases include IL-1,
IL-6, and TNF-α (Limongi and Fallahi, 2017). On the other hand, type 4 hypersensitivity reactions
result in the activation of T cells upon antigen exposure, leading to the production of pro-inflammatory cytokines like interferon-gamma (IFN-γ), interleukin-2 (IL-2), and interleukin-12 (IL-12)
(Limongi and Fallahi, 2017; Selman et al., 2012; Vasakova et al., 2017). Although, neutrophils are
limited than the other immune cells, but they also contribute to the inflammatory response in HP
(Pardo et al., 2000).

TABLE 5.6
Stages of Hypersensitivity Pneumonitis

		Stages of Hypersensitivity Pneumonitis			
S. No.	Conditions	Acute HP	Subacute HP	Chronic HP	References
1.	Appearance of symptoms after exposure	Within 2–23 h	Few weeks to months	Months to years	Alberti et al., 2021; Leone and Richeldi, 2020
2.	Clinical manifestation	Fever, chills, and cough	Cough and dyspnea	Chronic cough	Leone and Richeldi, 2020
3.	High-resolution computed tomography scans	Patchy alveoli	Nodule formation in alveoli	Lung fibrosis	Cherniaev et al., 2021; Zhang et al., 2022
4.	Immune response	Neutrophils activation	Type III and IV hypersensitivity	Type III and IV hypersensitivity	Zhang et al., 2022; Pardo et al., 2000

5.4.3.3   Clinical Manifestations of HP

The exaggerated immune response and resulting inflammation leads to a range of clinical symptoms
in HP patients like fever, cough, dyspnea, chest discomfort, and malaise (Zhang et al., 2022). Imaging
studies, such as chest X-rays and high-resolution computed tomography (HRCT), often reveal characteristic patterns of lung infiltrates and ground-glass opacities, which are indicative of inflammation

(Miyazaki et al., 2016; Ojanguren et al., 2019). Diagnosis of HP requires a comprehensive evaluation, including clinical history of antigen exposure, radiological data, and immunological markers like elevated levels of specific antibodies (precipitating antibodies or IgG) directed against relevant antigens (Cherniaev et al., 2021; Zhang et al., 2022). Understanding the immunopathogenesis, key immune cells, cytokines, and clinical implications of this condition is essential for accurate diagnosis and effective management (Table 5.6). A multidisciplinary approach, involving clinicians, radiologists, and immunologists, is often necessary to provide comprehensive care to individuals with HP.

5.5 DIAGNOSIS AND MANAGEMENT OF FUNGAL ALLERGIES

Diagnosis of fungal allergies present challenges due to limited awareness, overlapping symptoms, and heterogeneous nature of these conditions. Addressing these challenges and fulfilling current needs is crucial for timely intervention and effective management. Advancements in this field include standardization of diagnosis criteria, improved diagnostic tools, patient education, and interdisciplinary collaboration. Additionally, identification of biomarkers and new preventive measures can further enhance our ability to identify and manage fungal allergies. Increasing awareness among healthcare providers regarding the prevalence and diverse manifestations of fungal allergies will go a long way in improving the qualtity of patient life (Bachert et al., 2003). Educational initiatives and clinical guidelines should emphasize the importance of considering fungal allergies in differential diagnoses (Richardson and Warnock, 2012). Establishing standardized diagnostic criteria for fungal allergies is crucial. This includes defining clear clinical and immunological parameters that can assist in the accurate identification of sensitized individuals (Zhang et al., 2022). There is a need for more accurate and accessible diagnostic tests for fungal allergies. Skin prick tests, specific IgE assays, and serum precipitating antibodies are commonly used, but further advancements, such as molecular diagnostic techniques, could greatly improve sensitivity and specificity (Cherniaev et al., 2021). Similarly, identification of specific biomarkers associated with fungal allergies can improve diagnostic accuracy tremendously. Research into the unique immune responses and genetic factors involved in fungal sensitization is ongoing. Public health measures to reduce exposure to allergens should be promoted side by side. This includes strategies for controlling indoor humidity, reducing mold growth, and limiting outdoor exposure during seasons when fungal spores are high in the environment (Soumagne and Dalphin, 2018). Collaborative efforts of allergists, pulmonologists, otolaryngologists, and other healthcare specialists is crucial for providing comprehensive care for individuals with fungal allergies. An interdisciplinary approach can aid in accurate diagnosis and management.

5.6 CONCLUSION

Optimal management of allergic fungal infections, such as allergic bronchopulmonary aspergillosis (ABPA), allergic fungal sinusitis (AFS), and hypersensitivity pneumonitis (HP), is a multifaceted process that necessitates a comprehensive and patient-centered approach. Accurate diagnosis, interdisciplinary collaboration, and patient education are the foundational pillars of effective management. Allergen avoidance strategies play a pivotal role in reducing host exposure to fungal spores, minimizing the risk of exacerbations. Pharmacological treatments, including corticosteroids and, in select cases, antifungal medications, help control airway inflammation and alleviate symptoms. In more severe cases, immunomodulatory therapies have shown promise in reducing exacerbations and corticosteroid dependency. Optimal asthma management is essential for individuals with coexisting allergies, enhancing overall respiratory health. Long-term monitoring through regular checkups, lung function assessment, scans and imaging studies is vital for tracking treatment efficacy and detecting any disease progression. Additionally, surgical intervention by otolaryngologists may be necessary in AFS cases with severe nasal polyps or sinus obstruction to improve sinus drainage and facilitate delivery of medicines. As ongoing research continues to shed light on the immunopathogenesis of these infections, the prospect of personalized medicine and targeted therapies is on the horizon. Identifying specific biomarkers and tailoring treatments to individual patient profiles may revolutionize the management of ABPA, AFRS and HP in the future. Hence, management of allergic fungal infections require a holistic approach that considers not only the immediate symptomatic relief but also the long-term well-being of affected individuals. With a combination of medical advancements, patient education, and healthcare provider collaboration, these complex conditions can achieve improved symptom control and management of allergic fungal infections.

REFERENCES

Abel-Fernández, E., Martínez, M. J., Galán, T., & Pineda, F. (2023). Going over fungal allergy: Alternaria alternata and its allergens. *Journal of Fungi*, *9*(5), 582.

Agarwal, R. (2009). Allergic bronchopulmonary aspergillosis. *Chest*, *135*(3), 805–826.

Agarwal, R., Chakrabarti, A., Shah, A., Gupta, D., Meis, J. F., & Guleria, R., & ABPA Complicating Asthma ISHAM Working Group (2013). Allergic bronchopulmonary aspergillosis: Review of literature and proposal of new diagnostic and classification criteria. *Clinical & Experimental Allergy*, *43*(8), 850–873.

Agarwal, R., Khan, A., Garg, M., Aggarwal, A. N., & Gupta, D. (2011). Pictorial essay: Allergic bronchopulmonary aspergillosis. *Indian Journal of Radiology and Imaging*, *21*(04), 242–252.

Ahmad, Z., Shah, T. A., Reddy, K. P., Ghosh, S., Panpatil, V., Kottoru, S. K., & Rao, D. R. (2022). Immunology in Medical Biotechnology. In *Fundamentals and advances in medical biotechnology*. Springer International Publishing. 179–207.

Alberti, M. L., Rincon-Alvarez, E., Buendia-Roldan, I., & Selman, M. (2021). Hypersensitivity pneumonitis: Diagnostic and therapeutic challenges. *Frontiers in Medicine*, *8*, 718299.

Amado, M., & Barnes, C. (2016). Allergenic Microfungi and Human Health: A Review on Exposure, Sensitization, and Sequencing Allergenic Proteins. In Biology of microfungi. Springer, Vol. 1: 429–449.

Bachert, C., Hörmann, K., Mösges, R., Rasp, G., Riechelmann, H., Müller, R., & Rudack, C. (2003). An update on the diagnosis and treatment of sinusitis and nasal polyposis. *Allergy*, *58*(3), 176–191.

Barac, A., Ong, D. S., Jovancevic, L., Peric, A., Surda, P., Tomic Spiric, V., & Rubino, S. (2018). Fungi-induced upper and lower respiratory tract allergic diseases: One entity. *Frontiers in Microbiology*, *9*, 583.

Bartemes, K. R., & Kita, H. (2018). Innate and adaptive immune responses to fungi in the airway. *Journal of Allergy and Clinical Immunology*, *142*(2), 353–363.

Blanco, J. L., & Garcia, M. E. (2008). Immune response to fungal infections. *Veterinary Immunology and Immunopathology*, *125*(1–2), 47–70.

Bowyer, P., Fraczek, M., & Denning, D. W. (2006). Comparative genomics of fungal allergens and epitopes shows widespread distribution of closely related allergen and epitope orthologues. *BMC Genomics*, *7*, 1–14.

Cameron, B. H., & Luong, A. U. (2023). New developments in allergic fungal rhinosinusitis pathophysiology and treatment. *American Journal of Rhinology & Allergy*, *37*(2), 214–220.

Caraballo, L., Valenta, R., Puerta, L., Pomés, A., Zakzuk, J., Fernandez-Caldas, E., & Karaulov, A. (2020). The allergenic activity and clinical impact of individual IgE-antibody binding molecules from indoor allergen sources. *World Allergy Organization Journal*, *13*(5), 100118.

Chakrabarti, A., & Kaur, H. (2016). Allergic *Aspergillus* rhinosinusitis. *Journal of Fungi*, *2*(4), 32.

Cherniaev, A. L., Kusraeva, E. V., Samsonova, M. V., Avdeev, S. N., Trushenko, N. V., & Tumanova, E. L. (2021). Clinical, radiologic, and morphological diagnosis of hypersensitivity pneumonitis. *Bulletin of Siberian Medicine*, *20*(4), 93–102.

Choidis, P., Kraniotis, D., Lehtonen, I., & Hellum, B. (2021). A modelling approach for the assessment of climate change impact on the fungal colonization of historic timber structures. *Forests*, *12*(7), 819.

Chou, H., Wu, K. G., Yeh, C. C., Tai, H. Y., Tam, M. F., Chen, Y. S., & Shen, H. D. (2014). The transaldolase, a novel allergen of Fusarium proliferatum, demonstrates IgE cross-reactivity with its human analogue. *PLOS One*, *9*(7), e103488.

Crameri, R., Garbani, M., Rhyner, C., & Huitema, C. (2014). Fungi: The neglected allergenic sources. *Allergy*, *69*(2), 176–185.

Crameri, R., Limacher, A., Weichel, M., Glaser, A. G., Zeller, S., & Rhyner, C. (2006). Structural aspects and clinical relevance of *Aspergillus fumigatus* antigens/allergens. *Medical Mycology*, *44*(Supplement_1), S261–S267.

DeYoung, K., Wentzel, J. L., Schlosser, R. J., Nguyen, S. A., & Soler, Z. M. (2014). Systematic review of immunotherapy for chronic rhinosinusitis. *American Journal of Rhinology & Allergy*, *28*(2), 145–150.

Dutre, T., Al Dousary, S., Zhang, N., & Bachert, C. (2013). Allergic fungal rhinosinusitis—More than a fungal disease? *Journal of Allergy and Clinical Immunology, 132*(2), 487–489.

Dykewicz, M. S., Rodrigues, J. M., & Slavin, R. G. (2018). Allergic fungal rhinosinusitis. *Journal of Allergy and Clinical Immunology, 142*(2), 341–351.

Fernández-Soto, R., Navarrete-Rodríguez, E. M., Del-Rio-Navarro, B. E., Sienra-Monge, J. L., Meneses-Sánchez, N. A., & Saucedo-Ramírez, O. J. (2018). Fungal allergy: Pattern of sensitization over the past 11 years. *Allergologia et immunopathologia, 46*(6), 557–564.

Fukutomi, Y., & Taniguchi, M. (2015). Sensitization to fungal allergens: Resolved and unresolved issues. *Allergology International, 64*(4), 321–331.

Gabriel, M. F., Postigo, I., Tomaz, C. T., & Martínez, J. (2016). Alternaria alternata allergens: Markers of exposure, phylogeny and risk of fungi-induced respiratory allergy. *Environment International, 89*, 71–80.

Gilley, S. K., Goldblatt, M. R., & Judson, M. A. (2010). The treatment of ABPA. In *Aspergillosis: From diagnosis to prevention.* Springer, Vol. 1: 747–759.

Glass, D., & Amedee, R. G. (2011). Allergic fungal rhinosinusitis: A review. *Ochsner Journal, 11*(3), 271–275.

Greenberger, P. A., Bush, R. K., Demain, J. G., Luong, A., Slavin, R. G., & Knutsen, A. P. (2014). Allergic bronchopulmonary aspergillosis. *Journal of Allergy and Clinical Immunology: In Practice, 2*(6), 703–708.

Hedayati, M. T., Pasqualotto, A. C., Warn, P. A., Bowyer, P., & Denning, D. W. (2007). *Aspergillus flavus*: Human pathogen, allergen and mycotoxin producer. *Microbiology, 153*(6), 1677–1692.

Houser, S. M., & Corey, J. P. (2000). Allergic fungal rhinosinusitis: Pathophysiology, epidemiology, and diagnosis. *Otolaryngologic Clinics of North America, 33*(2), 399–408.

Hoyt, A. E., Borish, L., Gurrola, J., & Payne, S. (2016). Allergic fungal rhinosinusitis. *Journal of Allergy and Clinical Immunology: In Practice, 4*(4), 599–604.

Hutcheson, P. S., Schubert, M. S., & Slavin, R. G. (2010). Distinctions between allergic fungal rhinosinusitis and chronic rhinosinusitis. *American Journal of Rhinology & Allergy, 24*(6), 405–408.

Knutsen, A., & Kariuki, B. (2019). Allergic bronchopulmonary aspergillosis and fungal allergy in severe asthma. *Archives of Applied Medicine, 1*(2), 1–19.

Kokoszka, M., Stryjewska-Makuch, G., Kantczak, A., Górny, D., & Glück, J. (2023). Allergic fungal rhinosinusitis in Europe: Literature review and own experience. *International Archives of Allergy and Immunology, 184*(9), 856–865.

Kurup, V. P., Shen, H. D., & Vijay, H. (2002). Immunobiology of fungal allergens. *International Archives of Allergy and Immunology, 129*(3), 181–188.

Kwong, K., Robinson, M., Sullivan, A., Letovsky, S., Liu, A. H., & Valcour, A. (2023). Fungal allergen sensitization: Prevalence, risk factors, and geographic variation in the United States. *Journal of Allergy and Clinical Immunology. 152*(6), 1–35.

Laury, A. M., & Wise, S. K. (2013). Allergic fungal rhinosinusitis. *American Journal of Rhinology & Allergy, 27*(3_suppl), S26–S27.

Leone, P. M., & Richeldi, L. (2020). Current diagnosis and management of hypersensitivity pneumonitis. *Tuberculosis and Respiratory Diseases, 83*(2), 122.

Levetin, E., Horner, W. E., Scott, J. A., Barnes, C., Baxi, S., Chew, G. L., & Williams, P. B. (2016). Taxonomy of allergenic fungi. *Journal of Allergy and Clinical Immunology: In Practice, 4*(3), 375–385.

Limongi, F., & Fallahi, P. (2017). Hypersensitivity pneumonitis and alpha-chemokines. *La Clinica Terapeutica, 168*(2), e140–e145.

Lokhande, N. (2018). *Bioaerosols.* RUT Printer and Publisher. Vol 2, 1–79.

Lord, A. K., & Vyas, J. M. (2019). Host defenses to fungal pathogens. In *Clinical immunology*, Elsevier. 413–424.

Marglani, O. (2014). Update in the management of allergic fungal sinusitis. *Saudi Medical Journal, 35*(8), 791–795.

Mari, A., Schneider, P., Wally, V., Breitenbach, M., & Simon-Nobbe, B. (2003). Sensitization to fungi: Epidemiology, comparative skin tests, and IgE reactivity of fungal extracts. *Clinical & Experimental Allergy, 33*(10), 1429–1438.

Marple, B. F. (2006). Allergic fungal rhinosinusitis: A review of clinical manifestations and current treatment strategies. *Medical Mycology, 44*(Supplement_1), S277–S284.

Miller, R., Allen, T. C., Barrios, R. J., Beasley, M. B., Burke, L., Cagle, P. T., & Smith, M. L. (2018). Hypersensitivity pneumonitis a perspective from members of the pulmonary pathology society. *Archives of Pathology & Laboratory Medicine, 142*(1), 120–126.

Miyazaki, Y., Tsutsui, T., & Inase, N. (2016). Treatment and monitoring of hypersensitivity pneumonitis. *Expert Review of Clinical Immunology, 12*(9), 953–962.

Moss, R. B. (2005). Pathophysiology and immunology of allergic bronchopulmonary aspergillosis. *Medical Mycology, 43*(sup1), 203–206.

Mouyna, I., & Fontaine, T. (2008). Cell wall of *Aspergillus fumigatus*: A dynamic structure. *Aspergillus fumigatus and Aspergillosis, 2*, 169–183.

Muthu, V., Sehgal, I. S., Prasad, K. T., Dhooria, S., Aggarwal, A. N., Garg, M., & Agarwal, R. (2020). Allergic bronchopulmonary aspergillosis (ABPA) sans asthma: A distinct subset of ABPA with A lesser risk of exacerbation. *Medical Mycology, 58*(2), 260–263.

Nevalainen, A., Täubel, M., & Hyvärinen, A. (2015). Indoor fungi: Companions and contaminants. *Indoor Air, 25*(2), 125–156.

O'Reilly, A., & Dunican, E. (2021). The use of targeted monoclonal antibodies in the treatment of ABPA—A case series. *Medicina, 58*(1), 53.

Ogórek, R., Lejman, A., Pusz, W., Miłuch, A., & Miodyńska, P. (2012). Characteristics and taxonomy of Cladosporium fungi. *Mikologia Lekarska, 19*(2), 80–85.

Ojanguren, I., Morell, F., Ramón, M. A., Villar, A., Romero, C., Cruz, M. J., & Muñoz, X. (2019). Long-term outcomes in chronic hypersensitivity pneumonitis. *Allergy, 74*(5), 944–952.

Oliveira, M., Oliveira, D., Lisboa, C., Boechat, J. L., & Delgado, L. (2023). Clinical manifestations of human exposure to fungi. *Journal of Fungi, 9*(3), 381.

Pant, H., Schembri, M. A., Wormald, P. J., & Macardle, P. J. (2009). IgE-mediated fungal allergy in allergic fungal sinusitis. *Laryngoscope, 119*(6), 1046–1052.

Pardo, A., Barrios, R., Gaxiola, M., Segura-Valdez, L., Carrillo, G., Estrada, A., & Selman, M. (2000). Increase of lung neutrophils in hypersensitivity pneumonitis is associated with lung fibrosis. *American Journal of Respiratory and Critical Care Medicine, 161*(5), 1698–1704.

Patel, A. R., Patel, A. R., Singh, S., Singh, S., & Khawaja, I. (2019). Treating allergic bronchopulmonary aspergillosis: A review. *Cureus, 11*(4), 1–8.

Pathakumari, B., Liang, G., & Liu, W. (2020). Immune defence to invasive fungal infections: A comprehensive review. *Biomedicine & Pharmacotherapy, 130*, 110550.

Patterson, K., & Strek, M. E. (2010). Allergic bronchopulmonary aspergillosis. *Proceedings of the American Thoracic Society, 7*(3), 237–244.

Platts-Mills, T. A., & Woodfolk, J. A. (2011). Allergens and their role in the allergic immune response. *Immunological Reviews, 242*(1), 51–68.

Plonk, D. P., & Luong, A. (2014). Current understanding of allergic fungal rhinosinusitis and treatment implications. *current. Opinion in Otolaryngology & Head and Neck Surgery, 22*(3), 221–226.

Portnoy, J. M., Williams, P. B., & Barnes, C. S. (2016). Innate immune responses to fungal allergens. *Current Allergy and Asthma Reports, 16*, 1–6.

Reboux, G., Rocchi, S., Vacheyrou, M., & Millon, L. (2019). Identifying indoor air *Penicillium* species: A challenge for allergic patients. *Journal of Medical Microbiology, 68*(5), 812–821.

Richardson, M. D., & Warnock, D. W. (2012). *Fungal infection: Diagnosis and management*. John Wiley & Sons. Vol *1*(4): 10–33.

Rizzetto, L., De Filippo, C., & Cavalieri, D. (2014). Richness and diversity of mammalian fungal communities shape innate and adaptive immunity in health and disease. *European Journal of Immunology, 44*(11), 3166–3181.

Rudert, A., & Portnoy, J. (2017). Mold Allergy: Is it real and what do we do about it? *Expert Review of Clinical Immunology, 13*(8), 823–835.

Ryan, M. W., & Clark, C. M. (2015). Allergic fungal rhinosinusitis and The unified airway: The role of antifungal therapy in AFRS. *Current Allergy and Asthma Reports, 15*, 1–6.

Salamah, M. A., Alsarraj, M., Alsolami, N., Hanbazazah, K., Alharbi, A. M., & Khalifah, W., Sr (2020). Clinical, radiological, and histopathological patterns of allergic fungal sinusitis: A single-center retrospective study. *Cureus, 12*(7), e9233.

Saravanan, K., Panda, N. K., Chakrabarti, A., Das, A., & Bapuraj, R. J. (2006). Allergic fungal rhinosinusitis: An attempt to resolve the diagnostic dilemma. *Archives of Otolaryngology–Head & Neck Surgery, 132*(2), 173–178.

Schubert, M. S. (2004). Allergic fungal sinusitis. *Otolaryngologic Clinics of North America, 37*(2), 301–326.

Selman, M., Pardo, A., & King, T. E. Jr. (2012). Hypersensitivity pneumonitis: Insights in diagnosis and pathobiology. *American Journal of Respiratory and Critical Care Medicine, 186*(4), 314–324.

Shah, A., & Panjabi, C. (2016). Allergic bronchopulmonary aspergillosis: A perplexing clinical entity. *Allergy, Asthma & Immunology Research, 8*(4), 282–297.

Simon-Nobbe, B., Denk, U., Pöll, V., Rid, R., & Breitenbach, M. (2007). The spectrum of fungal allergy. *International Archives of Allergy and Immunology, 145*(1), 58–86.

Singh, A., Garg, S., & Upadhyay, A. K. (2023). Identification and analysis of allergens in edible mushroom (*Agaricus bisporus*). *Materials Today: Proceedings*, *1*, 1–32.

Singh, A. B., & Mathur, C. (2021). Fungal aerobiology and allergies in India: An overview. *Progress in mycology: An Indian perspective*, Springer, 397–417.

Soeria-Atmadja, D., Önell, A., & Borgå, Å. (2010). IgE sensitization to fungi mirrors fungal phylogenetic systematics. *Journal of Allergy and Clinical Immunology*, *125*(6), 1379–1386.

Soumagne, T., & Dalphin, J. C. (2018). Current and emerging techniques for the diagnosis of hypersensitivity pneumonitis. *Expert Review of Respiratory Medicine*, *12*(6), 493–507.

Steels, S., Proesmans, M., Bossuyt, X., Dupont, L., & Frans, G. (2023). Laboratory biomarkers in the diagnosis and follow-up of treatment of allergic bronchopulmonary aspergillosis in cystic fibrosis. *Critical Reviews in Clinical Laboratory Sciences*, *60*(1), 1–24.

Sussman, A. S., & Airworth, G. C. (2013). Longevity and survivability of fungi. *Fungi*, *3*, 447–486.

Tracy, M. C., Okorie, C. U., Foley, E. A., & Moss, R. B. (2016). Allergic bronchopulmonary aspergillosis. *Journal of Fungi*, *2*(2), 17.

Twaroch, T. E., Curin, M., Valenta, R., & Swoboda, I. (2015). Mold allergens in respiratory allergy: From structure to therapy. *Allergy, Asthma & Immunology Research*, *7*(3), 205–220.

Tyler, M. A., & Luong, A. U. (2018). Current understanding of allergic fungal rhinosinusitis. *World Journal of Otorhinolaryngology—Head and Neck Surgery*, *4*(03), 179–185.

Tyler, M. A., & Luong, A. U. (2020). Current concepts in the management of allergic fungal rhinosinusitis. *Immunology and Allergy Clinics*, *40*(2), 345–359.

Vasakova, M., Morell, F., Walsh, S., Leslie, K., & Raghu, G. (2017). Hypersensitivity pneumonitis: Perspectives in diagnosis and management. *American Journal of Respiratory and Critical Care Medicine*, *196*(6), 680–689.

Viegas, C., Pinheiro, A. C., Sabino, R., Viegas, S., Brandão, J., & Veríssimo, C. (Eds.). (2015). *Environmental mycology in public health: Fungi and mycotoxins risk assessment and management*. Academic Press. Vol. 1: 223–337.

Vijay, H. M., & Kurup, V. P. (2008). Fungal allergens. *Allergens and Allergen Immunotherapy*, *4*, 159–178.

Wang, L., Hu, Q., Pei, F., Mugambi, M. A., & Yang, W. (2020). Detection and identification of fungal growth on freeze-dried *Agaricus bisporus* using spectra and olfactory sensors. *Journal of the Science of Food and Agriculture*, *100*(7), 3136–3146.

Wark, P. A., & Gibson, P. (2010). Pathogenesis of ABPA. In *Aspergillosis: From diagnosis to prevention*. Springer, Vol. 1, 695–706.

Watts, M. M., & Grammer, L. C. (2019). Hypersensitivity pneumonitis. *Allergy & Asthma Proceedings*, *40*(6), 425–436.

Wheeler, M. L., Limon, J. J., & Underhill, D. M. (2017). Immunity to commensal fungi: Detente and disease. *Annual Review of Pathology: Mechanisms of Disease*, *12*, 359–385.

Wüthrich, M., Deepe, G. S. Jr., & Klein, B. (2012). Adaptive immunity to fungi. *Annual Review of Immunology*, *30*, 115–148.

Yamamoto, N., Bibby, K., Qian, J., Hospodsky, D., Rismani-Yazdi, H., Nazaroff, W. W., & Peccia, J. (2012). Particle-size distributions and seasonal diversity of allergenic and pathogenic fungi in outdoor air. *ISME Journal*, *6*(10), 1801–1811.

Zhang, C., Jiang, Z., & Shao, C. (2020). Clinical characteristics of allergic bronchopulmonary aspergillosis. *Clinical Respiratory Journal*, *14*(5), 440–446.

Zhang, F., Yang, T., Liu, Z., Jia, X., Yang, L., Wu, L., & Tang, L. (2022). Clinical features of hypersensitivity pneumonitis in children: A single center study. *Frontiers in Pediatrics*, *9*, 789183.

6 Therapeutic Challenge of Blastomycosis

*Heba S. Abbas**

6.1 INTRODUCTION

There are several species of dimorphic fungi that cause blastomycosis. There are several species of dimorphic fungi that cause blastomycosis. for example, *Blastomyces dermatitidis* and *B. gilchristii* cause an endemic mycosis. The description of a third species has been determined by analysis of typical specimens, *B. helices*. An in-depth understanding of this agent's role is needed (Schwartz et al., 2017; Vyas and Bradsher, 2020). North America has reported the highest number of cases of blastomycosis, occurring mainly in the Mississippi, Ohio, and St. Lawrence River basins, as well as the Great Lakes region (Bradsher et al., 2003). A few very rare cases of dogs with oral lesions transmitting infection to humans via bites have occurred. Dogs, cats, cows, and other mammals have been infected, but humans have not been infected. It is rare for people to transmit the disease from person to person. Only one case has been reported where a man with a genitourinary disease transmitted pelvic blastomycosis to his wife and there are a few examples of mother-to-child transmission during pregnancy (Craig et al., 1970; Maxson et al., 1992).

Blastomycosis has only been recognized as a major pathogen in AIDS patients on few occasions. When it does occur in this situation, it is usually in the latter stages of AIDS with extremely low CD4 lymphocyte levels. The disease is typically deadly, swiftly progressing, and globally distributed (Pappas et al., 1994). If an antifungal therapy response occurs, the affected individual should be kept on continuous secondary prophylaxis until a reactivation of the immune system begins (Chapman et al., 2008). Blastomycosis has been documented in individuals with other immunosuppressive sickness, as a result of organ transplantation blood vessel malignancies, or corticosteroid use (Vyas and Bradsher, 2020).

6.2 GEOGRAPHIC DISTRIBUTION

Blastomyces thrives in damp, acidic ground, particularly in densely vegetated places near waterways or lakes. Initial attempts to isolate Blastomyces in soil samples relied heavily cultured or recovery from injections of soil samples into animals, which was typically a laborious and unsuccessful operation. Blastomyces is usually regarded as endemic in the states located along the Ohio and Mississippi rivers, and the region around the Great Lakes of the United States, and the eastern provinces of Canada, and has been noticed in both previous and more current epidemiologic investigations (Brown et al., 2018; Seitz et al., 2015). Although the majority of data on epidemiology predates the discovery of *B. gilchristii*, which has been incorporated in the *B. dermatitidis* complex classification, some research implies that *B. gilchristii* is more geographically confined to Canada and the northern part of the United States (McTaggart et al., 2016). Though less frequently stated, blastomycosis has been observed in countries of the Middle East and the nation of India; however, variations in patterns of illness and an up-to-date molecular analysis of several cases indicate that two distinct species (*Blastomyces percusus* and *Blastomyces emzantsi*) may account for an important part of the burden of infection in these areas (Randhawa et al., 2013;

---

* Corresponding Author: heba181179@yahoo.com

DOI: 10.1201/9781032642864-6

Schwartz et al., 2020). The non-identified state of blastomycosis in the majority of the world, as well as a scarcity of larger epidemiological investigations of this illness, makes it difficult to accurately identify Blastomyces' expanding global distribution. When this is combined with the potential extension of endemic fungus domains as a result of climate change, it is clear that additional research on the real prevalence and geographic extent of fungal illnesses is required (Silversides et al., 2008).

6.3 PATHOGENESIS

Infection enters the lungs via inhaling of infective conidia, which are picked up by bronchopulmonary phagocytes. The fungus undergoes metamorphosis into the yeast phase here. Asymptomatic pulmonary infection, acute bacterial pneumonia, or more with lung nodules or perforation can all occur. Infection symptoms after spreading through the blood to the tissues of the skin, or others tend to be non-specific. The immune system's response is a hybrid of granulation tissue and pyogenic components. Pseudoepitheliomatous hypertrophy is observed in skin and mucous membrane infections, which has led to some individuals being misdiagnosed (Bradsher et al., 2003).

6.3.1 LUNG LESIONS

Acute pulmonary blastomycosis can be diagnosed as an undetected radiographs infiltrate or as a pneumonia with a high body temperature, chills, and persistent coughing with or without blood spitting. Another common sign of chronic pulmonary blastomycosis is a prolonged period of excessive sweating, a persistent cough, and severe pain in the chest. Respiratory distress syndrome, or RD syndrome, is associated with a significant death rate in individuals with endobronchial infection dissemination. Radiologically, pulmonary abnormalities are evident in half of the patients. In patients who show more intense symptoms, the radiologic image may look like bacterial pneumonia (Halvorsen et al., 1984). Patients usually claim a decrease in weight, irregular low-grade fever, discomfort in the chest, fatigue/malaise, breathing difficulties and coughing (often with limited production of sputum sometimes associated with bleeding). Patients who present with (or evolve to) a severe type of lung disease are less prevalent. In its severe form, pulmonary blastomycosis can cause symptoms ranging from asymptomatic pneumonia to acute respiratory distress syndrome (McBride et al., 2017; Fanella et al., 2011). An estimated 25%–40% of patients who have symptomatic infection develop extrapulmonary illness, which is most typically spread from lung disease (though direct inoculation is unusual). The skin (40%–80%), bones (5%–25%), and genitourinary structure (less than 10%) are the most typically affected extrapulmonary locations for the spread of the illnesses. Infection of the brain and spinal cord is uncommon, occurring in 5%–10% of disseminated illness, with immunocompromised patients at an increased risk (Bariola et al., 2010; McBride et al., 2017). Figure 6.1 highlights the most prevalent locations and frequency of dispersion.

6.3.2 SKIN PATCHES

The skin is the most prevalent location of infection, and affecting 40%–80% of patients. They are often erythematous, hyperkeratotic, encrusted nodules or patches that expand over a span of many weeks. The patches may ulcerate, leaving an exposed edge. In chronic situations, some central healing may occur, resulting in a hypopigmented, atrophic, and fibrous region. These patches can also occur in the nose, mouth, larynx, and so on (Vyas and Bradsher, 2020).

6.3.3 OSTEOMYELITIS

Distinctive features include osteolytic lesions and an associated cold abscess. For bone infections, the treatment regimen is frequently lengthier (Chapman et al., 2008).

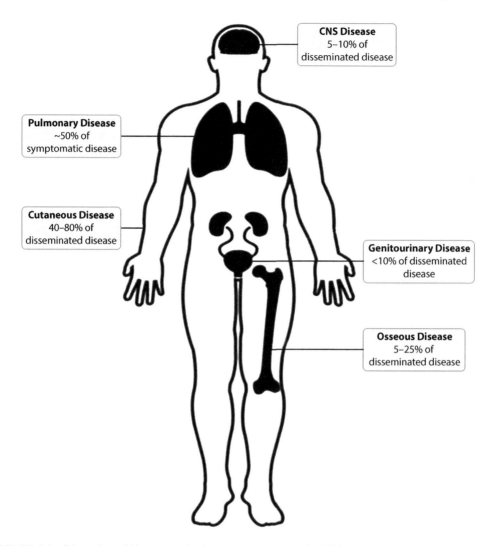

FIGURE 6.1   Disseminated blastomycosis sites. (From Pullen et al., 2022.)

6.3.4   INFECTION OF ENTIRE BODY

Nearly every part of the body can be infected with *B. dermatitidis*. Systemic symptoms such as fever and a decrease in body weight are minor at first but grow to be more serious as the disease progresses. Chronic infection, if left untreated, can spread to several organs and cause death over a period of weeks to years. Once a respiratory infection has spread throughout the lung, a relapse of infection is uncommon.

6.4   DIAGNOSIS

The microscopic analysis of stained histopathological or cytogenetic samples, as well as the culturing of discharges or tissues, is the primary method of blastomycosis identification. Thorough microscopic inspection of the sputum, skin, abscesses fluids, or liquid from the body after disintegration of the cells with 10% potassium hydroxide or stained calcofluor white or Papanicolaou's stain is the most immediate and reliable method of detection when clinical indication is high. The microbe shows as a yeast with only one broad-based budding in biopsy specimens stained with stained methenamine silver or periodic acid-Schiff (Saccente and Woods, 2010).

Almost always, the test is positive. The organism is expected to develop as a mold in a period of two to four weeks at a normal temperature on a variety of growth media, including Sabouraud dextrose agar. Skin testing is neither useful nor readily available. Serology, which employs a wide range of antigens and procedures, is ineffective for diagnosis (Bradsher and Pappas, 1995).

6.5 BLASTOMYCOSIS TREATMENT

The great majority of people who seek medical care will require treatment with antifungal medications. Blastomycosis has a wide spectrum of signs and symptoms, from asymptomatic infection to bronchitis to acute lung injury. Due to this clinical unpredictability, Blastomycosis has been dubbed as "the great deceiver." After soil disturbance, the principal entry point for dispersed conidia is the lung. Skin traumatization is uncommon but has been recorded (Larson et al., 1983). Symptoms appear 3 weeks to 3.5 months after inhaling mycelial shreds or spores (Klein et al., 1986). Extrapulmonary spreading affects roughly 25%–40% of individuals who become symptomatic (Chapman et al., 2008). The layers of skin, bones, genitourinary path, and the brain and spinal cord are common locations for disseminated illness; however, Blastomyces may infect nearly every organ in the human body (Gauthier and Klein, 2014).

The Infectious Diseases Society of America and the American Thoracic Society have established criteria for the identification and management of blastomycosis. For instance, in case of mild or moderate pulmonary blastomycosis, the patient should be given Itraconazole 200 mg orally 3 times per day for 3 days, then 1 or 2 times per day for 6–12 months as preliminary therapy. However, in case of severe pulmonary blastomycosis, Amphotericin B lipid formulation 3–5 mg/kg should be given every day or Amphotericin B deoxycholate 0.7–1 mg/kg every day for 1–2 weeks or until progress is seen (Limper et al., 2011). Before starting treatment, a baseline assessment of hematologic, liver, and renal health should be performed. To reduce interactions between medications usually linked with azole antifungals, a thorough examination of all drugs is essential. Itraconazole, voriconazole, posaconazole, and fluconazole can all extend the time between heartbeats, especially when used with other drugs that do. Isavuconazole, on the other hand, can reduce the QT period and is thus recommended in individuals with related short QT disorder. Itraconazole has a significant inotropic action and has the potential to aggravate pre-existing congestive heart failure (CHF), hence it should be administered with cautiously to individuals with ventricular failure (Ahmad et al., 2001). Azole antifungals raise serum levels of inhibitors of the enzyme HMG-CoA reductase processed by the enzyme cytochrome P450 3A4, thereby raising the possibility of statin-induced rhabdomyolysis. Because pravastatin is not processed by P450 3A4, it can be taken successfully with the azoles (Chapman et al., 2008). Immune-suppressive drugs, dihydropyridine blockers of calcium channels, sulfonylureas, and anticonvulsants are among the other important azole interactions between drugs. All females of sexual maturity should be checked for gestation due to the adverse effects of azole exposure on childbirth, including toxicity (De Santis et al., 2009; Pursley et al., 1996). Amphotericin B is also used as a first-line treatment for newborns and pregnant women (Chapman et al., 2008). Amphotericin B deoxycholate has a lengthy history of clinical effectiveness and excellent cure rates 15,42. Despite its well-established effectiveness, Amphotericin B usage is accompanied with severe chronic toxicity (Chapman et al., 1997). The most prevalent treatment-limiting complication is nephrotoxicity, which affects over 30% of treated individuals. Infusion responses (e.g., fever, rigors, a lack of oxygen, vomiting, nausea, high blood pressure, hypotension) and electrolyte changes are other possible side effects (Girois et al., 2005). Because of its decreased nephrotoxicity, lipid Amphotericin B preparations (e.g., liposomal amphotericin, Amphotericin B lipid complex, and Amphotericin B colloidal dispersion) are recommended over AmB deoxycholate. Liposomal amphotericin is the chosen polyene for CNS blastomycosis due to how it penetrates the blood–brain barrier most effectively of the lipid compositions (Chapman et al., 2008). Although proper antifungal treatment, blastomycosis-induced lung injury remains fatal (Lemos et al., 2001; Meyer et al., 1993; Vasquez et al., 1998). Case studies have shown that supplementary steroids may enhance survival; however, a current retrospective investigation of 43 patients with acute respiratory distress syndrome related to blastomycosis (1992–2014) found no reduction in death in individuals who received steroids (Lahm et al., 2008; Plamondon et al., 2010; Schwartz et al., 2016). Nonetheless, more study on the dosage, duration, and effectiveness of supplemental steroids in acute respiratory distress syndrome is required (McBride et al., 2017).

Case deaths in a large group of cases from Wisconsin and Manitoba range between 4.3% and 6.3% (Seitz et al., 2014). Death rates have been linked to a shorter period of symptoms, implying a more active presentation and a reduced patient immunological system. Even in patients getting proper antifungal therapy, blastomycosis-induced ARDS is linked with a significant fatality rate (Gauthier et al., 2007; Lemos et al., 2001; Schwartz et al., 2016). In the absence of immunological reconstitution, the mortality rate of blastomycosis in people with HIV is approximately 40%, with the majority of fatalities occurring within three weeks after diagnosis (Pappas et al., 1992). Similarly, individuals immunocompromised by transplantation of solid organs have a mortality rate of 33%–38%, which rises in the presence of respiratory distress (Grim et al., 2012).

6.6   CONCLUSION

Blastomycosis is frequently difficult to diagnose and treat. Even in endemic locations, blastomycosis's unpredictability clinical symptoms may cause an extended delay in detection. Concurrent pulmonary/cutaneous infection, acute respiratory distress syndrome, and the recognizable risk factors for Blastomyces exposure, should raise alarm among physicians practicing in Blastomyces-endemic regions. It is critical to understand that *Blastomyces* spp. may infect and spread in both immunocompromised and immunocompetent individuals alike. Finally, being aware of frequent concerns that doctors face while utilizing polyene and azole antifungals will reduce treatment risks.

REFERENCES

Ahmad SR, Singer SJ, Leissa BG. Congestive heart failure associated with itraconazole. Lancet 2001;357(9270):1766–1777.
Bariola JR, Perry P, Pappas PG, Proia L, Shealey W, Wright PW, Sizemore JM, Robinson M, Bradsher JRW. Blastomycosis of the Central nervous system: A multicenter review of diagnosis and treatment in the modern era. Clin Infect Dis 2010;50:797–804.
Bradsher RW, Chapman SW, Pappas PG. Blastomycosis. Infect Dis Clin North Am 2003;17:21–40.
Bradsher RW, Pappas PG. Detection of specific antibodies in human blastomycosis by enzyme immunoassay. South Med J 1995;88:1256–1259.
Brown E, McTaggart LR, Dunn D, Pszczolko E, Tsui KG, Morris SK, Stephens D, Kus JV, Richardson SE. Epidemiology and geographic distribution of blastomycosis, histoplasmosis, and coccidioidomycosis, Ontario, Canada, 1990–2015. Emerg Infect Dis 2018;24:1257–1266.
Chapman SW, Dismukes WI, Proia LA, et al. Infectious Diseases Society of America Clinical practice guidelines for the management of blastomycosis: 2008 updated by the Infectious Disease Society of America. Clin Infect Dis 2008;46(12): 1801–1811.
Chapman SW, Lin AC, Hendricks KA, et al. Endemic blastomycosis in Mississippi: Epidemiologic and clinical studies. Semin Respir Infect 1997;12(3): 219–228.
Craig MW, Davey WN, Green RA. Conjugal blastomycosis. Am Rev Respir Dis 1970;102:86–90.
De Santis M, Di Gianantonio E, Cesari E, Ambrosini G, Straface G, Clementi doneM. First-trimester itraconazole exposure and pregnancy outcome: A prospective cohort study of women contacting teratology information services in Italy. Drug Saf 2009;32(3):239–244.
Fanella S, Skinner S, Trepman E, Embil JM. Blastomycosis in children and adolescents: A 30-year experience from Manitoba. Med Mycol 2011;49:627–632.
Gauthier G, Klein BS. Blastomycosis. In: Cherry JD, Harrison GJ, editors, Feigin and Cherry's Textbook of Pediatric Infectious Disease. Seventh Edition. Philadelphia (PA): Elsevier Saunders; 2014. pp. 2723–2743.
Gauthier GM, Safdar N, Klein BS, Andes DR. Blastomycosis in solid organ transplant recipients. Transpl Infect Dis 2007;9(4):310–317.

Girois SB, Chapuis F, Decullier E, Revol GB. Adverse effects of antifungal therapies in invasive fungal infections: Review and meta-analysis. Eur J Clinc Microbiol Infect Dis 2005;24(2):119–130.

Grim SA, Proria L, Miller R, et al. A multicenter study of histoplasmosis and blastomycosis after solid organ transplantation. Transpl Infect Dis 2012;14(1): 17–23.

Halvorsen RA, Duncan JD, Merten DF, et al. Pulmonary blastomycosis: Radiologic manifestations. Radiology 1984;150:1–5.

Klein BS, Vergeront JM, Weeks RJ, et al. Isolation of *Blastomyces dermatitidis* in soil associated with a large outbreak of blastomycosis in Wisconsin. N Engl J Med 1986;314(9): 529–534.

Lahm T, Neese S, Thornburg AT, Ober MD, Sarosi GA, Hage CA. Corticosteroids for blastomycosis-induced ARDS: A report of two patients and review of the literature. Chest 2008;133(6):1478–1480.

Larson DM, Eckman MR, Alber RL, Goldschmidt VG. Primary cutaneous (inoculation) blastomycosis: An occupational hazard to pathologist. Am J Clinc Pathol 1983;79(2):253–255.

Lemos LB, Baliga M, Guo M. Acute respiratory distress syndrome and blastomycosis: Presentation of nine cases and review of the literature. Ann Diagn Pathol 2001;5(1):1–9.

Limper AH, Knox KS, Sarosi GA, et al. An official American thoracic society statement: Treatment of fungal infections in adult pulmonary and critical care patients. Am J Respir Crit Care Med 2011; 183:96–128.

Maxson S, Miller SF, Tryka AF, Schutze GE. Perinatal blastomycosis: A review. Pediatr Infect Dis J 1992;11:760–763.

McBride JA, Gauthier GM, Klein BS. Clinical manifestations and treatment of blastomycosis. Clin Chest Med 2017;38(3):435–449. https://doi.org/10.1016/j.ccm.2017.04.006

McTaggart LR, Brown E, Richardson SE. Phylogeographic analysis of blastomyces dermatitidis and blastomyces gilchristii reveals an association with North American freshwater drainage basins. PLOS ONE 2016;11:e0159396.

Meyer KC, McManus EJ, Maki DG. Overwhelming pulmonary blastomycosis associated with the adult respiratory distress syndrome. N Engl J Med 1993;329(17):1231–1236.

Pappas PG, Pottage JC, Powderly WG, et al. Blastomycosis in patients with the acquired immunodeficiency syndrome. Ann Intern Med 1994;116:847–853.

Plamondon M, Lamontagne F, Allard C, Pepin J. Corticosteroids as adjunctive therapy in severe blastomycosis-induced acute respiratory distress syndrome in an immunosuppressed patient. Clin Infect Dis 2010;51(1):1–3.

Pullen MF, Alpern JD, Bahr NC. Blastomycosis—Some progress but still much to learn. J Fungi 2022;8(8):824. https://doi.org/10.3390/jof8080824

Pursley TJ, Blomquist IK, Abraham, Andersen HF, Bartley JA. Fluconazole-induced congential anomalies in three infants. Clin Infect Dis 1996;22(2):336–340.

Randhawa HS, Chowdhary A, Kathuria S, Roy P, Misra DS, Jain S, Chugh TD. Blastomycosis in India: Report of an imported case and current status. Med Mycol 2013;51:185–192.

Saccente M, Woods GL. Clinical and laboratory update on blastomycosis. Clin Microbiol Rev 2010; 23:367–381.

Schwartz IS, Embil JM, Sharma A, Goulet S, Light RB. Management and outcome of acute respiratory distress syndrome caused by blastomyosis: A retrospective case series. Medicine 2016;95(18):3538.

Schwartz IS, Muñoz JF, Kenyon CR, Govender NP, McTaggart L, Maphanga TG, Richardson S, Becker P, Cuomo CA, McEwen JG, Sigler L. Blastomycosis in Africa and the middle East: A comprehensive review of reported cases and reanalysis of historical isolates based on molecular data. Clin Infect Dis 2020;73:e1560–e1569.

Schwartz IS, Weiderhold NP, Rofael M, et al. *Blastomyces helices*, an emerging systemic pathogen in western Canada and United States. Open Forum Infect Dis 2017;4:S83–S84.

Seitz AE, Adjemian J, Steiner CA, Prevots DR. Spatial epidemiology of blastomycosis hospitalizations: Detecting clusters and identifying environmental risk factors. Med Mycol 2015;53:447–454.

Seitz AE, Younes N, Steiner CA, Prevots DR. Incidence and trends of blastomycosis-associated hospitalizations in the United States. PLOS ONE 2014;9(8):e105466.

Silversides A, Greer A, Ng V, Fisman D. CMA opposes gender discrimination against doctors. Can Med Assoc J 2008;178:715–722.

Vasquez JE, Mehta JB, Agrawal R, et al. Blastomycosis in northeast Tennessee. Chest. 1998;114(2):436–443.

Vyas KS, Bradsher RW. Blastomycosis in northeast Tennessee. In Hunter's Tropical Medicine and Emerging Infectious Diseases. London: Elsevier; 2020. pp. 671–673. https://doi.org/10.1016/b978-0-323-55512-8.00087-9

7 Fungal Invasion
Deciphering the Pathogenesis of Infection

*Deepika Kumari, Pammi Kumari, Preeti Sharma,
Antresh Kumar, and Ritu Pasrija*

7.1 INTRODUCTION

Fungal species found in the human environment can cause various diseases that pose a serious public health problem. These infections are linked with fatal mycoses and a high mortality rate in people with a variety of diseases, especially those with a weakened immune response. Fungal infections of varying severity include superficial, mucosal, cutaneous, subcutaneous, and systemic infections. Among the disease-causing fungi, only a few are true pathogens (*Histoplasma*, *Paracoccidioides*, etc.) and infect healthy people. Most fungi are opportunistic pathogens (e.g., *Candida* and *Cryptococcus*) that cause disease, especially in immunocompromised individuals (Brown et al. 2012). High-ranking human pathogenic fungi are *Candida*, *Cryptococcus*, *Aspergillus*, etc. *Candida* is a commensal fungus commonly found on the skin, gastrointestinal tract, and other mucosal surfaces. It is the cause of three forms of candidiasis: Vaginal candidiasis, thrush, and invasive candidiasis (IC). In addition, IC is a life-threatening infection that can damage the brain, heart, eyes, lungs, blood, bones, and other organs and accounts for 46%–75% of all invasive fungal infections (IFI) (Brown et al. 2012). Aspergillosis, caused by *Aspergillus fumigatus*, is another important invasive fungal infection. *Aspergillus* is a saprophytic fungus that causes approximately 2,00,000 cases of aspergillosis annually, with a mortality rate ranging from 30%–95% (Brown et al. 2012). Another common infectious disease is caused by *Cryptococcus* spp., which affects more than one million people annually and has a mortality rate of about 20%–70% (Park et al. 2009). Despite these known fungal pathogens, *Candida auris*, a new multidrug-resistant fungus, poses an emerging new public health threat (Chow et al. 2018).

Fungal pathogens are mainly transmitted by inhalation or direct contact. Dermatophytes from the genera *Trichophyton, Epidermophyton, Microsporum, Malassezia* spp., and *Sporothrix* infect injured skin through direct contact. The secretion of various proteolytic enzymes triggers superficial mycosis in keratinized tissue. Another transmission route is the inhalation of spores/conidia, which cause lung infections. *Paracoccidioides brasiliensis, Blastomyces dermatitidis* (blastomycosis), *Pneumocystis jirovecii* (Pneumocystis pneumonia), *Histoplasma capsulatum* (histoplasmosis), *Aspergillus flavus* (aspergillosis), *Cryptococcus gattii*, and *Cryptococcus neoformans* (cryptococcosis) are mainly transmitted by inhalation (Reddy et al. 2022).

Fungi's capability to induce disease arises due to numerous virulence factors that promote the survival and persistence of fungal cells in the host, leading to damage in tissues and infection. These factors encompass the capacity to bind to host tissue, the secretion of enzymes causing harm to host tissue, and the direct impairment of host defenses. In addition, capsule formation, dimorphism, biofilm formation, melanin production, and the formation of the asteroid bodies and titan cells increase the virulence of the fungi (Odds 2017).

The studies from different laboratories have shown that the interactions between hosts and pathogens are complicated and many signal transduction pathways in fungi are crucial in regulating pathogenicity. Thus, a comprehensive knowledge of these interactions can likely lead to new

DOI: 10.1201/9781032642864-7

effective therapeutic options. This chapter discusses the virulence factors of the fungal pathogen, and essential features of host-fungal interactions, including the host immune response.

7.2 VIRULENCE FACTORS IN PATHOGENIC FUNGI

The genera *Candida*, *Cryptococcus*, and *Aspergillus* are the main causes of most fungal infections in humans. The interaction between fungi and their host is complex and many factors contribute to establishment of fungal disease. This interaction must involve capability of fungal pathogens to thrive at body temperatures and resist irritants such as free radicals (Zaragoza 2019). Like this, other events result in adaptation of fungi to their host's environment and during this course pathogen adopt different strategies, which contribute to their virulence. In the following sections, several important virulence factors used by pathogenic fungi are discussed and summarized in Figure 7.1.

7.2.1 THE CELL WALL AND ITS COMPONENTS

The cell wall constitutes the outermost layer of the cell, and maintains its viability and gives it its characteristic shape. Many cell wall components bestow the virulence of fungi, including different carbohydrates (mannan, glucan, and chitin) (Garcia-Rubio et al. 2020). The cell wall plays an important role during host interaction by facilitating adhesion and later interacting with phagocytic cells. In this interaction, cell wall components pathogen-associated molecular patterns (PAMPs) may connect with pattern recognition receptors (PRRs) of host cells, like Toll-like receptors (TLRs)

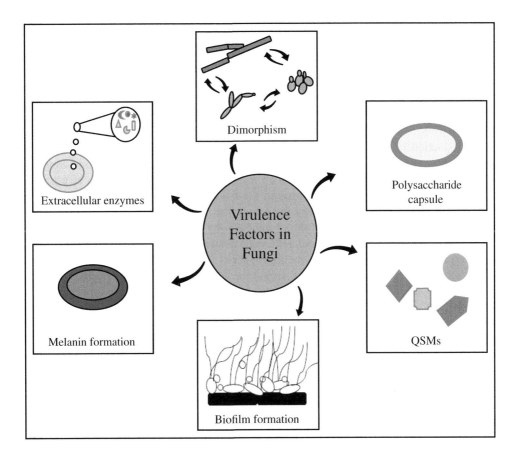

FIGURE 7.1 Diagrammatic representation of the virulence factors in fungi.

and C-type lectins and activate an immune response The response is generally initiated by phagocytic cells like macrophages and neutrophils. This process involves the production of reactive oxygen and nitrogen molecules as integral components of their defensive antimicrobial response (Arana et al. 2009). Given the significance of this contact, it is imperative to understand the role of different cell wall components contributing to virulence.

7.2.1.1 Glucans

Glucans constitute the principal component (50%–60%) of the dry weight. Although this component includes different bonds (α and β) in various fungi, β-glucans are easily recognized by the host immune system. The common β-1,3-glucan bond is a potent pro-inflammatory molecule, synthesized by glucan synthases, a product of *FKS* genes. The numbers of FKS genes vary from species to species, but fungi cannot tolerate the simultaneous deletion of all genes. The cell wall contains mannoproteins, which are covalently bound to the core of the polysaccharides and shield the β-glucan layer (Arana et al. 2009). This shielding helps to evade the host immunity (*via* the dectin-1 receptor on myeloid cells), and confers protection from complement activation (Klis et al. 2009). Unlike other polysaccharides, β-1,3-glucan is highly immunogenic and has a specific PRR, dectin-1 on the surface of the host immune cells (Brown and Gordon 2001). Disruption of *CEK1* (*C. albicans ERK*-like Kinase) (involved in the mitogen-activated protein kinase (MAPK) pathway) leads to increased β-glucan exposure, which triggers an immune response (Román et al. 2007, Román et al. 2016).

7.2.1.2 Chitin

Chitin exists inside of the cell wall and holds a crucial function in the interaction of *Candida* species with the host. It blocks the identification of *C. albicans* by murine macrophages and peripheral blood mononuclear cells (PBMCs), resulting in a noteworthy reduction in cytokine production. Since maneuvering the host immune response is crucial for infection, chitin in *C. albicans* reduces production of nitric oxide in host macrophages by inducing arginase-1 (Wagener et al. 2017). Chitin fragments in *C. neoformans* promote Th2 responses and inhibits the influx of neutrophils (Garcia-Rubio et al. 2020).

7.2.1.3 Capsule

Capsule production in certain species protects the fungi. The capsule of *C. neoformans* consists of two polysaccharides, glucuronoxylomannan (GXM) and glucuronoxylomannogalactan (GXMGal) (Ding et al. 2016). GXM makes up to 88% of the capsule and acts as a protective shield against dehydration and phagocytosis in the host (Zaragoza et al. 2009). It also absorbs microbicidal oxidative bursts released by phagocytic cells (Casadevall et al. 2018). It therefore makes sense that the size or content of the *C. neoformans* capsule increases during infection. In particular, GXM has been shown to promote fungal survival by triggering T cell and macrophage death, altering leukocyte cytokine production patterns and preventing leukocyte migration to infected sites (Previato et al. 2017). The other polysaccharide, GXMGal, although accounting for only ~8% of the capsule mass, has an important immunomodulatory role. It stimulates IFN-γ, I interleukin-17 (IL-17) and IL-10 production, activates dendritic cells, decreases peripheral blood mononuclear cell (PBMC) proliferation and triggers caspase-8-mediated T cell death, resulting in lower T cell numbers. Injecting mice with GXMGal preceding infection with *C. neoformans* provides protection from later infection. Uge1p, a presumed UDP-glucose epimerase, and Ugt1p, a presumed UDP-galactose transporter, play a crucial role in capsule formation and are indispensable for the virulence of *C. neoformans*. Strains lacking Uge1 ($\Delta uge1$) and Ugt1 ($\Delta ugt1$) do not construct GXMGal but instead synthesize a big capsule. The mutants, characterized by being GXMGal negative and GXM positive, exhibit an inability to colonize the brain. In contrast, strains lacking GXM (GXMGal positive) can still colonize the brain, albeit with reduced efficiency (Moyrand et al. 2007). In addition, GXMGal induces the FAS/FAS-L interaction supporting macrophage apoptosis (Previato et al. 2017).

7.2.1.4 Melanin

Melanin in fungi is either enclosed within the cell wall or released extracellularly. Many species, including *C. neoformans*, *Neurospora crassa*, *Aspergillus nidulans*, *C. albicans*, and other fungi produce various melanin pigments (Fling et al. 1963, Bull and Carter 1973, Walker et al. 2010, Smith and Casadevall 2019). The different types of melanin in fungi and the metabolic pathways involved are summarized in Table 7.1.

Melanin protects the fungus and increases the virulence of fungi by reducing the susceptibility of the pathogen to the host's antimicrobial systems and impairing the host's immunological response. *C. neoformans CnLAC1* (encoding laccase involved in melanin synthesis) mutants show reduced virulence in both *Galleria mellonella* and mice infection models (Zhu and Williamson 2004). Fungal melanins have been shown to influence host immunological responses through several mechanisms. Essentially, the melanin of *C. neoformans* provides electrons, that reduce ferric (Fe^{3+}) ions to ferrous (Fe^{2+}) ions, which have a redox buffering capacity in the host and protect the fungus (Nyhus et al. 1997; Jacobson and Hong 1997). *Exophiala dermatitidis* and *Aletrnaria alternata* DHN melanin protect them from hypochlorite and permanganate (Jacobson et al. 1995). In *C. auris*, melanization has been shown to protect cells from H_2O_2 stress (Smith et al. 2022).

Melanin also inhibits the deposition of C3 factor during complement activation in *A. fumigatus*. Compared to the melanized cells, non-pigmented *A. fumigatus* mutants (lacking alb1, arp1, or pksP) exhibit higher C3 deposition on their conidial surface (Tsai et al. 1997). Reduced C3 deposition inhibits recognition of fungus, subsequent binding, and clearance by the host. In addition to the innate response, melanin also interacts with components of the adaptive immune system. Immunization with cryptococcal melanin leads to a strong antibody response in mice and significantly increased levels of IgM and IgG are observed (Nosanchuk et al. 1998). The different roles of melanin in the pathogenesis of fungi are shown in Figure 7.2. Antibodies induced by melanin production have been evidenced in various fungal species such as *A. fumigatus C. neoformans*, *Fonsecaea pedrosoi* and *H. capsulatum* (Nosanchuk et al. 1998, Alviano et al. 2004, Nosanchuk et al. 2002, Youngchim et al. 2004).

7.2.2 DIMORPHISM

In dimorphic fungi, the cells have the ability to undergo a reversible transition from the unicellular yeast form to the multicellular hyphal form (true hyphae). In addition, some fungi also have a

---

TABLE 7.1
Types of Fungal Melanin and the Pathways Involved in Different Fungi

Species	Pigment Type	Pathway Involved	References
C. neoformans	DOPA, Pyo	Laccase	(Shaw and Kapica 1972; Frases et al. 2007)
C. albicans	DOPA	Laccase	(Morris-Jones et al. 2005)
A. flavus	DOPA	Tyrosinase	(Inamdar et al. 2014)
A. niger	DOPA, DHN	Unknown, PKS	(Pal et al. 2014)
A. fumigatus	DOPA, DHN, Pyo	Laccase, TDP, PKS	Tsai et al. 1999; Youngchim et al. 2004
A. nidulans	DOPA	Laccase and tyrosinase	(Bull and Carter 1973; Gonçalves et al. 2012)
C. gattii	DOPA	Laccase	(Chan and Tay 2010)
H. capsulatum	DOPA, DHN	TDP and laccase	(Almeida-Paes et al. 2018)
N. crassa	DHN	PKS, Laccase	(Ao et al. 2019)
C. auris	DOPA	?	(Smith et al. 2022)

Abbreviations: DOPA, DOPA-melanin; DHN, DHN-melanin; Pyo, pheomelanin; PKS, polyketide synthase pathway; TDP, tyrosine degradation pathway.

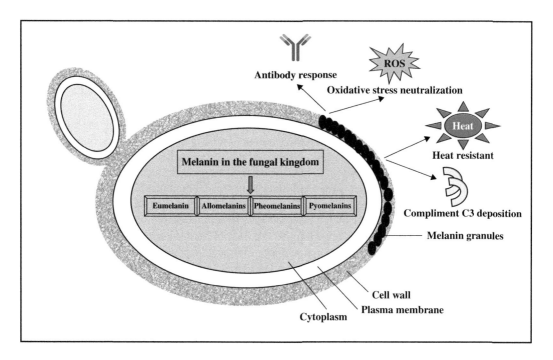

FIGURE 7.2   The complex role of melanin in the pathogenesis of fungi.

pseudohyphae form. The dimorphic change occurs in a few species of the Ascomycota phylum, including *C. albicans* (Figure 7.3). The dimorphism is dependent on the environmental conditions of the fungus such as 5% CO_2, a neutral pH at 37°C, serum, nitric oxide, hydrogen peroxide, and nitrogen deficiency, etc. (Mayer et al. 2013). It is assumed that both forms are required for infection and dissemination. The small yeast form spreads rapidly through the bloodstream, but the hyphae penetrate the epithelial barrier more effectively. It therefore makes sense that both forms occur in different parts of the infected host. *C. albicans* mutants arrested in one form are essentially avirulent (Lo et al. 1997).

However, understanding switching and filamentation is not an easy task, as many genes are involved. The main regulator genes include *EFG1*, *CPH1*, and *TUP1*, which control the change of

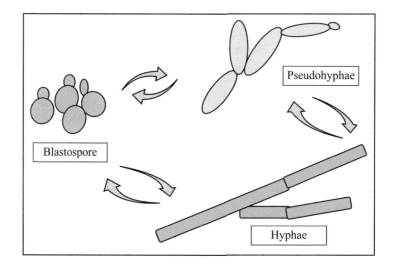

FIGURE 7.3   Yeast to hyphae transition in *Candida albicans*.

morphological forms *via* different pathways in *C. albicans*. In general, Efg1 and Cph1 are positive regulators of filamentation, while Tup1 is a negative regulator (Braun and Johnson 2000). Efg1 is the master regulator of filamentation and is linked to various regulatory networks, including biofilm formation, intestinal colonization, white/opaque switching, and virulence. *EFG1*-driven switching depends on the fungal environment and the availability of nutrients. Its reduced expression during the initiation of filament formation under serum-inducing conditions suppresses the formation of true hyphae in *C. albicans* and allows the formation of pseudohyphae (Braun and Johnson 2000). During initiation, it activates the adenylate cyclase Cyr1, converting ATP to cAMP. The interaction of cAMP with the regulatory subunit of the protein kinase A complex (PKA; Efg1-mediated cyclic AMP/protein kinase A) initiates a conformational change, resulting in liberation of the Tpk1 and Tpk2 catalytic subunits (Kim et al. 2023). Tpk1/2 phosphorylates target proteins, including Efg1, which binds at the promoter of hyphal-associated genes, including *EED1*, *TEC1*, and *CRZ1*. After initiation of filamentation, Efg1 negatively regulates its own expression. However, under hypoxic conditions, the deletion of *EFG1* results in an augmentation of filamentation.

Cph1 is essential for hyphal growth on agar, but its necessity is not observed in liquid media, through the mitogen-activated protein kinase pathway (Liu et al. 1994). A double mutation of Δ*cph1* and Δ*efg1* results in complete abrogation of filamentous growth indicating that Efg1 and Cph1 jointly govern the pathways that induce filamentous growth in *C. albicans*. However, homozygous mutants of *TUP1* in *C. albicans* produce cells arrested in the filamentous form (Lewis et al. 2002).

7.2.3 QUORUM SENSING (QS) AND BIOFILM FORMATION

Quorum sensing (QS) is a communication method among microorganisms that is dependent on cell density, allowing cells to function as multicellular organisms. This process is marked by the coordinated release of diverse quorum-signaling molecules (QSMs), which govern microbial metabolism and gene expression in various fungi (Albuquerque et al. 2013). QSMs are autoinducers and once the threshold level of QSMs is reached, it binds to its corresponding receptor and induces synchronized gene expression (Tian et al. 2021). In fungi, QSMs support fungal infection by forming biofilms and releasing virulence factors, etc., leading to multidrug resistance. Some of the QSMs are shown in Figure 7.4 and Table 7.2.

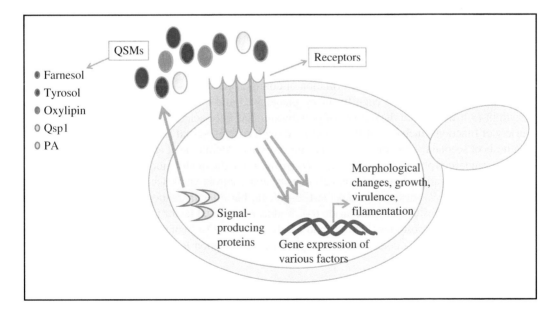

FIGURE 7.4 Quorum-sensing molecules.

TABLE 7.2

Role of Different Fungal Quorum-Sensing Molecules

QSM	Fungus	Role	References
Farnesol	*C. albicans*	Inhibits the transition from yeast to filamentous form, cell wall maintenance, influences genes linked to resistance against drugs	(Alem et al. 2006; Lindsay et al. 2012)
	A. nidulans	Induces apoptosis	(Semighini et al. 2006)
Tyrosol	*C. albicans*	Stimulates hyphae production, germ tube formation	(Alem et al. 2006)
Volatile compounds (CO$_2$)	*Trichoderma* spp.	Conidia formation	(Majumdar and Mondal 2015)
	C. neoformans	Capsule formation	(Majumdar and Mondal 2015)
	C. albicans	White-opaque switching	(Huang et al. 2009)
Farnesoic acid	*C. albicans*	Suppresses the growth of hyphae	(Davis-Hanna et al. 2008)
Oxylipin	*A. nidulans*	Alters sporulation	(Majumdar and Mondal 2015)
α-1,3-glucan	*H. capsulatum*	Protection in phagolysosomes	(Rappleye et al. 2007)
Pantothenic acid	*C. neoformans*	Increases growth and melanization	(Albuquerque et al. 2013)

Farnesol and tyrosol are known quorum-sensing molecules in *C. albicans*. Farnesol inhibits the induction of hyphal growth (Lindsay et al. 2012) and tyrosol induces the opposite effect of transition from stimulating the development of filamentous forms (Kovács and Majoros 2020). Farnesol considerably reduces the IL-12 and the Th1 cytokines interferon-γ (IFN-γ) production and thus inhibits the activation of the immune defense against systemic candidiasis when administered exogenously. Furthermore, farnesol impedes the ability of monocytes to differentiate into underdeveloped dendritic cells and the capability of these cells to properly activate T-cell response (Tian et al. 2021). Following the morphological characteristics of the cells, QSMs also regulate the development of *C. albicans* biofilms in conjunction with a decrease in the production of Hwp1, a hyphal-specific wall protein (Wongsuk et al. 2016). Tyrosol is an activator of hyphal growth and control of the cell cycle, chromosomal segregation and DNA duplication, as shown by transcriptional profiling studies (Wongsuk et al. 2016; Padder et al. 2018).

Within *C. neoformans*, the presence of multiple quorum-sensing systems imparts increased adaptability to various environmental conditions (Tian et al. 2021). A secreted peptide, quorum-sensing peptide 1 (Qsp1), pantothenic acid (PA) and oxylipin are QSM in *C. neoformans* (Lee et al. 2007; Albuquerque et al. 2013). In *C. neoformans*, virulence-induced changes in cell size have been linked to QS regulation. This transformation of common yeast cells into giant cells (titans), makes the fungal cells resistant to engulfment by phagocytes and nitrosative stress (Tian et al. 2021). Oxylipin (3-hydroxylated derivatives of polyunsaturated fatty acids (PUFA), a lipid QSM serve a variety of functions, including differentiation, growth control, sexual and asexual reproduction and synthesis of secondary metabolites (such as mycotoxins) (Mehmood et al. 2019).

In thermodimorphic pathogenic fungus *H. capsulatum*, the modulation of α-1, 3-glucan production in the cell wall is dependent on cell density and controls virulence. Free-living saprophytic mycelial form of *H. capsulatum* lacks α-1, 3-glucan, but after inhalation it transitions to a yeast form and synthesizes the polysaccharide α-1, 3-glucan, a critical factor for virulence, and provides protection in phagolysosomes (Klimpel and Goldman 1988). Qsp1 controls the production of melanin in *C. neoformans* by controlling binding of the transcription factor at the laccase gene, *LAC1* (Summers et al. 2020).

7.2.3.1 Biofilm Formation

Biofilms are important virulence factors and refer to a dense network of pseudohyphae embedded in the host cells' surface. They prevent the penetration of antifungal drugs inside, control the expression and extrusion by drug pumps and shield the fungus from host defense mechanisms (Martinez and Casadevall 2015). This recurrent phenomenon may be the reason why many infections persist despite effective antimicrobial therapy (Jabra-Rizk et al. 2004). Most *Candida* species associated with clinical candidiasis develop biofilms, posing a significant risk and complicating treatment (Ajetunmobi et al. 2023). However, infections with *C. auris* are of particular concern as the fungus itself is notoriously resistant, forming robust biofilms and increasing the challenges of treating infection (Smith et al. 2022).

Biofilm formation consists of several phases. In the first phase, the so-called 'colonization phase', fungal cells attach to the surface and in the second phase, which is the so-called 'proliferation phase', the fungal cells multiply and form an organized cell structure. During the 'maturation phase', the formation of an extracellular matrix begins, which envelops the entire cell network. The detachment of yeast cells from a mature biofilm during the 'dispersion phase' enables the re-formation of a new biofilm at a different location (Ajetunmobi et al. 2023) (Figure 7.5). The spread of biofilm-associated cells leads to widespread invasive infection and candidemia (Tsui et al. 2016). Adhesins, product of the *HWP1*, *ALS3*, and *ALS1* genes of the ALS family (agglutinin-like sequence) are hyphal wall proteins that serve a crucial function in biofilm formation and adhesion of *Candida species*. In hyphal form, Hwp1 is expressed and an Δ*hwp1* is avirulent in murine models of disseminated and oesophageal candidiasis (Staab et al. 2013). Similarly, deletion of *ALS1* in *Candida* delays germ-tube formation and fungal dissemination in the mouse model (Zhao et al. 2022).

Colonization with *Aspergillus* and the formation of biofilms is more likely in patients with hereditary lung diseases. Under conditions of biofilm formation in *Aspergillus*, all antifungal drugs lose efficacy, so either high doses or combination therapy must be used to treat the infection (Sardi et al. 2014). The ECM of the fungus *A. fumigatus* biofilms consists of lipids, proteins, and exopolysaccharides, including galactosaminogalactan (GAG), and strains lacking GAG production are unable to form a matrix and adherent biofilms in invasive aspergillosis models (Speth et al. 2019, Liu et al. 2022). In addition to adhesion, GAGs reduce host recognition and activate endothelial cells, contributing to formation of blood clots during the course of infection. They reduce the antifungal ability

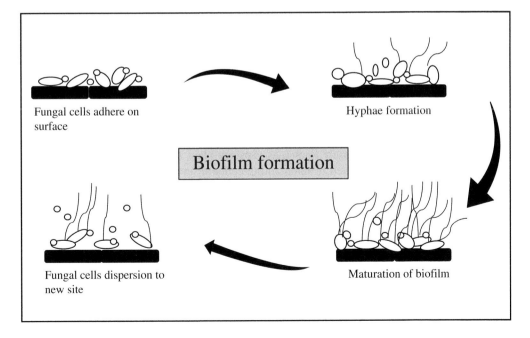

Fungal cells adhere on surface

Hyphae formation

Biofilm formation

Fungal cells dispersion to new site

Maturation of biofilm

FIGURE 7.5   Biofilm formation in fungus.

of neutrophils by inducing their apoptosis and reducing the cytokine response of Th1 and Th17 cells (Speth et al. 2019).

7.2.4 EXTRACELLULAR ENZYMES

Extracellular enzyme production and secretion are related to virulence in fungal pathogens. Mostly hydrolytic enzymes, such as proteases, lipases, and phospholipases are important virulence factors and are described in fungi such as *Candida, Aspergillus*, and the *Cryptococcus* species (Ghannoum 2000).

7.2.4.1 Proteases

Proteases break down various host proteins, including collagen, mucin, and keratin. These enzymes also degrade the antibodies, complementing proteins and cytokines. These enzymes are either membrane bound or can be secreted in the surrounding extracellular space and their location varies between species. The most extensively studied proteases are secreted aspartyl proteases (SAPs). Besides mucosal membranes and the breakdown of immune defense proteins, aspartic proteases also contribute to adherence, phenotypic switching (W/O transition), conversion of yeast cells into hyphae, and biofilm formation (Mroczyńska and Brillowska-Dąbrowska 2021, Kulshrestha and Gupta 2023). *SAP* genes are found in different *Candida* spp., including *C. albicans, Candida dubliniensis, Candida tropicalis, Candida glabrata, Candida parapsilosis*, etc., which enable the production of active extracellular proteinases *in vitro* (Naglik et al. 2003). *C. albicans* pathogenicity is associated with extracellular proteolytic activity of Sap proteins, a family of 10 proteins, and a null mutant study is summarized by Naglik et al. (2003). *sap1, sap2*, and *sap3* null mutants' mucosal infection studies show reduced adherence to buccal epithelium, cause less tissue damage, and are less virulent in different models of *C. albicans. sap4-sap6* triple mutants showed the most significant reduction in systemic model and their clearance was much faster than wild-type cells. *C. parapsilosis* possesses three genes encoding aspartyl acid proteases, Namely Sapp1, Sapp2, and Sapp3. Among these, Sapp1 and Sapp2 (excluding Sapp3) also impact processes such as adhesion, phagosome-lysosome maturation, phagocytosis, and cytokine secretion by human macrophages (Singh et al. 2020). Studies suggest *SAPP1* duplication in the genome (*SAPP1a, SAPP1b*). Δ/Δ*sapp1a*, Δ/Δ*sapp1b* mutant show attenuated damage to the host cell, exhibit enhanced susceptibility to phagocytosis and are efficiently eliminated by macrophages and monocytes more than are the wild-type (Horváth et al. 2012). In a separate investigation involving reconstituted human oral epithelium (RHOE), the addition of the Sap inhibitor pepstatin leads to a decrease in tissue damage caused by *C. parapsilosis* (Silva et al. 2009). Out of the seven Saps of *C. auris* (named Sapa1 to Sapa7), Sapa3 acts as a key molecule and its removal results in a sizeable decrease in Sap activity. Additionally, it also plays a role in *C. auris'* biofilm formation, and its deletion significantly reduced virulence (Kim et al. 2023). In *C. glabrata*, a family of 11 cell surface-bound aspartyl proteases, referred to as CgYps1–11 (commonly known as yapsins), plays a pivotal role as a virulence factor. Yapsins are reportedly involved in maintaining cell wall integrity, adherence to mammalian cells, survival in macrophages, and virulence (Kaur et al. 2007). *A. fumigatus* possesses two genes encoding aspartyl proteases, namely *PEP1* and *PEP2*, with *PEP1* being expressed during infection in various animal models. Studies involving mutants indicate that Pep1 can break down proteins of extracellular matrix within the host; however, its absence does not impact host mortality. *C. neoformans*, major aspartyl peptidase 1 (May1) is also involved in virulence, as the *may1Δ* mutant shows a remarkable reduction in virulence in the mouse model (Clarke et al. 2016).

7.2.4.2 Lipases

Lipases facilitate the breakdown of ester bonds in triacylglycerols, releasing fatty acids. Lipases are integral to lipid metabolism and fungal lipases digest host lipids to acquire nutrients, and affect the inflammatory process by lysing competing microflora. In *C. albicans*, ten different lipases

(*LIP1–10*) have been recognized, while *C. parapsilosis* has two *LIP* genes, *CpLIP1* and *CpLIP2* (Tóth et al. 2014). In the murine model, lipase-deficient mutants of both *C. albicans* and *C. parapsilosis* results in loss of virulence (Mroczyńska and Brillowska-Dąbrowska 2021). *C. albicans LIP8* mutants (Δ*lip8*) in an intravenous infection model (Gácser et al. 2007) and *C. parapsilosis CpLIP1* and *CpLIP2* double mutant (*lip−/−*), display significantly reduced virulence across diverse infection models and are easily eliminated by mouse macrophages (Tóth et al. 2014). Similar results are reported in the skin-associated fungal pathogen *Malassezia* (DeAngelis et al. 2007). However, limited details exist regarding lipases of *C. neoformans*. Its genomic sequencing revealed three genes, and there is no evidence of extracellular lipase activity from clinically isolated *C. neoformans* (Park et al. 2013).

7.2.4.3 Phospholipases

Phospholipases are ubiquitous and hydrolyze phospholipids. Phospholipases are broadly divided into four types: Phospholipase A, Phospholipase B (known as *PLB1*), Phospholipase C, and Phospholipase D (known as PLD), which in turn are divided into subtypes based on the hydrolysis of ester bonds in their substrate phospholipids (Barman et al. 2018). Although phospholipases belong to a multi gene family, *PLB* is important in the virulence of pathogenic fungi. Three genes (*PLB1*, *PLB2*, and *PLB5*) exist in *C. albicans* but one in *C. neoformans* (*CnPLB1*) (Leidich et al. 1998, Merkel et al. 1999). The phospholipases may be produced intracellularly but are secreted in pathogenic fungi; however, limited studies are available highlighting their role in virulence. During infection, they disrupt host membranes and/or modulate fungal cell signaling to produce immunomodulators and increase fungal survival (Alspaugh 2015). *C. albicans* isolates exhibiting strong adherence to buccal epithelium, were pathogenic in mice with higher phospholipase activity, while few pathogenic isolates displayed lower phospholipase activity (Barrett-Bee et al. 1985). The null mutant, *plb1*Δ of *C. albicans*, although not impaired in adherence to epithelial cells, its infiltration into deeper tissue sites is greatly reduced in animal studies (Leidich et al. 1998). *CaPLB5* inactivation results in impaired organ colonization (Theiss et al. 2006). *C. neoformans'* Plb1 increases fungal pathogenesis in the central nervous system and causes brain damage. Its *PLB1* mutant show altered capsular polysaccharide with impaired ability to stimulate glial cells in mice (Hamed et al. 2023). Additionally, *PLC1* of *C. neoformans* is also implicated in capsule formation, cell wall integrity, and adaptation for growth in host (Djordjevic 2010). Δ*plc1* mutants show reduced growth and melanin levels at host's standard temperature (37°C). This mutant has a compromised cell wall and an inability to trigger mitogen-activated protein kinase (MAPK) with cell wall-disrupting agents. Unlike Δ*plc2*, which has comparable virulence to wild type, Δ*plc1* lacks virulence in mice and demonstrates weakened lethality in the *Caenorhabditis elegans* model (Chayakulkeeree et al. 2008). However, *PLD* of *A. fumigatus* holds significant importance in its virulence factor and its release during infection elevates reactive oxygen species (ROS) levels during signaling pathways in lung infection (Yu and Huo 2018). *A. fumigatus* Δ*pld* has attenuated virulence in mice, suggesting a role in internalization in lung epithelium cells (Li et al. 2012).

7.2.4.4 Urease

Urease helps in nitrogen assimilation under nutrient-limited conditions in both fungal and bacterial pathogens (Xiong et al. 2023). Xiong and coworkers studied the *A. fumigatus* urease deficient mutant (Δ*ureB*), which was found to have reduced virulence in the murine model, and spores were more susceptible to macrophage-mediated killing. Δ*ureB* exhibits 50% greater survival than wild-type cells, which submit entirely to infection within four days in the lung infection model. The lungs and livers of the mice that survived were nearly free of infection (Xiong et al. 2023). Moreover, urease plays a role in multiple stages of cryptococcal pathogenesis, including facilitating the crossing of the blood–brain barrier. Δ*ure1* of *C. neoformans* exhibit reduced growth *in vitro*, and mice injected with macrophages having urease-deficient strain exhibit a reduced pathogen load in the brain compared to wild-type strain. Urease influences fungal fitness within the mammalian phagosome. It encourages non-destructive exocytosis, while impeding intracellular replication, consequently minimizing damage to the phagolysosomal membrane. These mechanisms may facilitate the spread of crypto-coccal infection with transport inside macrophages (Fu et al. 2018). *C. albicans* reports suggest a lack of urease activity, but most strains grow well with urea as the sole nitrogen source (Navarathna et al. 2010).

7.2.5 MANNITOL

Mannitol, a six-carbon cyclic sterol alcohol is produced in large quantities by various fungi, includ-ing *C. neoformans* and *Aspergillus* spp., suggesting its suitability as a biomarker (Nguyen et al. 2020). Mannitol plays a role in *C. neoformans'* pathogenesis and survival by increasing the toler-ance to osmotic thermal and oxidative stress. It has been hypothesized that mannitol helps to pro-mote an increase in CSF osmolality, leading to significant brain damage in patients. It safeguards *Aspergillus niger*'s spores from any harm faced during stressful situations. In their research, Ruijter and colleagues investigated the significance of the gene encoding mannitol 1-phosphate dehydroge-nase (*mpdA*) and found that conidiospores of Δ*mpdA* strain are highly sensitive to various stress fac-tors, including elevated temperatures and oxidative stress, and lyophilization (Ruijter et al. 2003). However, there is currently no convincing evidence regarding the role of mannitol in the virulence of *Candida* spp.

7.2.6 GENERATION OF TITAN CELLS AND ASTEROID BODIES

C. neoformans can change its cell size and are transformed into titan cells (Figure 7.6). These cells are exceptionally enlarged, reaching sizes that are 14 to 20 times larger than typical yeasts grown *in vitro*. Reportedly, the fungal cell body can grow when confronted with amoebae and macrophages, suggesting that the increase in size serves as a protective mechanism (Chrisman et al. 2011). Crabtree and co-workers studied the G protein-coupled receptors encoding mutant (*gpr4Δ*, *gpr5Δ*), having minimal *in vivo* titan cell production and show attenuated virulence, a lower colony forming unit (CFU) in tissue and reduced dissemination, indicating that the titan cell is a distinc-tive morphology boosts the virulence of *C. neoformans* and possibly implicated in latent infections (Crabtree et al. 2012).

Asteroid bodies (AB) have been identified in various fungi, such as *Candida*, *Histoplasma*, *Sporothrix*, and *Aspergillus* (Daniel Da Rosa et al. 2008). These ABs act as protective shields for the yeast, safeguarding them from the external environment, while simultaneously trapping antibody fragments in the outer crown, indicating potential interference from the immune system (Hernández-Chávez et al. 2017). Further, Daniel Da Rosa and co-workers hypothesized that AB might function as resilient structures. (Daniel Da Rosa et al. 2008). Contradictorily, there is limited research on *Candida* and *Aspergillus* in relation to AB, highlighting the urgent need for compre-hensive study in this area.

7.2.7 THERMOTOLERANCE

A crucial element contributing to fungal virulence in systemic infections is their capacity to thrive at temperatures exceeding 37°C. The thermotolerant fungi can endure temperatures beyond 37°C, and are essential in triggering invasive fungal diseases in mammalian hosts. *A. fumigatus* exhib-its notable thermophily, thriving in temperatures ranging from 55°C to 77°C. Under high tempera-ture, opportunistic fungal pathogens employ an adaptive strategy referred to as the heat shock (HS) response, regulated by heat shock transcription factors (HSFs). The ability to grow at elevated tem-peratures in fungi depends on heat shock proteins (HSPs), as *C. albicans* and *S. cerevisiae* require Hsp104 and Hsp70, while *C. neoformans* and *P. brasiliensis* need Hsp90 (Berman and Krysan 2020, Chatterjee and Tatu 2017).

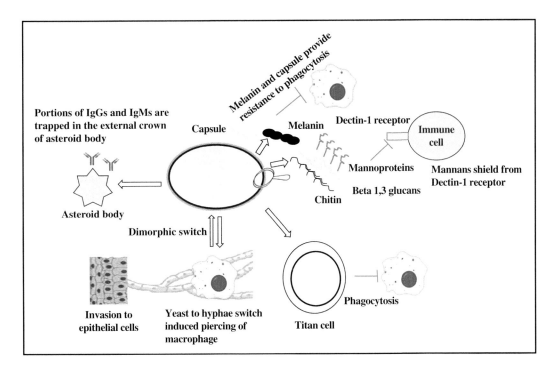

FIGURE 7.6 Components of the fungal cell wall involved in evasion of the immune system.

Hsp104 serves as a pro-survival intermediary in response to temperature elevation, indicating a crucial role for this protein in enhancing thermotolerance (Gong et al. 2017). The expression of Hsp104 rises in *C. albicans*, when cells are exposed at high temperatures. In *C. neoformans*, Hsp90 is implicated in the regulation of thermotolerance, as evidenced by increased sensitivity to growth at 37°C when there is a slight impairment of Hsp90 function compared to 25°C (Chatterjee and Tatu 2017). The cell wall ultra structure undergoes significant alterations after *A. fumigatus* is subjected to heat shock, and the transcription factor HsfA plays a crucial role in ensuring the viability, thermotolerance, and integrity of the cell wall in *A. fumigatus* (Fabri et al. 2021). Moreover, in *P. brasiliensis*, Hsp90 functions to safeguard cells from cellular and molecular damage during exposure to heat stress. Its role involves the regulation of reactive oxygen species (ROS) levels. In yeast cells, inhibiting Hsp90 at an elevated temperature (42°C) resulted in an increase in ROS levels, while no significant effect on ROS levels was observed at the optimal temperature of 37°C (Matos et al. 2013).

7.2.8 Fungal Toxins

Toxins form a unique class of fungal factors, contributing to the range of tools for virulence that pathogens use during infection. Candidalysin, a fungal toxin is a cytolytic peptide produced by *C. albicans*, and is vital in both initiating infections and triggering the immune response (Moyes et al. 2016). During mucosal infections, candidalysin plays a vital role and is generated through targeted proteolytic modification of the protein Ece1p by kexin proteases. Upon secretion, candidalysin triggers disruption of epithelial cell membranes, initiating cellular stress that results in cell death and facilitates fungal invasion (Richardson et al. 2022). *A. fumigatus* toxins act as defense molecules against predators and competitors in an ecological habitat, but might also play a role in its pathogenicity by directly assaulting the host. One such toxin is gliotoxin, identified in the lungs of patients of invasive aspergillosis (Lewis et al. 2005). In laboratory settings, it prompts cytoskeleton restructuring in human alveolar epithelial cells, thereby facilitating conidial uptake (Brown et al. 2021).

7.3 HOST-PATHOGEN INTERACTIONS

The relationship between fungi constantly evolves with time, as physiological states can alter fungal phenotype and promote its pathogenicity.

Infection starts with the pathogen's attachment to the cell surface and invasion inside the host cell. During this interaction, the pathogen binds to a variety of surface molecules of ECM (Mendes-Giannini et al. 2005). After successful invasion, phagocytic cells are the first defense against fungal pathogens. Phagocytes recognize fungi with their cell wall component, PAMPs *via* several multiple PRR (Díaz-Jiménez et al. 2012). Here, fungi initially resorted to evading the immune system, hiding their PRR by structure and assembly of the cell wall, capsule, dimorphism, release of virulence factors, generation of resistance and tolerance (Brown 2023). In response, the host immune system also has vital strategies to combat these pathogens, which results in a tug of war situation. However, the outcome depends on multiple factors like pathogenicity of the species/strain, health of the host, and environmental conditions for topical infections. To understand this, we must discuss the series of events, which essentially starts with adhesion.

7.3.1 ADHESION

The capability of *Candida* and other fungal species to securely adhere to diverse surfaces, such as human skin, endothelial, epithelial, and mucosa, is regarded as the initial stage in the onset of fungal infection. Basic studies on *C. albicans* were later extended to *A. fumigatus*, *H. capsulatum*, *Coccidioides immitis*, *B. dermatitidis*, and *C. neoformans* (Mendes-Giannini et al. 2005). Aminolevulinate synthase, Als1p, Ala1p/Als5p, hyphal wall protein (Hwp1) of *C. albicans*, and Epa1p (epithelial adhesion) of *C. glabrata* have been studied already. They belong to class of cell wall protein called glycosylphosphatidylinositol-reliant (GPI-CWP) with a signaling peptide (N-terminal) and another (C-terminal) for attachment of GPI-anchors, and lead to binding of glucans to the cell wall. Epa1 is a major contributor of virulence and adherence to host cells and *EPA1* deletion in *C. glabrata* reduces adhesion (Mendes-Giannini et al. 2005). In the case of *C. albicans*, another adhesion gene known as *EAP1* (enhanced adherence to polystyrene) has been identified to code for a putative cell wall adhesin, and deleting *EAP* reduces cell adhesion to polystyrene and epithelial cells. Sequence analysis has revealed that it also contains a GPI anchor site, has a signal peptide, and is homologous with several other genes, which encode cell-wall proteins (Li and Palecek 2003).

An alternative form of communication that allows the fungus to attach to the host involves the precise identification of glycoconjugates (with lectins) on the epithelium located on the fungal wall surface. Adhesion is followed by fungal cell uptake inside host cells.

7.3.2 INTERNALIZATION OF FUNGI

Invasion can occur in endothelial or epithelial cells and in professional phagocytes (Finlay and Falkow 1997). Fungi invade the host by polymerizing microfilaments and microtubules of endothelial cells (Filler et al. 1995). *In vitro* experiments conducted using the human umbilical vein endothelium, J774 mouse macrophages, and an A549 lung epithelial cell investigates the uptake of two strains of *A. fumigatus*. Conidial uptake by A549 requires reorganization of the cytoskeleton in the host and colchicine, cytochalasin D treatment inhibited fungal uptake. Similarly, in the human brain microvascular endothelial cells (HBMEC), *C. neoformans* induces reorganization of the actin cytoskeleton. It is reported that a dephosphorylated form of cofilin (an acting binding protein) increases, along with the actin rearrangement (Chen et al. 2003). Newman and colleagues discovered that the cytochalasin D treatment, which targets actin microfilaments, impedes both the macrophage attachment and the *H. capsulatum* digestion (Newman et al. 1990). Successful internalization led to a cascade of changes studied under signaling.

7.3.3 SIGNALING

Due to their key role in the control of pathogenicity, signal transduction pathways are of utmost importance in research on fungal pathogenesis. Fungi undergo adaptation and activation of signaling pathways, enabling its survival in the hostile setting within the host cell. Diverse types of MAP (Mitogen-Activated Protein) kinase pathway are involved: Hog1 (High-Osmolarity Glycerol) and Cek1 (*C. albicans* Extracellular Signal-Regulated Kinase 1)-mediated pathways, the cyclic adenosine monophosphate (cAMP)-dependent MAPK pathways (Singh et al. 2020).

7.3.3.1 Mitogen-Activated Protein Kinases (MAPKs)

During colonization, *C. albicans* cells induce MAPK signaling. MAPKs participate in the activation of transcription factors involved in stress responses, hyphal development, cell cycle regulation, and alterations in glycosylation, among other functions (Herrero de Dios et al. 2013). The MAP kinase responds to environmental stimuli, and MAPKKKs phosphorylate MAPKKs, which, in turn, phosphorylate MAPKs. In *C. albicans*, three different MAPK pathways are involved during colonization (Figure 7.7)

7.3.3.1.1 CEK1-Mediated Pathway

It is implicated in filamentation, mating, cell wall formation, and the interaction with the host. CEK1-mediated MAPK pathway controls β-glucan display to PRR. Cell wall damage also initiates this pathway and is involved in unmasking of β-glucan, which results in fungus recognition by an innate immune response; thus, *CEK1* protects from recognition and evasion of immune response and *cek1Δ* mutants expose more β-1,3 glucan on their surface. Mutants of the *CEK1* gene are defective in hyphal growth and demonstrate diminished virulence in the murine mastitis model (Csank et al. 1998).

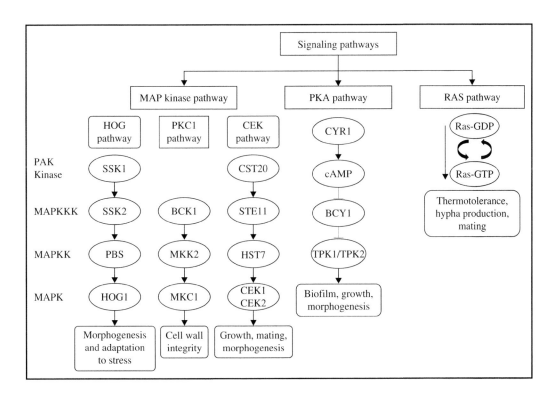

FIGURE 7.7 Signal transduction pathways in *C. albicans*.

7.3.3.1.2 HOG Pathway

C. albicans accrues glycerol through the Hog1-dependent pathway. Hog1 undergoes phosphorylation in response to hyperosmotic stress and subsequently translocates to the nucleus during osmotic stress. It represses filamentation in *C. albicans* independent of *Efg1* and *Cph1* (Chen et al. 2020). There exists a cross talk linking the Cek1 and HOG signaling and control of filamentation. Under basal conditions, Hog1 inhibits the Cek1 MAP kinase, and the activation of Cek1 is associated with resistance to specific cell wall-inhibiting agents. The HOG pathway is also involved in cell wall biogenesis of *C. albicans*. In *C. neoformans*, the HOG pathway is associated with stress response and virulence.

The stress-sensitive two-component arrangement, which senses the external stress, includes the response regulator Ssk1 and two additional components, Tco1 and Tco2. Deleting *TCO1* results in a decrease of pathogenicity in an intranasal animal model. The condition of *A. fumigatus* remains uncertain, as *sakA* mutants demonstrates sensitivity to oxidants; however, the importance of *sakA* remains unexplored in virulence. Nonetheless, mutants lacking *hog1* (MAPK) and *pbs2* (MAPKK) display bigger capsules in *C. neoformans*. Despite increased capsular formation, *hog1* and *pbs2* mutants exhibit diminished virulence in animal models (Bahn et al. 2005). The conidia germination of *A. fumigatus* also depends on the HOG signaling, as the *sakA* strain (an ortholog of HOG1) displays failure to thrive in elevated osmotic conditions and their germination in limiting nutrients is hindered. Kirkland et al. (2021) reported that SskA regulates *A. fumigatus* germination in the airways via the SakA-MAPK pathway, contributing to increased disease initiation and inflammation in the lungs.

7.3.3.1.3 MKC-1 Pathway

In *C. albicans*, the Mkc1-MAPK pathway regulates the cell wall stability. Deletion of Mkc1 results in increased sensitivity to cell wall-targeting enzymes and antifungals, accompanied by surface changes at higher temperature. These *mkc1* cells display diminished virulence in the mouse model and display inefficient colonization of target organs (Diez-Orejas et al. 1997). In *C. neoformans*, *mpk1* mutants do not have functional CWI, are avirulent in murine models, and do not survive the human body temperature (Hartland et al. 2016).

7.3.3.2 cAMP/PKA Signaling

cAMP levels play a role in regulating diverse cellular processes in fungi. *C. albicans* and *A. fumigatus* signaling pathways involve adenylyl cyclase and play a crucial role in shaping specific phenotypes necessary for virulence. This pathway involves cAMP generation from ATP, through adenylate cyclase (a Ras protein).

The PKA enzyme in *C. albicans* comprises two catalytic subunits (Tpk1 and Tpk2), both associated with a regulatory subunit (Bcy1), which remains in an inactive state. External stimulants such as glucose, amino acids, and serum initiate the cAMP-PKA signaling by activating adenylyl cyclase (Cyr1), resulting in the conversion of ATP to cAMP. cAMP is a secondary messenger which binds to the regulatory subunits of PKA, initiating a structural change, resulting in its dissociation and activation of catalytic subunits (Figure 7.7) (McDonough and Rodriguez 2011). Survival in *C. albicans* relies on both the regulatory subunit Bcy1 and the catalytic subunits Tpk1 and Tpk2. The *bcy1/bcy1* mutant is viable with a *tpk2/tpk2* mutant, displaying diminished PKA activity, than wild type. Furthermore, while the *tpk2/tpk2* strain shows decreased virulence in a murine model, the *tpk1/tpk1* strain does not exhibit such an effect. The signaling pathway involving cAMP/PKA plays a regulatory role in capsule size, melanin formation, and virulence in *C. neoformans*. Various components of this pathway have been identified, including genes responsible for adenylyl cyclase (Cac1), Gα protein (Gpa1), a potential G-protein–coupled receptor (Gpr4), and the catalytic (Pka1, Pka2) and regulatory (Pkr1) PKA subunits (Hu et al. 2007). A defect in *PKA1*, which encodes the catalytic subunit of cAMP-dependent PKA, is known for diminishing capsule size and reducing

C. neoformans virulence. The significance of the cAMP/PKA pathway in *C. neoformans* pathogenesis is highlighted by results that any defect in pathway cause altered virulence in a cryptococcosis mouse model. For instance, *pka1*, *gpa1*, and *cac1* mutants display lower melanin levels and decreased virulence (Alspaugh et al. 2002).

7.3.4 RAS-MEDIATED SIGNALING PATHWAY

The Ras superfamily comprises small G-bindings and like all G proteins, the activation of Ras is contingent upon the substitution of GDP with GTP and its subsequent hydrolysis (Boguski and McCormick 1993). *C. albicans* has two Ras genes, namely *RAS1* and *RAS2*, and mutants lacking *RAS1* display defects in hyphal transition and a reduced virulence in mouse infection models. On the other hand, hyphal growth of *ras2Δ* remains normal; however, deleting *RAS2* in the *ras1Δ* background significantly exacerbates the hyphal defect. This suggests that Ras2 also plays a role in hyphal development (Zhu et al. 2009). *C. neoformans* also has *RAS1* and *RAS2*. *RAS1* is indispensable for various processes in *C. neoformans*, such as growth at physiological temperature, filamentation, mating, and agar adherence. Given its pivotal function, especially in temperature tolerance, Ras1 is regarded as a significant virulence factor. Notably, the Δ*ras1* mutant exhibits avirulence in a cryptococcal meningitis model in rabbit (Alspaugh et al. 2000). In *C. elegans*, *Drosophila melanogaster*, and other models, the Δ*ras1* mutant exhibits reduced virulence at a low temperature. Thus, in non-mammalian model systems, Ras1 signaling is crucial in the pathogenicity of *C. neoformans* under reduced temperature conditions.

7.3.5 DAMAGE TO THE HOST

Due to different outcomes of host-pathogen interactions, host damage may be variable. Disease develops when damage increases over time and reaches a certain threshold (Casadevall and Pirofski 2003). Other consequences, such as latency (*Cryptococcus*) or commensalism, result in fungal survival in the host for a longer period with limited damage. Various fungi can cause chronic infections and have symbiotic relationships with their host, although the balance between the benefit and harm they provide may vary. The oral, vaginal, and gastrointestinal tracts are colonized by *C. albicans*, a well-studied commensal fungus that cause disease and damage in immune compromised hosts. There is no known environmental reservoir for *C. albicans*, and it is considered an obligate commensal fungus (Jabra-Rizk et al. 2016).

7.4 INNATE IMMUNITY AGAINST FUNGI: FROM DETECTION TO ELIMINATION

7.4.1 FUNGAL INVASION OBSTACLES AND RECOGNITION OF FUNGI BY HOSTS

The anatomical and physical barriers keep pathogens out, including skin and mucus membranes, which act as first lines of defense against fungal invasion. Skin colonization by *C. albicans* is influenced by skin-resident dendritic cells, whereas the epithelial tight junctions serve as a physical barrier to fungal invasion (Drummond et al. 2015). Inhalation of spores, especially *Aspergillus* spp., is one of the main routes of exposure to the lungs. Once in the airways, tight junctions of epithelial cells prevent fungi from entering the host. The mucus forms a layer that traps fungal spores and allows the cilia to transport the fungal particles out of the airways and into the stomach for digestion (Bartemes and Kita 2018).

Fungal infections can breach the physical barriers and enter the host. PAMPs and DAMPs recognition by PRRs of host triggers a succession, involving intracellular signaling events that lead to the attraction of innate immune cells (Marshall et al. 2018). Much work on PRRs participating in the immune response against fungi, including C-type lectin receptors (CLRs), as Mincle, and dectin-1 have shown PRRs role in fungal identification in the host. In particular, the fungal cell wall

1,3-glucan is recognized by dectin-1, and activate NF-B signaling, phagocytosis, inflammasome activation, and ROS generation (Vautier et al. 2012).

7.4.2 MACROPHAGES AND NEUTROPHILS

Macrophages are important immune cells in combating fungal infection. The process of attracting macrophages to the site of fungal infection is quite dynamic. *A. fumigatus*, which is surrounded by macrophage clusters in zebrafish is inhibited in transitioning from spore to hyphae, generally associated with pathogenicity (Rosowski et al. 2018). Phagocytosis is the main method of pathogen clearance utilized by macrophages. Their PRRs or Fc receptors attach with opsonizing antibodies or fungal PAMPs, respectively, results in fungal engulfment. Two modes of fungal engulfment have been identified, including coiling and zipper phagocytosis. IgG antibodies and Fcγ receptors bound to *C. neoformans* often cross-link and enhance phagocytosis of *C. neoformans*. IgG-mediated phagocytosis is prevented by the lack of lipid rafts formation, having densely localized Fcγ receptors (Bryan et al. 2021). Hatinguais et al. (2021) have shown that mitochondrial ROS contribute to the destruction of phagocytosed *A. fumigatus* conidia and result in the formation of TNF and IL-1 *in vitro*, which in turn triggers further antifungal responses. In a mouse model, suppression of mitochondrial H_2O_2 reduces the ability of alveolar macrophages to phagocytose *A. fumigatus* without affecting neutrophil activity, as well as survival and fungal burden (Shlezinger and Hohl 2021). Phagocytosis of *C. albicans* can lead to yeast-to-hyphae switching that kills macrophages by mechanically penetrating them with hyphae or inducing pyroptosis, allowing the fungus to escape. To neutralize this, macrophages can fold hyphae at the septal junction, which drastically slows down hyphal growth and damages the cell membrane. Although the extent to which hyphal folding contributes to the elimination of the fungus is not known, this reflects a macrophage function that has not yet been investigated and may apply to other hyphal pathogens (Bain et al. 2021).

Similar to macrophages, neutrophil migration to fungi is also stimulated by the recognition of PAMP/DAMPs by PRRs. Like macrophages, recognition of PAMP/DAMPs by PRRs induces neutrophil migration to fungi. The pro-inflammatory signaling and neutrophil recruitment are triggered by binding of Mincle to α-mannose and another component in fungal cell wall. In addition, the neutrophil-mediated solid antifungal response is triggered by the recognition of double-stranded RNA by the RIG-I-like receptor *via* MDA5/MAVS signaling (Burgess et al. 2022). Neutrophils manage their movement to areas of infection via a method of neutrophil swarming (Lämmermann et al. 2013). *In vitro*, swarming diminishes the growth of different types of fungi. Strains of *C. auris*, *C. albicans*, and *C. glabrata* that are incapable of forming hyphae, known as yeast-locked variants, exhibited smaller swarming patterns compared to the typical wild-type *Candida* spp. Swarming is also observed around hyphae in an *in vitro* model of *A. fumigatus* infection. This suggests a possible link between hyphal development and neutrophil swarming that needs to be further explored. Since swarming in fungi depends on *LTB4*, it works in the same way as in other infections or injuries. (Burgess et al. 2022). After migration to sites of infection, neutrophils can eliminate the fungus by a variety of strategies, e.g., degranulation, phagocytosis, and destruction in the phagolysosome, ROS production and release of neutrophil extracellular traps (NETs). A variety of bactericidal and fungicidal effectors, including cathepsins, lactoferrin, defensins, and myeloperoxidase, are produced by neutrophils as granules (Lacy 2006). Neutrophils also release ROS species such as superoxide (O_2^-) and H_2O_2 that can be used intracellularly to destroy phagocytized fungi or extracellularly to combat hypha. In addition, extracellular pathogens can be intercepted and killed by the ability of the neutrophils' to release chromatin coated with antimicrobial proteins (Urban et al. 2009).

7.5   FUNGAL STRATEGIES TO EVADE THE IMMUNE DEFENSE

Fungi employ many advanced strategies to evade the host's immune defense. Some of these strategies are discussed in the following sections.

7.5.1 Overcoming Phagocytosis

Some fungal pathogens have evolved a remarkable defense mechanism called volume expansion to prevent engulfment by phagocytic cells, either by cell morphogenesis or by an increase in cell size. This strategy has been reported to be adopted by titan cells in *C. neoformans* (Zaragoza and Nielsen 2013). Some other fungi, including *Coccidioides posadasii* and *C. immitis*, resort to the volume growth strategy by changing the morphotype. These fungi exist in in the form of hyphae in the environment, but when they invade the host, they transform into giant, rounded spheres that can grow up 120 μm in diameter and are large enough for macrophages to engulf (Gonzalez et al. 2011). In a recent study, it was shown that the hyphal length of *C. albicans* determines the rate of phagocytosis when macrophages engulf these hyphae: Shorter hyphae (less than 20 μm) are engulfed at a rate comparable to yeasts, while longer hyphae have a slower rate of consumption (Lewis et al. 2012). The conidia of *A. fumigatus* are processed in the phagolysosome after being ingested by alveolar macrophages using ROS, as these macrophages do not produce NADPH oxidase-mediated ROS in mice so they cannot kill the conidia (Philippe et al. 2003). Numerous fungi, including *H. capsulatum* and *P. brasiliensis*, also have this strong antioxidant defense system. Moreover, the latter has been shown to increase CAT activity upon on exposure to H_2O_2 (Hernández-Chávez et al. 2017).

7.5.2 Deception of the Humoral Immune Response

Antimicrobial peptides are short peptides with 20–50 amino acid residues that can vary in sequence, structure, and length. In the humoral response to fungi, peptides and the complement system serve as the primary defense mechanisms. These peptides can penetrate the membranes of microbial cells causing leakage, thereby killing the pathogens (Lehrer and Ganz 1999). The production of proteinase that can inactivate the antimicrobial peptides is one of main ways *Candida* evade these immune molecules. The killing abilities of histatin 5 are lost due to cleavage by the secreted aspartyl proteases, Sap9 and Sap10 (Meiller et al. 2009).

On the other hand, lectin pathways, or classical complement activation led to opsonization and formation of the membrane attack complex. Since fungi have a cell wall, it is difficult for this complex to effectively insert inside fungal membrane. *A. fumigatus* secrete proteases such as Alp1, that degrade complement proteins C3, C5 and C1q, preventing complement activation (Behnsen et al. 2010). Using quantitative flow cytometry, Henwick et al. showed that less pathogenic species binds to more C3 molecules than highly pathogenic ones (Henwick et al. 1993). Furthermore, melanin is also crucial for protection against antimicrobial peptides, Magainin, Defensin, and Protegrin. These peptides are inhibited after binding to melanin (Doering et al. 1999).

7.6 CONCLUSION

Opportunistic commensals and environmental fungi possess the potential to induce infections, varying from surface level to systemic, potentially leading to fatality. The close relationship between fungal pathogens and their hosts necessitates that these pathogens possess various vital elements for their survival. Fungi employ an array of proteins, peptides, and secondary compounds, enabling them to infiltrate hosts by eluding the immune system, manipulating other host targets, and exploiting their resources similarly. Usually, these pathogens rely on numerous factors, each performing a pivotal role in the pathogen's virulence, facilitating it in navigating through the intricate defense mechanisms of hosts, regardless of host type. Currently, only a few antifungal drugs demonstrate effectiveness against human fungal infections. To enhance our ability to combat these infections, there is a need to devise new treatment methods that take into account the pathobiology of these fungi. This presents an opportunity to innovate novel diagnostic and treatment strategies.

ACKNOWLEDGMENTS

R. Pasrija thanks the Science & Engineering Research Board (CRG/2020/004986) under the Department of Science and Technology for funding. D. Kumari acknowledges the Council of Scientific and Industrial Research (CSIR), New Delhi, for their financial support through the Junior and Senior Research Fellowship (09/382(0264)/2020-EMR-I).

REFERENCES

Ajetunmobi OH, Badali, H, Romo JA, Ramage G, Lopez-Ribot JL (2023) Antifungal therapy of *Candida* biofilms: Past, present and future. Biofilm 5:100126. https://doi.org/10.1016/j.bioflm.2023.100126

Albuquerque P, Nicola AM, Nieves E, Paes HC, Williamson PR, Silva-Pereira I, Casadevall A (2013) Quorum sensing-mediated, cell density-dependent regulation of growth and virulence in *Cryptococcus neoformans*. mBio 5(1):e00986-13. https://doi.org/10.1128/mBio.00986-13

Alem MAS, Oteef MDY, Flowers TH, Douglas LJ (2006) Production of tyrosol by *Candida albicans* biofilms and its role in quorum sensing and biofilm development. Eukaryot Cell 5(10):1770–1779. https://doi.org/10.1128/EC.00219-06

Almeida-Paes, R, Almeida-Silva, F, Pinto GCM, de Abreu Almeida M, de Medeiros Muniz M, Pizzini CV, Gerfen GJ, Nosanchuk JD, Zancopé-Oliveira RM (2018) L-tyrosine induces the production of a pyomelanin-like pigment by the parasitic yeast-form of *Histoplasma capsulatum*. Med Mycol 56, 506–509. https://doi.org/10.1093/mmy/myx068

Alspaugh JA, Cavallo LM, Perfect JR, Heitman J (2000) RAS1 regulates filamentation, mating and growth at high temperature of *Cryptococcus neoformans*. Mol Microbiol 36(2):352–365. https://doi.org/10.1046/j.1365-2958.2000.01852

Alspaugh JA, Pukkila-Worley R, Harashima T, Cavallo LM, Funnell D, Cox GM, Perfect JR, Kronstad JW, Heitman J (2002) Adenylyl cyclase functions downstream of the Galpha protein Gpa1 and controls mating and pathogenicity of *Cryptococcus neoformans*. Eukaryot Cell 1(1):75–84. https://doi.org/10.1128/EC.1.1.75-84.2002

Alspaugh JA (2015) Virulence mechanisms and *Cryptococcus neoformans* pathogenesis. Fungal Genet Biol 78:55–58. https://doi.org/10.1016/j.fgb.2014.09.004

Alviano DS, Franzen AJ, Travassos LR, Holandino C, Rozental S, Ejzemberg R, Alviano CS, Rodrigues ML (2004) Melanin from *Fonsecaea pedrosoi* induces production of human antifungal antibodies and enhances the antimicrobial efficacy of phagocytes. Infect Immun 72(1):229–237. https://doi.org/10.1128/iai.72.1.229-237.2004

Ao J, Bandyopadhyay S, Free SJ, (2019) Characterization of the *Neurospora crassa* DHN melanin biosynthetic pathway in developing ascospores and peridium cells. Fungal Biol 123(1):1–9. https://doi.org/10.1016/j.funbio.2018.10.005

Arana DM, Prieto D, Román E, Nombela C, Alonso-Monge R, Pla J (2009) The role of the cell wall in fungal pathogenesis. Microb Biotechnol 2(3):308–320. https://doi.org/10.1111/j.1751-7915.2008.00070.x

Bahn Y-S, Kojima K, Cox GM, Heitman J (2005) Specialization of the HOG pathway and its impact on differentiation and virulence of *Cryptococcus neoformans*. Mol Biol Cell 16(5):2285–2300. https://doi.org/10.1091/mbc.E04-11-0987

Bain JM, Alonso MF, Childers DS, Walls CA, Mackenzie K, Pradhan A, Lewis LE, Louw J, Avelar GM, Larcombe DE, Netea MG, Gow NAR, Brown GD, Erwig LP, Brown AJP (2021) Immune cells fold and damage fungal hyphae. Proc Natl Acad Sci USA 118(15):e2020484118. https://doi.org/10.1073/pnas.2020484118

Barman A, Gohain D, Bora U, Tamuli R (2018) Phospholipases play multiple cellular roles including growth, stress tolerance, sexual development, and virulence in fungi. Microbiol Res 209:55–69. https://doi.org/10.1016/j.micres.2017.12.012

Barrett-Bee K, Hayes Y, Wilson RG, Ryley JF (1985) A comparison of phospholipase activity, cellular adherence and pathogenicity of yeasts. J Gen Microbiol 131(5):1217–1221. https://doi.org/10.1099/00221287-131-5-1217

Bartemes KR, Kita H (2018) Innate and adaptive immune responses to fungi in the airway. J Allergy Clin Immunol 142(2):353–363. https://doi.org/10.1016/j.jaci.2018.06.015

Behnsen J, Lessing F, Schindler S, Wartenberg D, Jacobsen ID, Thoen M, Zipfel PF, Brakhage AA (2010) Secreted *Aspergillus fumigatus* protease Alp1 degrades human complement proteins C3, C4, and C5. Infect Immun 78(8):3585–3594. https://doi.org/10.1128/IAI.01353-09

Berman J, Krysan DJ (2020) Drug resistance and tolerance in fungi. Nat Rev Microbiol 18(6):319–331. https://doi.org/10.1038/s41579-019-0322-2

Boguski MS, McCormick F (1993) Proteins regulating Ras and its relatives. Nature 366(6456):643–654. https://doi.org/10.1038/366643

Braun BR, Johnson AD (2000) TUP1, CPH1 and EFG1 make independent contributions to filamentation in *candida albicans*. Genetics 155(1):57–67. https://doi.org/10.1093/genetics/155.1.57

Brown AJP (2023) Fungal resilience and host–pathogen interactions: Future perspectives and opportunities. Parasite Immunol 45(2):e12946. https://doi.org/10.1111/pim.12946

Brown GD, Denning DW, Gow NAR, Levitz SM, Netea MG, White TC (2012) Hidden killers: Human fungal infections. Sci Transl Med 4(165):165rv13. https://doi.org/10.1126/scitranslmed.3004404

Brown GD, Gordon S (2001) Immune recognition. A new receptor for beta-glucans. Nature 413(6851):36–37. https://doi.org/10.1038/35092620

Brown R, Priest E, Naglik JR, Richardson JP (2021) Fungal toxins and host immune responses. Front Microbiol 12:643639. https://doi.org/10.3389/fmicb.2021.643639

Bryan AM, You JK, Li G, Kim J, Singh A, Morstein J, Trauner D, Pereira de Sá N, Normile TG, Farnoud AM, London E, Del Poeta M (2021) Cholesterol and sphingomyelin are critical for Fcγ receptor–mediated phagocytosis of *Cryptococcus neoformans* by macrophages. J Biol Chem 297(6):101411. https://doi.org/10.1016/j.jbc.2021.101411

Bull AT, Carter BLA (1973) The isolation of tyrosinase from *Aspergillus nidulans*, its kinetic and molecular properties and some consideration of its activity *in vivo*. Microbiol 75(1):61–73. https://doi.org/10.1099/00221287-75-1-61

Burgess TB, Condliffe AM, Elks PM (2022) A fun-guide to innate immune responses to fungal infections. J Fungi 8(8):805. https://doi.org/10.3390/jof8080805

Casadevall A, Coelho C, Cordero RJB, Dragotakes Q, Jung E, Vij R, Wear MP (2018) The capsule of *Cryptococcus neoformans*. Virulence 10(1):822–831. https://doi.org/10.1080/21505594.2018.1431087

Casadevall A, Pirofski L (2003) The damage-response framework of microbial pathogenesis. Nat Rev Microbiol 1(1):17–24. https://doi.org/10.1038/nrmicro732

Chan MY, Tay ST (2010) Enzymatic characterisation of clinical isolates of *Cryptococcus neoformans*, *Cryptococcus gattii* and other environmental *Cryptococcus* spp. Mycoses 53, 26–31. https://doi.org/10.1111/j.1439-0507.2008.01654.x

Chatterjee S, Tatu U (2017) Heat shock protein 90 localizes to the surface and augments virulence factors of *Cryptococcus neoformans*. PLOS Negl Trop Dis 11(8):e0005836. https://doi.org/10.1371/journal.pntd.0005836

Chayakulkeeree M, Sorrell TC, Siafakas AR, Wilson CF, Pantarat N, Gerik KJ, Boadle R, Djordjevic JT (2008) Role and mechanism of phosphatidylinositol-specific phospholipase C in survival and virulence of *Cryptococcus neoformans*. Mol Microbiol 69(4):809–826. https://doi.org/10.1111/j.1365-2958.2008.06310.x

Chen H, Zhou X, Ren B, Cheng L (2020) The regulation of hyphae growth in *Candida albicans*. Virulence 11(1):337–348. https://doi.org/10.1080/21505594.2020.1748930

Chen SHM, Stins MF, Huang S-H, Chen YH, Kwon-Chung KJ, Chang Y, Kim KS, Suzuki K, Jong AY (2003) *Cryptococcus neoformans* induces alterations in the cytoskeleton of human brain microvascular endothelial cells. J Med Microbiol 52(11):961–970. https://doi.org/10.1099/jmm.0.05230-0

Chow NA, Gade L, Tsay SV, Forsberg K, Greenko JA, Southwick KL, Barrett PM, Kerins JL, Lockhart SR, Chiller TM, Litvintseva AP, Adams E, Barton K, Beer KD, Bentz ML, Berkow EL, Black S, Bradley KK, Brooks R, Chaturvedi S, Clegg W, Cumming M, DeMaria A, Dotson N, Epson E, Fernandez R, Fulton T, Greeley R, Jackson B, Kallen A, Kemble S, Klevens M, Kuykendall R, Le NH, Leung V, Lutterloh E, Mcateer J, Pacilli M, Peterson J, Quinn M, Ross K, Rozwadowski F, Shannon DJ, Skrobarcek KA, Vallabhaneni S, Welsh R, Zhu Y (2018) Multiple introductions and subsequent transmission of multidrug-resistant *Candida auris* in the USA: A molecular epidemiological survey. Lancet Infect Dis 18(12):1377–1384. https://doi.org/10.1016/S1473-3099(18)30597-8

Chrisman CJ, Albuquerque P, Guimaraes AJ, Nieves E, Casadevall A (2011) Phospholipids trigger *Cryptococcus neoformans* capsular enlargement during interactions with amoebae and macrophages. PLOS Pathog 7(5):e1002047. https://doi.org/10.1371/journal.ppat.1002047

Clarke SC, Dumesic PA, Homer CM, O'Donoghue AJ, Greca FL, Pallova L, Majer P, Madhani HD, Craik CS (2016) Integrated activity and genetic profiling of secreted peptidases in *Cryptococcus neoformans* reveals an aspartyl peptidase required for low pH survival and virulence. PLOS Pathog 12(12):e1006051. https://doi.org/10.1371/journal.ppat.1006051

Crabtree JN, Okagaki LH, Wiesner DL, Strain AK, Nielsen JN, Nielsen K (2012) Titan cell production enhances the virulence of *Cryptococcuss neoformans*. Infect Immun 80(11):3776–3785. https://doi.org/10.1128/IAI.00507-12

Csank C, Schröppel K, Leberer E, Harcus D, Mohamed O, Meloche S, Thomas DY, Whiteway M (1998) Roles of the *Candida albicans* mitogen-activated protein kinase homolog, Cek1p, in hyphal development and systemic candidiasis. Infection and Immunity 66(6):2713–2721. https://doi.org/10.1128/iai.66.6.2713-2721.1998

Daniel Da Rosa W, Gezuele E, Calegari L, Goñi F (2008) Asteroid body in sporotrichosis. Yeast viability and biological significance within the host immune response. Med Mycol 46(5):443–448. https://doi.org/10.1080/13693780801914898

Davis-Hanna A, Piispanen AE, Stateva LI, Hogan DA (2008) Farnesol and dodecanol effects on the *Candida albicans* Ras1-cAMP signalling pathway and the regulation of morphogenesis. Mol Microbiol 67(1):47–62. https://doi.org/10.1111/j.1365-2958.2007.06013.x

DeAngelis YM, Saunders CW, Johnstone KR, Reeder NL, Coleman CG, Kaczvinsky JR, Gale C, Walter R, Mekel M, Lacey MP, Keough TW, Fieno A, Grant RA, Begley B, Sun Y, Fuentes G, Youngquist RS, Xu J, Dawson TL (2007) Isolation and expression of a *Malassezia globosa* lipase gene, LIP1. J Invest Dermatol 127(9):2138–2146. https://doi.org/10.1038/sj.jid.5700844

Díaz-Jiménez DF, Pérez-García LA, Martínez-Álvarez JA, Mora-Montes HM (2012) Role of the fungal cell wall in pathogenesis and antifungal resistance. Curr Fungal Infect Rep 6(4):275–282. https://doi.org/10.1007/s12281-012-0109-7

Diez-Orejas R, Molero G, Navarro-García F, Pla J, Nombela C, Sanchez-Pérez M (1997) Reduced virulence of *Candida albicans* MKC1 mutants: A role for mitogen-activated protein kinase in pathogenesis. Infect Immun 65(2):833–837. https://doi.org/10.1128/iai.65.2.833-837.1997

Ding H, Mayer FL, Sánchez-León E, de S. Araújo GR, Frases S, Kronstad JW (2016) Networks of fibers and factors: Regulation of capsule formation in *Cryptococcus neoformans*. F1000Research 5(F1000 Faculty Rev):1786. https://doi.org/10.12688/f1000research.8854.1

Djordjevic JT (2010) Role of phospholipases in fungal fitness, pathogenicity, and drug development – Lessons from *Cryptococcus neoformans*. Front Microbiol 1:125. https://doi.org/10.3389/fmicb.2010.00125

Doering TL, Nosanchuk JD, Roberts WK, Casadevall A (1999) Melanin as a potential cryptococcal defence against microbicidal proteins. Med Mycol 37(3):175–181. https://doi.org/10.1080/j.1365-280X.1999.00218.x

Drummond RA, Gaffen SL, Hise AG, Brown GD (2015) Innate defense against fungal pathogens. Cold Spring Harb Perspect Med 5(6):a019620. https://doi.org/10.1101/cshperspect.a019620

Fabri JHTM, Rocha MC, Fernandes CM, Persinoti GF, Ries LNA, da Cunha AF, Goldman GH, Del Poeta M, Malavazi I (2021) The heat shock transcription factor HsfA is essential for thermotolerance and regulates cell wall integrity in *Aspergillus fumigatus*. Front Microbiol 12:656548. https://doi.org/10.3389/fmicb.2021.656548

Filler SG, Swerdloff JN, Hobbs C, Luckett PM (1995) Penetration and damage of endothelial cells by *Candida albicans*. Infect Immun 63(3):976–983. https://doi.org/10.1128/iai.63.3.976-983.1995

Finlay BB, Falkow S (1997) Common themes in microbial pathogenicity revisited. Microbiol Mol Biol Rev 61(2):136–169. https://doi.org/10.1128/mmbr.61.2.136-169.1997

Fling M, Horowitz NH, Heinemann SF (1963) The isolation and properties of crystalline tyrosinase from *Neurospora*. J Biol Chem 238(6):2045–2053. https://doi.org/10.1016/S0021-9258(18)67939-6

Frases S, Salazar A, Dadachova E, Casadevall A (2007) *Cryptococcus neoformans* can utilize the bacterial melanin precursor homogentisic acid for fungal melanogenesis. Appl Environ Microbiol 73, 615–621. https://doi.org/10.1128/AEM.01947-06

Fu MS, Coelho C, De Leon-Rodriguez CM, Rossi DCP, Camacho E, Jung EH, Kulkarni M, Casadevall A (2018) *Cryptococcus neoformans* urease affects the outcome of intracellular pathogenesis by modulating phagolysosomal pH. PLOS Pathog 14(6):e1007144. https://doi.org/10.1371/journal.ppat.1007144

Gácser A, Stehr F, Kröger C, Kredics L, Schäfer W, Nosanchuk JD (2007) Lipase 8 affects the pathogenesis of *Candida albicans*. Infect Immun 75(10):4710–4718. https://doi.org/10.1128/IAI.00372-07

Garcia-Rubio R, de Oliveira HC, Rivera J, Trevijano-Contador N (2020) The fungal cell wall: *Candida*, *Cryptococcus*, and *Aspergillus* species. Front Microbiol 10:2993. https://doi.org/10.3389/fmicb.2019.02993

Ghannoum MA (2000) Potential role of phospholipases in virulence and fungal pathogenesis. Clin Microbiol Rev 13(1):122–143.

Gonçalves RCR, Lisboa HCF, Pombeiro-Sponchiado SR (2012) Characterization of melanin pigment produced by *Aspergillus nidulans*. World J Microbiol Biotechnol 28, 1467–1474. https://doi.org/10.1007/s11274-011-0948-3

Gong Y, Li T, Yu C, Sun S (2017) *Candida albicans* heat shock proteins and Hsps-associated signaling pathways as potential antifungal targets. Front Cell Infect Microbiol 7:520. https://doi.org/10.3389/fcimb.2017.00520

Gonzalez A, Hung CY, Cole GT (2011) Coccidioides releases a soluble factor that suppresses nitric oxide production by murine primary macrophages. Microb Pathog 50(2):100–108. https://doi.org/10.1016/j.micpath.2010.11.006

Hamed MF, Araújo GR de S, Munzen ME, Reguera-Gomez M, Epstein C, Lee HH, Frases S, Martinez LR (2023) Phospholipase B is critical for *Cryptococcus neoformans* survival in the central nervous system. mBio 14(2):e02640–22. https://doi.org/10.1128/mbio.02640-22

Hartland K, Pu J, Palmer M, Dandapani S, Moquist PN, Munoz B, DiDone L, Schreiber SL, Krysan DJ (2016) High-throughput screen in *Cryptococcus neoformans* identifies a novel molecular scaffold that inhibits cell wall integrity pathway signaling. ACS Infect Dis 2(1):93–102. https://doi.org/10.1021/acsinfecdis.5b00111

Hatinguais R, Pradhan A, Brown GD, Brown AJP, Warris A, Shekhova E (2021) Mitochondrial reactive oxygen species regulate immune responses of macrophages to *Aspergillus fumigatus*. Front Immunol 12:641495. https://doi.org/10.3389/fimmu.2021.641495

Henwick S, Hetherington SV, Patrick CC (1993) Complement binding to *Aspergillus* conidia correlates with pathogenicity. J Lab Clin Med 122(1):27–35. https://doi.org/10.5555/uri:pii:0022214393901999

Hernández-Chávez MJ, Pérez-García LA, Niño-Vega GA, Mora-Montes HM (2017) Fungal strategies to evade the host immune recognition. J Fungi (Basel) 3(4):51. https://doi.org/10.3390/jof3040051

Herrero de Dios C, Román E, Diez C, Alonso-Monge R, Pla J (2013) The transmembrane protein Opy2 mediates activation of the Cek1 MAP kinase in *Candida albicans*. Fungal Genet Biol 50:21–32. https://doi.org/10.1016/j.fgb.2012.11.001

Horváth P, Nosanchuk JD, Hamari Z, Vágvölgyi C, Gácser A (2012) The identification of gene duplication and the role of secreted aspartyl proteinase 1 in *Candida parapsilosis* virulence. J Infect Dis 205(6):923–933. https://doi.org/10.1093/infdis/jir873

Hu G, Steen BR, Lian T, Sham AP, Tam N, Tangen KL, Kronstad JW (2007) Transcriptional regulation by protein kinase A in *Cryptococcus neoformans*. PLoSPathog 3(3):e42. https://doi.org/10.1371/journal.ppat.0030042

Huang G, Srikantha T, Sahni N, Yi S, Soll DR (2009) CO_2 regulates white-to-opaque switching in *Candida albicans*. Curr Biol 19(4):330–334. https://doi.org/10.1016/j.cub.2009.01.018

Inamdar S, Joshi S, Bapat V, Jadhav J (2014) Purification and characterization of RNA allied extracellular tyrosinase from *Aspergillus* species. Appl Biochem Biotechnol 172, 1183–1193. https://doi.org/10.1007/s12010-013-0555-x

Jabra-Rizk MA, Falkler WA, Meiller TF (2004) Fungal biofilms and drug resistance. Emerg Infect Dis 10(1):14–19. https://doi.org/10.3201/eid1001.030119

Jabra-Rizk MA, Kong EF, Tsui C, Nguyen MH, Clancy CJ, Fidel PL, Noverr M (2016) *Candida albicans* pathogenesis: Fitting within the host-microbe damage response framework. Infect Immun 84(10):2724–2739. https://doi.org/10.1128/IAI.00469-16

Jacobson ES, Hong JD (1997) Redox buffering by melanin and Fe(II) in *Cryptococcus neoformans*. J Bacteriol 179(17):5340–5346. https://doi.org/10.1128/jb.179.17.5340-5346.1997

Jacobson ES, Hove E, Emery HS (1995) Antioxidant function of melanin in black fungi. Infect Immun 63(12):4944–4945. https://doi.org/10.1128/iai.63.12.4944-4945.1995

Naglik JR, Challacombe SJ, Hube B (2003) *Candida albicans* secreted aspartyl proteinases in virulence and pathogenesis. Microbiol Mol Biol Rev MMBR 67(3). https://doi.org/10.1128/MMBR.67.3.400-428.2003

Kaur R, Ma B, Cormack BP (2007) A family of glycosylphosphatidylinositol-linked aspartyl proteases is required for virulence of *Candida glabrata*. Proc Natl Acad Sci 104(18):7628–7633. https://doi.org/10.1073/pnas.0611195104

Kim J-S, Lee K-T, Bahn Y-S (2023) Deciphering the regulatory mechanisms of the cAMP/protein kinase a pathway and their roles in the pathogenicity of *Candida auris*. Microbiol Spectr 11(5):e02152–23. https://doi.org/10.1128/spectrum.02152-23

Kirkland ME, Stannard M, Kowalski CH, Mould D, Caffrey-Carr A, Temple RM, Ross BS, Lofgren LA, Stajich JE, Cramer RA, Obar JJ (2021) Host lung environment limits *Aspergillus fumigatus* Germination through an SskA-dependent signaling response. mSphere 6(6):e00922–21. https://doi.org/10.1128/msphere.00922-21

Klimpel KR, Goldman WE (1988) Cell walls from avirulent variants of *Histoplasma capsulatum* lack alpha-(1,3)-glucan. Infect Immun 56(11):2997–3000. https://doi.org/10.1128/iai.56.11.2997-3000.1988.

Klis FM, Sosinska GJ, de Groot PWJ, Brul S (2009) Covalently linked cell wall proteins of *Candida albicans* and their role in fitness and virulence. FEMS Yeast Res 9(7):1013–1028. https://doi.org/10.1111/j.1567-1364.2009.00541.x

Kovács R, Majoros L (2020) Fungal quorum-sensing molecules: A review of their antifungal effect against *Candida* biofilms. J Fungi 6(3):99. https://doi.org/10.3390/jof6030099

Kulshrestha A, Gupta P (2023) Secreted aspartyl proteases family: A perspective review on the regulation of fungal pathogenesis. Future Microbiol 18:295–309. https://doi.org/10.2217/fmb-2022-0143

Lacy P (2006) Mechanisms of degranulation in neutrophils. Allergy Asthma Clin Immunol 2(3):98–108. https://doi.org/10.1186/1710-1492-2-3-98

Lämmermann T, Afonso PV, Angermann BR, Wang JM, Kastenmüller W, Parent CA, Germain RN (2013) Neutrophil swarms require LTB4 and integrins at sites of cell death *in vivo*. Nature 498(7454). https://doi.org/10.1038/nature12175

Lee H, Chang YC, Nardone G, Kwon-Chung KJ (2007) TUP1 disruption in *Cryptococcus neoformans* uncovers a peptide-mediated density-dependent growth phenomenon that mimics quorum sensing. Mol Microbiol 64(3):591–601. https://doi.org/10.1111/j.1365-2958.2007.05666.x

Lehrer RI, Ganz T (1999) Antimicrobial peptides in mammalian and insect host defence. Curr Opin Immunol 11(1):23–27. https://doi.org/10.1016/S0952-7915(99)80005-3

Leidich SD, Ibrahim AS, Fu Y, Koul A, Jessup C, Vitullo J, Fonzi W, Mirbod F, Nakashima S, Nozawa Y, Ghannoum MA (1998) Cloning and disruption of caPLB1, a phospholipase B gene involved in the pathogenicity of *Candida albicans*. J Biol Chem 273(40):26078–26086. https://doi.org/10.1074/jbc.273.40.26078

Lewis LE, Bain JM, Lowes C, Gillespie C, Rudkin FM, Gow NAR, Erwig LP (2012) Stage specific assessment of *Candida albicans* phagocytosis by macrophages identifies cell wall composition and morphogenesis as key determinants. PLOS Pathog 8(3):e1002578. https://doi.org/10.1371/journal.ppat.1002578

Lewis RE, Lo H-J, Raad II, Kontoyiannis DP (2002) Lack of catheter infection by the efg1/efg1 cph1/cph1 double-null mutant, a Candida albicans strain that is defective in filamentous growth. Antimicrob Agents Chemother 46(4):1153–1155. https://doi.org/10.1128/AAC.46.4.1153-1155.2002

Lewis RE, Wiederhold NP, Chi J, Han XY, Komanduri KV, Kontoyiannis DP, Prince RA (2005) Detection of gliotoxin in experimental and human aspergillosis. Infect Immun 73(1):635–637. https://doi.org/10.1128/IAI.73.1.635-637.2005

Li F, Palecek SP (2003) EAP1, a *Candida albicans* gene involved in binding human epithelial cells. Eukaryot Cell 2(6):1266–1273. https://doi.org/10.1128/ec.2.6.1266-1273.2003

Li X, Gao M, Han X, Tao S, Zheng D, Cheng Y, Yu R, Han G, Schmidt M, Han L (2012) Disruption of the phospholipase D Gene attenuates the virulence of *Aspergillus fumigatus*. Infect Immun 80(1):429–440. https://doi.org/10.1128/iai.05830-11

Lindsay AK, Deveau A, Piispanen AE, Hogan DA (2012) Farnesol and cyclic AMP signaling effects on the hypha-to-yeast transition in *Candida albicans*. Eukaryot Cell 11(10):1219–1225. https://doi.org/10.1128/EC.00144-12

Liu H, Köhler J, Fink GR (1994) Suppression of hyphal formation in *Candida albicans* by mutation of a STE12 homolog. Science 266(5191):1723–1726. https://doi.org/10.1126/science.7992058

Liu S, Le Mauff F, Sheppard DC, Zhang S (2022) Filamentous fungal biofilms: Conserved and unique aspects of extracellular matrix composition, mechanisms of drug resistance and regulatory networks in *Aspergillus fumigatus*. NPJ Biofilms Microbi 8(1):1–8. https://doi.org/10.1038/s41522-022-00347-3

Lo HJ, Köhler JR, DiDomenico B, Loebenberg D, Cacciapuoti A, Fink GR (1997) Nonfilamentous *C. albicans* mutants are avirulent. Cell 90(5):939–949. https://doi.org/10.1016/s0092-8674(00)80358-x

Majumdar S, Mondal S (2015) Perspectives on quorum sensing in fungi. Biophysics. Int J Modern Biol Med 6(3):170–180. https://doi.org/10.1101/019034

Marshall JS, Warrington R, Watson W, Kim HL (2018) An introduction to immunology and immunopathology. Allergy Asthma Clin Immunol 14(Suppl 2):49. https://doi.org/10.1186/s13223-018-0278-1

Martinez LR, Casadevall A (2015) Biofilm formation by *Cryptococcus neoformans*. Microbiol Spect 3(3):. https://doi.org/10.1128/microbiolspec.mb-0006-2014

Matos TGF, Morais FV, Campos CBL (2013) Hsp90 regulates *Paracoccidioides brasiliensis* proliferation and ROS levels under thermal stress and cooperates with calcineurin to control yeast to mycelium dimorphism. Med Mycol 51(4):413–421. https://doi.org/10.3109/13693786.2012.725481

Mayer FL, Wilson D, Hube B (2013) *Candida albicans* pathogenicity mechanisms. Virulence 4(2):119–128. https://doi.org/10.4161/viru.22913

McDonough KA, Rodriguez A (2011) The myriad roles of cyclic AMP in microbial pathogens: From signal to sword. Nat Rev Microbiol 10(1):27–38. https://doi.org/10.1038/nrmicro2688

Mehmood A, Liu G, Wang X, Meng G, Wang C, Liu Y (2019) Fungal quorum-sensing molecules and inhibitors with potential antifungal activity: A review. Molecules 24(10):1950. https://doi.org/10.3390/molecules24101950

Meiller TF, Hube B, Schild L, Shirtliff ME, Scheper MA, Winkler R, Ton A, Jabra-Rizk MA (2009) A novel immune evasion strategy of *Candida albicans*: Proteolytic cleavage of a salivary antimicrobial peptide. PLOS One 4(4):e5039. https://doi.org/10.1371/journal.pone.0005039

Mendes-Giannini MJS, Soares CP, da Silva JLM, Andreotti PF (2005) Interaction of pathogenic fungi with host cells: Molecular and cellular approaches. FEMS Immunol Med Microbiol 45(3):383–394. https://doi.org/10.1016/j.femsim.2005.05.014

Merkel O, Fido M, Mayr JA, Prüger H, Raab F, Zandonella G, Kohlwein SD, Paltauf F (1999) Characterization and function *in vivo* of two novel phospholipases B/lysophospholipases from *Saccharomyces cerevisiae*. J Biol Chem 274(40):28121–28127. https://doi.org/10.1074/jbc.274.40.28121

Morris-Jones R, Gomez BL, Diez S, Uran M, Morris-Jones SD, Casadevall A, Nosanchuk JD, Hamilton AJ 2005. Synthesis of melanin pigment by *Candida albicans in vitro* and during infection. Infect Immun 73, 6147–6150. https://doi.org/10.1128/IAI.73.9.6147-6150.2005

Moyes DL, Wilson D, Richardson JP, Mogavero S, Tang SX, Wernecke J, Höfs S, Gratacap RL, Robbins J, Runglall M, Murciano C, Blagojevic M, Thavaraj S, Förster TM, Hebecker B, Kasper L, Vizcay G, Iancu SI, Kichik N, Häder A, Kurzai O, Luo T, Krüger T, Kniemeyer O, Cota E, Bader O, Wheeler RT, Gutsmann T, Hube B, Naglik JR (2016) Candidalysin is a fungal peptide toxin critical for mucosal infection. Nature 532(7597):64–68. https://doi.org/10.1038/nature17625

Moyrand F, Fontaine T, Janbon G (2007) Systematic capsule gene disruption reveals the central role of galactose metabolism on *Cryptococcus neoformans* virulence. Molecular Microbiology 64(3):771–781. https://doi.org/10.1111/j.1365-2958.2007.05695.x

Mroczyńska M, Brillowska-Dąbrowska A (2021) Virulence of clinical *Candida* isolates. Pathogens 10(4):466. https://doi.org/10.3390/pathogens10040466

Navarathna DHMLP, Harris SD, Roberts DD, Nickerson KW (2010) Evolutionary aspects of urea utilization by fungi. FEMS Yeast Res 10(2):209–213. https://doi.org/10.1111/j.1567-1364.2010.00602.x

Newman SL, Bucher C, Rhodes J, Bullock WE (1990) Phagocytosis of *Histoplasma capsulatum* yeasts and microconidia by human cultured macrophages and alveolar macrophages. Cellular cytoskeleton requirement for attachment and ingestion. J Clin Invest 85(1):223–230. https://doi.org/10.1172/JCI114416

Nguyen S, Truong JQ, Bruning JB (2020) Targeting unconventional pathways in pursuit of novel antifungals. Front Mol Biosci 7. https://doi.org/10.3389/fmolb.2020.621366

Nosanchuk JD, Gómez BL, Youngchim S, Díez S, Aisen P, Zancopé-Oliveira RM, Restrepo A, Casadevall A, Hamilton AJ (2002) *Histoplasma capsulatum* synthesizes melanin-like pigments *in vitro* and during mammalian infection. Infect Immun 70(9):5124–5131. https://doi.org/10.1128/iai.70.9.5124-5131.2002

Nosanchuk JD, Rosas AL, Casadevall A (1998) The antibody response to fungal melanin in Mice1. J Immunol 160(12):6026–6031. https://doi.org/10.4049/jimmunol.160.12.6026

Nyhus KJ, Wilborn AT, Jacobson ES (1997) Ferric iron reduction by *Cryptococcus neoformans*. Infect Immun 65(2):434–438. https://doi.org/10.1128/iai.65.2.434-438.1997

Odds FC (2017) Pathogenesis of fungal disease. In: Kibbler CC, Barton R, Gow NAR, Howell S, MacCallum DM, Manuel RJ, Kibbler CC, Barton R, Gow NAR, Howell S, MacCallum DM, Manuel RJ (eds) Oxford Textbook of Medical Mycology. Oxford University Press. https://doi.org/10.1093/med/9780198755388.003.0008

Padder SA, Prasad R, Shah AH (2018) Quorum sensing: A less known mode of communication among fungi. Microbiol Res 210:51–58. https://doi.org/10.1016/j.micres.2018.03.007

Pal AK, Gajjar DU, Vasavada AR (2014) DOPA and DHN pathway orchestrate melanin synthesis in *Aspergillus* species. Med Mycol 52:10–18. https://doi.org/10.3109/13693786.2013.826879

Park BJ, Wannemuehler KA, Marston BJ, Govender N, Pappas PG, Chiller TM (2009) Estimation of the current global burden of cryptococcal meningitis among persons living with HIV/AIDS. AIDS 23(4):525–530. https://doi.org/10.1097/QAD.0b013e328322ffac

Park M, Do E, Jung WH (2013) Lipolytic enzymes involved in the virulence of human pathogenic fungi. Mycobiology 41(2):67–72. https://doi.org/10.5941/MYCO.2013.41.2.67

Philippe B, Ibrahim-Granet O, Prévost MC, Gougerot-Pocidalo MA, Sanchez Perez M, Van der Meeren A, Latgé JP (2003) Killing of *Aspergillus fumigatus* by alveolar macrophages is mediated by reactive oxidant intermediates. Infect Immun 71(6):3034–3042. https://doi.org/10.1128/IAI.71.6.3034-3042.2003

Previato JO, Vinogradov E, Maes E, Fonseca LM, Guerardel Y, Oliveira PAV, Mendonça-Previato L (2017) Distribution of the O-acetyl groups and β-galactofuranose units in galactoxylomannans of the opportunistic fungus *Cryptococcus neoformans*. Glycobiol 27(6):582–592. https://doi.org/10.1093/glycob/cww127

Rappleye CA, Eissenberg LG, Goldman WE (2007) *Histoplasma capsulatum* alpha-(1,3)-glucan blocks innate immune recognition by the beta-glucan receptor. Proc Natl Acad Sci USA 104(4):1366–1370. https://doi.org/10.1073/pnas.0609848104

Reddy GKK, Padmavathi AR, Nancharaiah YV (2022) Fungal infections: Pathogenesis, antifungals and alternate treatment approaches. Curr Res Microb Sci 3:100137. https://doi.org/10.1016/j.crmicr.2022.100137

Richardson JP, Brown R, Kichik N, Lee S, Priest E, Mogavero S, Maufrais C, Wickramasinghe DN, Tsavou A, Kotowicz NK, Hepworth OW, Gallego-Cortés A, Ponde NO, Ho J, Moyes DL, Wilson D, D'Enfert C, Hube B, Naglik JR (2022) Candidalysins are a new family of cytolytic fungal peptide toxins. mBio 13(1):e03510–21. https://doi.org/10.1128/mbio.03510-21

Román E, Arana DM, Nombela C, Alonso-Monge R, Pla J (2007) MAP kinase pathways as regulators of fungal virulence. Trends Microbiol 15(4):181–190. https://doi.org/10.1016/j.tim.2007.02.001

Román E, Correia I, Salazin A, Fradin C, Jouault T, Poulain D, Liu F-T, Pla J (2016) The Cek1-mediated MAP kinase pathway regulates exposure of α-1,2 and β-1,2-mannosides in the cell wall of Candida albicans modulating immune recognition. Virulence 7(5):558–577. https://doi.org/10.1080/21505594.2016.1163458

Rosowski EE, Raffa N, Knox BP, Golenberg N, Keller NP, Huttenlocher A (2018) Macrophages inhibit *Aspergillus fumigatus* germination and neutrophil-mediated fungal killing. PLOS Pathog 14(8):e1007229. https://doi.org/10.1371/journal.ppat.1007229

Ruijter GJG, Bax M, Patel H, Flitter SJ, van de Vondervoort PJI, de Vries RP, vanKuyk PA, Visser J (2003) Mannitol is required for stress tolerance in *Aspergillus niger* conidiospores. Eukaryot Cell 2(4):690–698. https://doi.org/10.1128/EC.2.4.690-698.2003

Sardi JDCO, Pitangui NDS, Rodríguez-Arellanes G, Taylor ML, Fusco-Almeida AM, Mendes-Giannini MJS (2014) Highlights in pathogenic fungal biofilms. Rev Iberoam Micol 31(1):22–29. https://doi.org/10.1016/j.riam.2013.09.014

Semighini CP, Hornby JM, Dumitru R, Nickerson KW, Harris SD (2006) Farnesol-induced apoptosis in *Aspergillus nidulans* reveals a possible mechanism for antagonistic interactions between fungi. Mol Microbiol 59(3):753–764. https://doi.org/10.1111/j.1365-2958.2005.04976.x

Shaw CE, Kapica L (1972) Production of diagnostic pigment by phenoloxidase activity of *Cryptococcus neoformans*. Appl Microbiol 24, 824–830. https://doi.org/10.1128/am.24.5.824-830.1972

Shlezinger N, Hohl TM (2021) Mitochondrial reactive oxygen species enhance alveolar macrophage activity against *Aspergillus fumigatus* but are dispensable for host protection. mSphere 6(3):e00260–21. https://doi.org/10.1128/mSphere.00260-21

Silva S, Henriques M, Oliveira R, Azeredo J, Malic S, Hooper SJ, Williams DW (2009) Characterization of *Candida parapsilosis* infection of an *in vitro* reconstituted human oral epithelium. Eur J Oral Sci 117(6):669–675. https://doi.org/10.1111/j.1600-0722.2009.00677

Singh S, Rehman S, Fatima Z, Hameed S (2020) Protein kinases as potential anticandidal drug targets. FBL 25(8):1412–1432. https://doi.org/10.2741/4862

Smith DFQ, Casadevall A (2019) The role of melanin in fungal pathogenesis for animal hosts. In: Rodrigues ML (ed) Fungal Physiology and Immunopathogenesis. Springer International Publishing, pp 1–30. https://doi.org/10.1007/82_2019_173

Smith DFQ, Mudrak NJ, Zamith-Miranda D, Honorato L, Nimrichter L, Chrissian C, Smith B, Gerfen G, Stark RE, Nosanchuk JD, Casadevall A (2022) Melanization of *Candida auris* is associated with alteration of extracellular pH. J Fungi (Basel) 8(10):1068. https://doi.org/10.3390/jof8101068

Speth C, Rambach G, Lass-Flörl C, Howell PL, Sheppard DC (2019) Galactosaminogalactan (GAG) and its multiple roles in *Aspergillus* pathogenesis. Virulence 10(1):976–983. https://doi.org/10.1080/21505594.2019.1568174

Staab JF, Datta K, Rhee P (2013) Niche-specific requirement for hyphal wall protein 1 in virulence of *Candida albicans*. PLOS One 8(11):e80842. https://doi.org/10.1371/journal.pone.0080842

Summers DK, Perry DS, Rao B, Madhani HD (2020) Coordinate genomic association of transcription factors controlled by an imported quorum sensing peptide in *Cryptococcus neoformans*. PLOS Genet 16(9):e1008744. https://doi.org/10.1371/journal.pgen.1008744

Theiss S, Ishdorj G, Brenot A, Kretschmar M, Lan C-Y, Nichterlein T, Hacker J, Nigam S, Agabian N, Köhler GA (2006) Inactivation of the phospholipase B gene PLB5 in wild-type *Candida albicans* reduces cell-associated phospholipase A2 activity and attenuates virulence. International Journal of Medical Microbiology 296(6):405–420. https://doi.org/10.1016/j.ijmm.2006.03.003

Tian X, Ding H, Ke W, Wang L (2021) Quorum sensing in fungal species. Annu Rev Microbiol 75(1):449–469. https://doi.org/10.1146/annurev-micro-060321-045510

Tóth A, Németh T, Csonka K, Horváth P, Vágvölgyi C, Vizler C, Nosanchuk JD, Gácser A (2014) Secreted *Candida parapsilosis* lipase modulates the immune response of primary human macrophages. Virulence 5(4):555–562. https://doi.org/10.4161/viru.28509

Tsai HF, Washburn RG, Chang YC, Kwon-Chung KJ (1997) *Aspergillus fumigatus* arp1 modulates conidial pigmentation and complement deposition. Mol Microbiol 26(1):175–183. https://doi.org/10.1046/j.1365-2958.1997.5681921.x

Tsai H-F, Wheeler MH, Chang YC, Kwon-Chung KJ (1999) A developmentally regulated gene cluster involved in conidial pigment biosynthesis in *Aspergillus fumigatus*. J Bacteriol 181, 6469–6477. https://doi.org/10.1128/JB.181.20.6469-6477.1999

Tsui C, Kong EF, Jabra-Rizk MA (2016) Pathogenesis of *Candida albicans* biofilm. Pathog Dis 74(4):ftw018. https://doi.org/10.1093/femspd/ftw018

Urban CF, Ermert D, Schmid M, Abu-Abed U, Goosmann C, Nacken W, Brinkmann V, Jungblut PR, Zychlinsky A (2009) Neutrophil extracellular traps contain calprotectin, a cytosolic protein complex involved in host defense against *Candida albicans*. PLOS Pathog 5(10):e1000639. https://doi.org/10.1371/journal.ppat.1000639

Vautier S, MacCallum DM, Brown GD (2012) C-type lectin receptors and cytokines in fungal immunity. Cytokine 58(1):89–99. https://doi.org/10.1016/j.cyto.2011.08.031

Wagener J, MacCallum DM, Brown GD, Gow NAR (2017) *Candida albicans* chitin increases arginase-1 activity in human macrophages, with an impact on macrophage antimicrobial functions. mBio 8(1):e01820–16. https://doi.org/10.1128/mBio.01820-16

Walker CA, Gómez BL, Mora-Montes HM, Mackenzie KS, Munro CA, Brown AJP, Gow NAR, Kibbler CC, Odds FC (2010) Melanin externalization in *Candida albicans* depends on cell wall chitin structures. Eukaryot Cell 9(9):1329–1342. https://doi.org/10.1128/ec.00051-10

Wongsuk T, Pumeesat P, Luplertlop N (2016) Fungal quorum sensing molecules: Role in fungal morphogenesis and pathogenicity. J Basic Microbiol 56(5):440–447. https://doi.org/10.1002/jobm.201500759

Youngchim S, Morris-Jones R, Hay RJ, Hamilton AJ (2004) Production of melanin by *Aspergillus fumigatus*. J Med Microbiol 53(3):175–181. https://doi.org/10.1099/jmm.0.05421-0

Xiong Z, Zhang N, Xu L, Deng Z, Limwachiranon J, Guo Y, Han Y, Yang W, Scharf DH (2023) Urease of aspergillus fumigatus is required for survival in macrophages and virulence. Microbiol Spectr 11(2):e0350822. https://doi.org/10.1128/spectrum.03508-22

Yu S, Huo K (2018) *Aspergillus fumigatus* phospholipase D may enhance reactive oxygen species production by accumulation of histone deacetylase 6. Biochem Biophys Res Commun 505(3):651–656. https://doi.org/10.1016/j.bbrc.2018.09.157

Zaragoza O (2019) Basic principles of the virulence of *Cryptococcus*. Virulence 10(1):490–501. https://doi.org/10.1080/21505594.2019.1614383

Zaragoza O, Nielsen K (2013) Titan cells in *Cryptococcus neoformans*: Cells with a giant impact. Curr Opin Microbiol 16(4):409–413. https://doi.org/10.1016/j.mib.2013.03.006

Zaragoza O, Rodrigues ML, De Jesus M, Frases S, Dadachova E, Casadevall A (2009) The capsule of the fungal pathogen *Cryptococcus neoformans*. Adv Appl Microbiol 68:133–216. https://doi.org/10.1016/S0065-2164(09)01204-0

Zhao X, Oh S-H, Coleman DA, Hoyer LL (2022) ALS1 deletion increases the proportion of small cells in a *Candida albicans* culture population: Hypothesizing a novel role for Als1. Front Cell Infect Microbiol 12:895068.

Zhu X, Williamson PR (2004) Role of laccase in the biology and virulence of *Cryptococcus neoformans*. FEMS Yeast Res 5(1):1–10. https://doi.org/10.1016/j.femsyr.2004.04.004

Zhu Y, Fang H-M, Wang Y-M, Zeng G-S, Zheng X-D, Wang Y (2009) Ras1 and Ras2 play antagonistic roles in regulating cellular cAMP level, stationary-phase entry and stress response in *Candida albicans*. Mol Microbiol 74(4):862–875. https://doi.org/10.1111/j.1365-2958.2009.06898.

8 Diagnostic and Therapeutic Advancements in the Field of Antifungal Management

Ashok Kumar Yadav and Pranava Prakash

8.1 MOLECULAR DIAGNOSTICS

Molecular diagnostics have revolutionized antifungal management by providing rapid and accurate methods for identifying fungal infections, determining drug susceptibility, and monitoring treatment response. These techniques have significantly improved patient outcomes by enabling timely and precise interventions.

Traditional methods for diagnosing fungal infections, such as culture-based techniques, are slow and often fail to provide a definitive diagnosis. Molecular diagnostics, on the other hand, offer several advantages. Polymerase chain reaction (PCR) is a commonly used technique that amplifies specific fungal DNA sequences, allowing for the rapid and sensitive detection of fungal pathogens in clinical samples. This technology can identify a wide range of fungal species, enabling healthcare providers to tailor antifungal therapy to the specific pathogen, which is crucial given the diversity in fungal susceptibility to different antifungal agents.

Furthermore, molecular diagnostics have expanded our understanding of antifungal resistance. By identifying genetic markers associated with resistance, clinicians can predict which antifungal drugs are likely to be effective, avoiding ineffective treatments and the development of resistance. This information is invaluable in guiding therapy decisions, especially in critically ill patients who cannot afford treatment delays.

In addition to diagnosis and susceptibility testing, molecular methods are essential for monitoring treatment response. Real-time PCR and other quantitative assays can measure the fungal load in patient samples over time, allowing clinicians to assess whether the antifungal therapy is effective. This approach helps detect treatment failures early, preventing the progression of infections and improving patient outcomes.

One notable example of the impact of molecular diagnostics in antifungal management is the management of invasive aspergillosis, a life-threatening infection often affecting immunocompromised patients. Traditional diagnostic methods were often unreliable, leading to delayed treatment initiation. Molecular assays, such as galactomannan and β-D-glucan tests, have greatly improved early detection, and PCR-based techniques provide rapid species identification.

While molecular diagnostics have transformed antifungal management, challenges remain. These methods require specialized equipment and expertise, limiting their availability in resource-constrained settings. Additionally, ongoing research is needed to expand the range of detectable fungal species and resistance markers.

8.1.1 MASS SPECTROMETRY

Mass spectrometry is a reliable and fast technique to detect fungi in the sample with accurate species identification. Matrix-assisted laser desorption ionization–time-of-flight mass spectrometry (MALDI-TOF MS) detects fungal species using a laser to ionize biomolecules on the basis of their

DOI: 10.1201/9781032642864-8

mass-to-charge ratio. The time-of-flight (TOF) component produces a unique peptide mass finger-print that matches with a database. It requires minimum samples and can be analyzed in few minutes. MALDI-TOF has an extended database for microbes like fungi and bacteria.

8.1.2 DNA Sequencing

The "gold standard" technique for rare fungal species is DNA sequencing. It is common DNA barcode region, which lies between small subunit (18S) and large subunit (28S) ribosomal RNA genes in the fungal genome. In this technique, a clinical sample is collected and DNA is extracted followed by PCR amplification using ITS1 and ITS4 primers. Sequence analysis is done and compared with a database such as GenBank.

In conclusion, molecular diagnostics have significantly improved antifungal management by providing rapid and accurate tools for diagnosis, susceptibility testing, and treatment monitoring. These advancements have led to more precise and personalized therapy, ultimately saving lives and reducing the burden of fungal infections. However, efforts to make these technologies more accessible and comprehensive continue to be crucial for maximizing their impact on patient care.

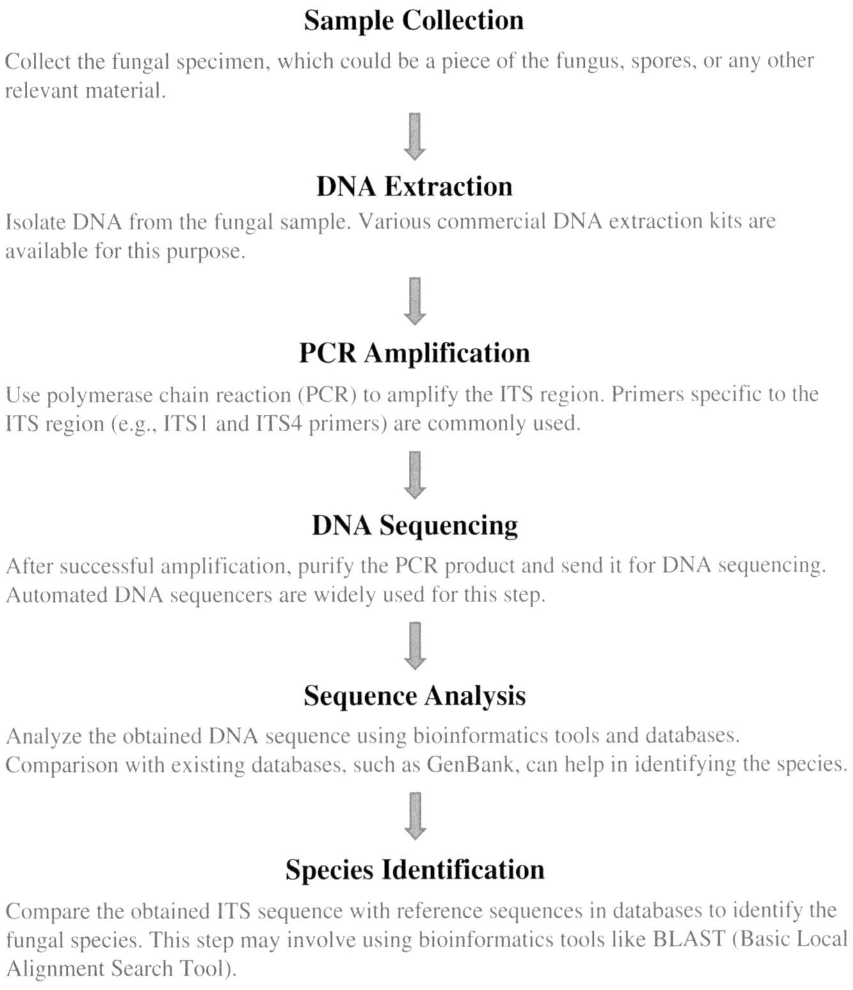

Sample Collection

Collect the fungal specimen, which could be a piece of the fungus, spores, or any other relevant material.

DNA Extraction

Isolate DNA from the fungal sample. Various commercial DNA extraction kits are available for this purpose.

PCR Amplification

Use polymerase chain reaction (PCR) to amplify the ITS region. Primers specific to the ITS region (e.g., ITS1 and ITS4 primers) are commonly used.

DNA Sequencing

After successful amplification, purify the PCR product and send it for DNA sequencing. Automated DNA sequencers are widely used for this step.

Sequence Analysis

Analyze the obtained DNA sequence using bioinformatics tools and databases. Comparison with existing databases, such as GenBank, can help in identifying the species.

Species Identification

Compare the obtained ITS sequence with reference sequences in databases to identify the fungal species. This step may involve using bioinformatics tools like BLAST (Basic Local Alignment Search Tool).

FIGURE 8.1 Flowchart representation of fungal diagnosis by ITS sequencing.

8.2 ANTIFUNGAL RESISTANCE

Fungal infections are becoming a major problem, and to make matters worse, some fungi are becoming resistant to the drugs we use to treat them (1). This resistance can be divided into two types: Microbiologic and clinical resistance. Microbiologic resistance means that the fungus is no longer affected by the drug at normal doses. This resistance is like the fungus building a shield that the drug can't penetrate. Just like with antibiotics, antifungal resistance is a big threat to public health and can make once-easily treatable fungal infections much harder to control (2).

8.2.1 How Fungi Resist Treatment

Different antifungal drugs work in different ways to stop the growth of fungal pathogens. The resistance mechanisms can be primary (inherent in the fungus) or secondary (acquired during treatment). Some fungi can develop resistance by changing their drug targets or by lowering the drug's effectiveness. Resistance can also happen when environmental factors lead to the growth of resistant fungal species (3).

8.2.2 Antifungal Therapy Challenges

Antifungal therapy is crucial for managing fungal infections, but our options are limited because there aren't many classes of antifungal drugs. The emergence of antifungal resistance is making things even more difficult (2). This resistance used to be rare, but now it's on the rise, especially in high-risk medical centers. One major concern is the development of multidrug-resistant fungi that don't respond to multiple classes of antifungal drugs. The reasons behind this resistance are shared among strains that have always been less susceptible and those that become resistant during treatment. These mechanisms include changes in how the drug interacts with the fungus, efflux pumps that remove the drug from the fungus, and the formation of biofilms.

In conclusion, antifungal resistance is a growing problem that threatens our ability to treat and control fungal infections. To address this issue, we need a multi-pronged approach. This includes infection control measures to prevent the spread of resistant fungi, responsible use of antifungal drugs to reduce the risk of resistance, better diagnostic tools to identify resistant strains, and research to develop new antifungal drugs. By tackling antifungal resistance head-on, we can protect the effectiveness of our antifungal treatments and ensure better outcomes for patients. It's a challenge, but one that we must face to keep fungal infections in check.

8.3 IMMUNOTHERAPY

Fungal infections, also known as mycoses, are a significant global health concern, affecting millions of individuals every year. They can range from superficial skin infections to life-threatening systemic diseases. Traditional antifungal treatments include drugs like azoles, polyenes, and echinocandins. While these medications have been effective, they are not without limitations, including drug resistance and adverse effects. Immunotherapy has emerged as a promising adjunctive or alternative approach to manage fungal infections, harnessing the power of the immune system to combat these pathogens.

Immunotherapy for fungal infections primarily relies on enhancing the host's immune response against the invading fungi. This approach involves various strategies, such as monoclonal antibodies, adoptive T-cell therapy, and vaccines, to improve the body's ability to recognize and eliminate fungal pathogens.

One of the most notable developments in antifungal immunotherapy is the use of monoclonal antibodies. These antibodies are designed to target specific fungal antigens, helping the immune system in recognizing and neutralizing the pathogen. For example, monoclonal antibodies like caspofungin and anidulafungin are used to treat invasive fungal infections by blocking specific enzymes involved in fungal cell wall synthesis.

Adoptive T-cell therapy is another promising approach. It involves the isolation and expansion of patient-derived T cells that can specifically target fungal antigens. This therapy has shown potential in treating refractory or recurrent fungal infections, especially in immunocompromised individuals. However, it is a complex and resource-intensive method that requires precise patient-specific customization.

Vaccines have been explored as a preventive measure against fungal infections, especially in high-risk populations. Fungal vaccines work by priming the immune system to recognize and respond more effectively to fungal pathogens. The development of a vaccine for *Candida* species, such as *Candida albicans*, has shown promise in animal models and may offer protection against invasive candidiasis in the future.

Immunomodulatory agents, such as interferons, granulocyte colony-stimulating factor (G-CSF), and colony-stimulating factors (CSFs), are also being investigated in antifungal immunotherapy. These agents can enhance the immune response and improve the outcome of antifungal treatment, particularly in patients with compromised immune systems.

Combination therapy is a growing area of interest, where traditional antifungal drugs are used in conjunction with immunotherapy. This approach aims to maximize the effectiveness of treatment and potentially reduce the risk of developing drug resistance. For instance, combining antifungal drugs with monoclonal antibodies or adoptive T-cell therapy has shown promise in preclinical and clinical studies.

Despite the potential benefits of immunotherapy, several challenges need to be addressed. Immunotherapy can be expensive and labor-intensive, limiting its widespread use. Additionally, the effectiveness of immunotherapy can vary depending on the specific fungal species and the patient's immune status. Research is ongoing to better understand these variables and optimize immunotherapeutic strategies.

In conclusion, immunotherapy has emerged as a promising avenue in antifungal management. It offers new ways to enhance the immune response against fungal pathogens, complementing traditional antifungal treatments. While challenges remain, ongoing research and development in this field hold the potential to improve the outcome of fungal infections, particularly in high-risk patient populations. As our understanding of immunotherapy advances, it may become an integral part of the antifungal arsenal, offering more effective and personalized solutions for fungal infection management. Therefore, antifungal immunotherapy continues to be attractive as an adjunct to the currently available antifungal chemotherapy options for a number of reasons, including the fact that existing antifungal drugs, albeit largely effective, are not without limitations, and that morbidity and mortality associated with invasive mycoses are still unacceptably high. For several decades, intense basic research efforts have been directed at development of fungal immunotherapies. Nevertheless, this approach suffers from a severe bench-bedside disconnect owing to several reasons: The chemical and biological peculiarities of the fungal antigens, the complexities of host-pathogen interactions, an under-appreciation of the fungal disease landscape, the requirement of considerable financial investment to bring these therapies to clinical use, as well as practical problems associated with immunizations. In this general, non-exhaustive review, we summarize the features of ongoing research efforts directed toward devising safe and effective immunotherapeutic options for mycotic diseases, encompassing work on antifungal vaccines, adoptive cell transfers, cytokines, antimicrobial peptides (AMPs), monoclonal antibodies (mAbs), and other agents.

Antifungal immunotherapy and immunomodulation are a double-hitter approach to deal with invasive fungal infections.

8.4   COMBINING ANTIFUNGAL TREATMENTS

Invasive fungal infections have been on the rise, particularly in individuals who have undergone organ transplants, those with hematological malignancies, and patients with weakened immune systems. This increase is a cause for concern, but there's a problem, the rate of antifungal resistance is growing faster than the development of new antifungal drugs.

Given this mismatch, the idea of using a combination of antifungal medications to treat invasive fungal infections has become appealing to many doctors. This interest is fueled by *in vitro* studies showing that some antifungal drugs work better together. However, despite some promising findings, the actual results of combining antifungals in clinical practice remain uncertain (4).

8.4.1   THE CHALLENGES OF COMBINATION ANTIFUNGAL THERAPY

Several randomized controlled trials have been conducted to evaluate the effectiveness and safety of using combinations of antifungal drugs. However, there are challenges that make it difficult to draw clear conclusions from these studies. The limited number of cases, high costs, and numerous factors that can influence results sometimes lead to weak and conflicting findings (3).

One of the major issues in the field of combination antifungal therapy is the lack of consensus. In many clinical scenarios, it's unclear whether combining antifungal drugs is the best approach. This creates confusion for healthcare providers, making it difficult to determine the most effective treatment for patients with invasive fungal infections.

The current situation underscores the need for more research into combination antifungal therapies. While some studies have shown promise, there's a lack of definitive guidance on when and how to use these combinations effectively. Further research is necessary to better understand the circumstances in which combination therapy is most beneficial.

8.4.2   DEVELOPING GUIDELINES FOR PHYSICIANS

Infectious disease societies play a vital role in guiding clinicians in their approach to treating invasive fungal infections. As the research in this field advances, these societies should develop specific recommendations to help doctors make informed decisions.

8.4.3   CONCLUSION

The rise in invasive fungal infections, coupled with the growing issue of antifungal resistance, has led to an interest in combining antifungal drugs as a treatment strategy. While *in vitro* studies have shown promise, the clinical outcomes of such combinations remain uncertain due to challenges like limited case numbers and high costs. This has resulted in a lack of consensus among healthcare providers (4, 5).

More research is needed to determine when and how combination antifungal therapy is most effective. As the field progresses, infectious disease societies should work on developing clear guidelines to help clinicians make informed decisions when treating patients with invasive fungal infections (6).

In the end, the goal is to find the most effective and safe ways to combat these infections and improve the outcomes for patients at risk.

8.5   FUNGAL BIOFILMS

Fungal biofilms are complex communities of fungal cells that adhere to surfaces and are embedded in a matrix of extracellular polymeric substances (EPS). They are a subject of increasing interest in microbiology due to their relevance in various fields, including medicine, agriculture, and industry.

8.5.1   BIOFILM FORMATION

Fungi, such as *Candida* and *Aspergillus* species, form biofilms by adhering to surfaces using adhesins and other surface-associated proteins. Once attached, they produce EPS that provide structural support and protect the biofilm from environmental stressors (7, 8).

8.5.2   MEDICAL RELEVANCE

In healthcare, fungal biofilms have gained attention as they are associated with serious infections. For example, *Candida* biofilms on medical devices like catheters can lead to biofilm-associated candidiasis, which is challenging to treat due to increased resistance to antifungal agents (9–11).

8.5.3   AGRICULTURAL IMPACT

In agriculture, fungal biofilms can develop on plant surfaces, affecting crop health. Understanding these biofilms is essential for the management of plant diseases.

8.5.4   INDUSTRIAL APPLICATIONS

In the industrial sector, fungal biofilms have been explored for their potential in bioremediation and biotechnology. They can be employed to remove contaminants from water and soil or to produce valuable compounds through fermentation.

8.5.5   DRUG RESISTANCE

Fungal biofilms exhibit greater resistance to antifungal drugs than their planktonic counterparts. This is partly attributed to the reduced penetration of drugs through the EPS matrix and the altered physiology of biofilm-associated fungal cells.

8.5.6   RESEARCH CHALLENGES

Studying fungal biofilms presents challenges, including the need for specialized techniques to investigate biofilm structure and composition. Researchers also face difficulties in developing effective antifungal strategies.

Fungal biofilms represent a complex and intriguing area of research with far-reaching implications. Understanding their formation, impact on health, agriculture, and industry, as well as strategies to combat them, is crucial. Researchers and industries alike continue to explore the potential of fungal biofilms, and their study remains an emerging frontier in microbiology (12).

8.6   NATURAL ANTIFUNGAL AGENTS: THERAPEUTIC AND DIAGNOSTIC ADVANCEMENTS

The rise of antifungal resistance has prompted a renewed interest in natural antifungal agents, particularly those derived from plants. These compounds offer promising alternatives or adjuncts to traditional antifungal drugs, potentially providing effective and less toxic treatment options.

8.6.1   DEVELOPMENT OF NATURAL ANTIFUNGAL AGENTS

The discovery of natural antifungal agents dates back to ancient times, with various civilizations utilizing plant extracts and herbal remedies to treat fungal infections. Over time, scientific investigations have led to the identification and characterization of numerous plant-derived compounds with potent antifungal activity.

8.6.2   MECHANISMS OF ACTION

Natural antifungal agents exhibit a diverse range of mechanisms of action, disrupting various aspects of fungal growth and survival. Some common mechanisms are discussed next.

8.6.2.1 Membrane Disruption

Certain natural compounds can disrupt the integrity of fungal cell membranes, leading to leakage of cellular contents and cell death.

8.6.2.2 Enzyme Inhibition

Natural antifungal agents can target specific enzymes involved in fungal metabolism, such as those involved in cell wall synthesis or protein synthesis.

8.6.2.3 Antifungal Biosynthesis Inhibition

Some natural compounds can interfere with the synthesis of essential fungal cell components, such as ergosterol, a key component of fungal cell membranes.

8.6.3 ADVANTAGES OF NATURAL ANTIFUNGAL AGENTS

Natural antifungal agents offer several potential advantages over traditional antifungal drugs.

8.6.3.1 Reduced Toxicity

Natural compounds are often less toxic to host cells than synthetic antifungal drugs, minimizing the risk of side effects.

8.6.3.2 Broad-Spectrum Activity

Some natural antifungal agents exhibit broad-spectrum activity against a wide range of fungal species, including those resistant to traditional drugs.

8.6.3.3 Synergistic Effects

Natural compounds can interact synergistically with traditional antifungal drugs, enhancing their efficacy and reducing the risk of resistance development.

8.6.4 EXAMPLES OF NATURAL ANTIFUNGAL AGENTS

Numerous plant-derived compounds have demonstrated antifungal activity, including the following.

8.6.4.1 Tea Tree Oil

Tea tree oil, derived from the Australian Melaleuca alternifolia tree, contains a variety of antifungal compounds, including terpinenes and terpinols, which exhibit broad-spectrum activity against various fungal species.

8.6.4.2 Garlic

Garlic, a common culinary herb, contains allicin, a sulfur-containing compound that has been shown to have antifungal activity against a wide range of fungi, including *Candida albicans* and *Aspergillus fumigatus*.

8.6.4.3 Oregano Oil

Oregano oil, derived from the herb *Origanum vulgare*, contains carvacrol and thymol, phenolic compounds with potent antifungal activity. Oregano oil has been shown to be effective against various fungal species, including *Candida albicans* and *Aspergillus fumigatus*.

8.6.4.4 Curcumin

Curcumin, the main active compound in turmeric, has demonstrated antifungal activity against various fungi, including *Candida albicans* and *Aspergillus fumigatus*. Curcumin is thought to act by disrupting fungal cell membranes and inhibiting fungal enzymes.

8.6.5 CLINICAL APPLICATIONS OF NATURAL ANTIFUNGAL AGENTS

Natural antifungal agents are currently being investigated for their potential clinical applications in various settings.

8.6.5.1 Dermatological Infections

Natural antifungal agents are being explored as topical treatments for dermatological fungal infections, such as *tinea pedis* (athlete's foot) and *candidiasis.*

8.6.5.2 Oral Fungal Infections

Natural antifungal agents are being studied as potential alternatives to oral antifungals for the treatment of oral candidiasis.

8.6.5.3 Fungal Biofilms

Natural antifungal agents are being investigated for their ability to disrupt fungal biofilms, which are notoriously resistant to traditional antifungal drugs.

8.6.6 CHALLENGES AND FUTURE DIRECTIONS

Despite the promising potential of natural antifungal agents, several challenges need to be addressed before they can be widely adopted in clinical practice.

8.6.6.1 Standardization

The standardization of natural antifungal products is crucial to ensure their consistent quality and efficacy.

8.6.6.2 Mechanism of Action Elucidation

Further research is needed to fully elucidate the mechanisms of action of natural antifungal agents, enabling the development of more targeted and effective therapies.

8.6.6.3 Clinical Trials

Large-scale clinical trials are necessary to evaluate the safety and efficacy of natural antifungal agents in various patient populations.

Natural antifungal agents represent a promising avenue for therapeutic and diagnostic advancements in antifungal therapy. Their potential to provide effective, less toxic, and more sustainable treatment options for fungal infections merits further investigation and development.

8.7 DIAGNOSTIC IMAGING: ADVANCED IMAGING TECHNIQUES IN FUNGAL INFECTION DIAGNOSIS AND MONITORING

In the realm of fungal infections, advanced imaging techniques have emerged as invaluable tools for accurate diagnosis and monitoring disease progression. These techniques, particularly computed tomography (CT), magnetic resonance imaging (MRI), and positron emission tomography (PET) scans, offer comprehensive insights into the extent and severity of fungal infections, particularly in deep-seated or systemic cases.

8.7.1 COMPUTED TOMOGRAPHY (CT)

CT scans provide detailed cross-sectional images of the body, allowing for precise visualization of anatomical structures and abnormalities. In the context of fungal infections, CT scans are particularly useful for detecting fungal lesions in the lungs, sinuses, and bones.

8.7.1.1   Pulmonary Fungal Infections

CT scans are highly sensitive for detecting fungal infiltrates in the lungs, including nodules, masses, and consolidations. They can also reveal signs of pleural effusion, cavitation, and broncholithiasis, which are associated with fungal infections (13).

8.7.1.2   Sinus Fungal Infections

CT scans are invaluable for diagnosing fungal sinusitis, enabling visualization of fungal masses, sinus opacification, and bony erosion (14).

8.7.1.3   Bone Fungal Infections

CT scans are particularly useful for detecting osteomyelitis, a bone infection caused by fungi. They can reveal bone destruction, abscess formation, and periosteal thickening, characteristic signs of fungal osteomyelitis (15).

8.7.2   Magnetic Resonance Imaging (MRI)

MRI utilizes magnetic fields and radio waves to generate detailed images of the body's soft tissues and internal structures. While CT scans excel at visualizing bony structures, MRI offers superior soft tissue contrast, making it particularly useful for detecting fungal infections in brain, spinal cord, and joints.

8.7.2.1   Brain Fungal Infections

MRI is the preferred imaging modality for diagnosing fungal brain abscesses, which are often difficult to detect with CT scans. MRI can reveal the abscess location, extent, and presence of surrounding edema (16).

8.7.2.2   Spinal Cord Fungal Infections

MRI is crucial for diagnosing spinal cord fungal infections, including spondylodiscitis and epidural abscesses. It can visualize the extent of spinal cord involvement, bone destruction, and abscess formation (17).

8.7.2.3   Joint Fungal Infections

MRI is valuable for detecting fungal arthritis, particularly in cases where CT scans are inconclusive. It can reveal synovial effusion, joint capsule thickening, and bone erosions (18).

8.7.3   Positron Emission Tomography (PET)

PET scans employ radioactive tracers to visualize areas of increased metabolic activity in the body. In the context of fungal infections, PET scans can identify active fungal lesions and differentiate them from inactive or necrotic tissue.

8.7.3.1   PET Scan Utility

PET scans are particularly useful for monitoring the response to antifungal therapy, as they can detect changes in fungal activity as early as 24–48 hours after treatment initiation (19).

8.7.4   Conclusion

Advanced imaging techniques, including CT, MRI, and PET scans, have revolutionized the diagnosis and monitoring of fungal infections. Their ability to provide detailed anatomical and metabolic information has significantly improved the detection and characterization of fungal lesions, leading to more accurate diagnoses, better treatment planning, and improved patient outcomes.

8.8 TELEMEDICINE AND REMOTE MONITORING: BRIDGING THE GAP IN ANTIFUNGAL CARE FOR UNDERSERVED PATIENTS

The global burden of fungal infections is significant, with an estimated 1.7 billion people affected annually. Despite the availability of antifungal therapies, access to appropriate diagnosis, treatment, and follow-up remains a challenge, particularly for individuals residing in remote or underserved areas. Telemedicine and remote monitoring technologies offer promising solutions to bridge this gap and improve the quality of antifungal care for these patients.

8.8.1 CHALLENGES IN ACCESSING ANTIFUNGAL CARE

Several factors contribute to the challenges in accessing antifungal care for underserved populations.

8.8.1.1 Geographic Barriers

Remote or rural areas often lack access to specialized healthcare providers with expertise in fungal infections.

8.8.1.2 Limited Resources

Underserved communities may have limited access to diagnostic facilities and antifungal medications.

8.8.1.3 Socioeconomic Factors

Poverty, lack of transportation, and cultural barriers can further hinder access to antifungal care.

8.8.1.4 Telemedicine and Remote Monitoring as Solutions

Telemedicine and remote monitoring technologies have the potential to address these challenges and improve access to antifungal care for underserved patients.

8.8.1.5 Virtual Consultations

Telemedicine platforms allow patients to consult with specialists remotely, enabling timely diagnosis and treatment recommendations.

8.8.1.6 Real-Time Monitoring

Remote monitoring devices can track vital signs, medication adherence, and treatment response, providing valuable data for clinicians.

8.8.1.7 Patient Education

Telehealth platforms can be used to provide patient education on fungal infections, prevention strategies, and medication management.

8.8.2 BENEFITS OF TELEMEDICINE AND REMOTE MONITORING

Telemedicine and remote monitoring offer several benefits for antifungal care.

8.8.2.1 Improved Access

These technologies expand access to specialists and specialized care for patients in remote or underserved areas.

8.8.2.2 Timely Diagnosis and Treatment

Early diagnosis and initiation of appropriate therapy can improve patient outcomes and reduce the risk of complications.

8.8.2.3   Enhanced Patient Engagement

Telehealth platforms can empower patients to take an active role in their care, leading to improved medication adherence and self-management.

8.8.2.4   Cost-Effectiveness

Telemedicine and remote monitoring can reduce healthcare costs by minimizing travel expenses, unnecessary hospitalizations, and readmissions.

8.8.3   CASE STUDIES AND IMPLEMENTATION STRATEGIES

Several successful case studies demonstrate the effectiveness of telemedicine and remote monitoring in improving antifungal care.

8.8.3.1   Teledermatology Programs

Teledermatology programs have been shown to be effective in diagnosing and managing fungal skin infections in remote areas (19).

8.8.3.2   Remote Monitoring of Invasive Fungal Infections

Remote monitoring of patients with invasive fungal infections has allowed for early detection of complications and timely interventions (2).

8.8.3.3   Telehealth Interventions for Aspergillosis

Telehealth interventions have been shown to improve patient outcomes and reduce healthcare costs for patients with aspergillosis, a serious fungal infection (20).

To effectively implement telemedicine and remote monitoring for antifungal care, several strategies are crucial.

8.8.3.4   Infrastructure Development

Access to reliable internet connectivity and appropriate technology is essential for successful implementation.

8.8.3.5   Training and Education

Healthcare providers need training in telemedicine and remote monitoring techniques, while patients require education on using these technologies.

8.8.3.6   Integration with Existing Healthcare Systems

Telemedicine and remote monitoring should be integrated into existing healthcare systems to ensure seamless patient care.

8.8.3.7   Policy and Regulatory Support

Clear policies and regulations are needed to address issues such as reimbursement, patient privacy, and data security.

CONCLUSION

Telemedicine and remote monitoring have the potential to revolutionize antifungal care for underserved patients, particularly those residing in remote or resource-limited settings. These technologies can improve access to specialists, facilitate timely diagnosis and treatment, enhance patient engagement, and reduce healthcare costs. By addressing the challenges and implementing effective strategies, telemedicine and remote monitoring can play a pivotal role in bridging the gap in antifungal care and improving patient outcomes.

REFERENCES

1. Antonovics J, et al. Evolution by any other name: Antibiotic resistance and avoidance of the E-word. PLOS Biol. 2007;5:e30.
2. Arendrup MC, Mavridou E, Mortensen KL, Snelders E, Frimodt-Moller N, Khan H, Melchers WJ, Verweij PE. Development of azole resistance in *Aspergillus fumigatus* during azole therapy associated with change in virulence. PLOS ONE. 2010;5:e10080.
3. Hamad M. Current antifungal drugs and immunotherapeutic approaches as promising strategies to treatment of fungal diseases. Scand J Immunol. 2008.
4. Nami S, et al. Novel immunotherapeutic strategies for invasive fungal disease. Biomed Pharmacother. 2019;110:857–868.
5. van Spriel AB. Antifungal drugs: New insights in research & development. Curr Drug Targets Cardiovasc Haematol Disord. 2003;3(3):209–217.
6. Nicola AM, et al. Promising immunotherapy against fungal diseases. Pharmacol Ther. 2019.
7. Day JN, et al. Combination antifungal therapy for cryptococcal meningitis. N Engl J Med. 2013;368(14):1291–1302. doi: 10.1056/NEJMoa1110404.
8. Finkel JS, Mitchell AP. Genetic control of *Candida albicans* biofilm development. Nat Rev Microbiol. 2011;9:109–118.
9. Beauvais A, Muller FM. Biofilm formation in *Aspergillus fumigatus*. In: Latge JP, Steinbach WJ, editors. Aspergillus fumigatus and aspergillosis. ASM Press; 2009. pp. 149–157.
10. Martinez LR, Casadevall A. *Cryptococcus neoformans* biofilm formation depends on surface support and carbon source and reduces fungal cell susceptibility to heat, cold, and UV light. Appl Environ Microbiol. 2007;73:4592–4601.
11. Di Bonaventura G, et al. Biofilm formation by the emerging fungal pathogen *Trichosporon asahii*: Development, architecture, and antifungal resistance. Antimicrob Agents Chemother. 2006; 50:3269–3276.
12. Davis LE, Cook G, Costerton JW. Biofilm on ventriculo-peritoneal shunt tubing as a cause of treatment failure in coccidioidal meningitis. Emerg Infect Dis. 2002;8:376–379.
13. Denning DW. CT and MRI of pulmonary fungal infections. Clin Radiol. 2012;67(9):825–833.
14. Li K, Zhu XK. CT and MRI findings of fungal sinusitis. Eur J Radiol. 2015;84(8):1419–1427.
15. Kontoyiannis DP, Kontoyiannis MN. Fungal osteomyelitis: Imaging findings with clinical correlation. RadioGraphics. 2017;37(2):371–388.
16. Garg P, Kauffman RS. Fungal brain abscess: A review of contemporary imaging and treatment. Clin Infect Dis. 2017;65(10):1678–1685.
17. Ozturk A, Saribas, O.. Spinal cord fungal infections. In Handbook of Clinical Neurosurgery. Springer; 2017. pp. 1–16.
18. Lee WC, Chen WH. Imaging of fungal arthritis. 2017.
19. Azumah I, Chinn D. Teledermatology in underserved communities: A systematic review. Telemedicine and eHealth. 2017;23(8):693–701.
20. Hoenig H, et al. Telemonitoring in invasive fungal infections: Results of the pilot study 'Mycoses'. Crit Care Med. 2010;38(3):617–625.

9 Mucormycosis Infection and Methods of Detection
An Overview

Saumya Chaturvedi, Khushboo Arya,
Sana Akhtar Usmani, Shikha Chandra,
Deeksha, Nitin Bhardwaj, and Ashutosh Singh

9.1 INTRODUCTION

According to the global datasets, the chance of occurrence of mucormycosis on the Indian subcontinent is 70% higher than in the rest of the world (Prakash and Chakrabarti, 2021). Researchers in analyzing 434 papers with incidence/prevalence/proportion data have found that 57 million people (which is equivalent to 4.4% of 1.3 billion people) in India suffer from serious fungal diseases, 10% of whom have potentially dangerous fungal infections. According to these data, around 24 million women were found to have recurrent episodes of vulvovaginal candidiasis (Ray et al, 2022). Another prominent fungal infection, allergic bronchopulmonary aspergillosis, which occurs because of an immune reaction against *Aspergillus fumigates*, affects nearly 2 million people in India (Agarwal et al, 2022; Ray et al, 2022). The number of cases for *Tinia capitis* in school children between the ages of 3 and 7 is around 25 million (Leung et al, 2020; Ray et al, 2022). *Tinia* species causes fungal infection of the scalp and is very difficult to treat; *Trichophyton tonsurans* and *Microsporum canis* are the major causal organisms (Leung et al, 2020). According to study conducted in 2022, around 1.7 million people suffer from chronic pulmonary aspergillosis, 58,400 from pneumocystis pneumonia, and nearly 200,000 had mucormycosis annually (Ray et al, 2022).

Mucormycosis is a fungal infection also known as black fungus. It is a life-threatening fungal infection which occurs in immunocompromised individuals, patients suffering from uncontrolled diabetes, renal failure, cancer (mostly at the time of chemotherapy), or those having undergone solid organ transplant. (Ponnaiyan et al, 2022). Additionally, patients using corticosteroids, or fluconazole, or having had neutropenia are at high risk of mucormycosis infection (Acosta-España and Voigt, 2022). Interestingly, several reports have claimed that patients with AIDS are not at high risk of developing mucormycosis (Sugar, 2005; Ibrahim et al, 2012). The spectrum of agents causing mucormycosis in India is higher. Mucormycosis is caused by thermotolerant saprophytic mucormycetes fungi of order Mucorales, mostly found in a terrestrial environment. It is found in decaying organic matter and soil samples which spreads through rotting fruits and vegetables. Fungi spores enter in body by nose, directly from sinus or vascular occlusion (Prakash and Chakrabarti, 2021; Ponnaiyan et al, 2022; Sree Lakshmi et al, 2023) (Figure 9.1). The primary source for cutaneous mucormycosis has been found to be non-sterile adhesive tapes and dressings (Ibrahim et al, 2012). This infection can also spread gastrointestinally and cutaneously (Acosta-España and Voigt, 2022).

The order Mucorales comprises 261 species in 55 genera, at least 38 of which have been associated with human infections and the total causative agents of order mucorale include 11 genera and 27 species. Common causal species include *Mucor*, *Rhizopus*, *Rhizomucor*, *Cunninghamella*, *Syncephalastrum*, *Saksenaea*, and *Lichtheimia*. (Centers for Disease Control and Prevention).

Mucormycosis is often misdiagnosed because of its non-specific symptoms, and also because of similar symptoms as observed during any bacterial and fungal infection (Acosta-España and Voigt, 2022).

DOI: 10.1201/9781032642864-9

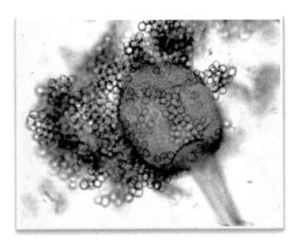

FIGURE 9.1 A representative picture of a mature sporangium of *Rhizopus* species. (Image taken from Flickr [CC BY-NC-SA 4.0]; Creator: George Shepherd.)

9.2 PREDISPOSING FACTORS

9.2.1 DIABETES

Diabetes is the most common underlying disease of patients prone to mucormycosis. In two separate meta-analyses, it has been shown that globally the predominance of mucormycosis in diabetic patients is around 36%–40% (Roden et al, 2005; Jeong et al, 2019), compared to around 50% in the Indian context (Chakrabarti et al, 2006, 2009; Prakash et al, 2019). Another study has concluded that the risk factor for mucormycosis in patients with diabetes mellitus was much higher in patients from North India than those from South India (Prakash et al, 2019).

In diabetic patients, hyperglycemia and diabetic keto acidosis (DKA), the intake of corticosteroid increases the risk of mucormycosis. This is probably because such conditions result in lower pH and an increase in free iron, which leads to the loss of phagocytic activity of white blood cells, and thereby increases the risk of mucormycosis (Sree Lakshmi et al, 2023). Hyperglycemia enhances the expression of the glucose-regulated 78 kDa protein (GRP78) on human endothelium cells, which is an important receptor for Mucorales vascular invasion via its spore coat protein (CotH). *Rhizopus* also interacts with GRP78 on nasal epithelial cells via CotH3 in order to penetrate and harm the cells. High glucose, iron, and ketones (a hallmark of DKA) dramatically increase the expression of GRP78 and CotH3, likely producing rhino-orbital mucormycosis (Muthu et al, 2021).

Mucormycosis is responsible for causing tissue necrosis but some other fungi such as *Aspergillus*, *Lomentospora*, and *Fusarium* species are also responsible for causing tissue necrosis (Gupta et al, 2022). During the COVID-19 pandemic, a concerning resurgence was seen in cases of mucormycosis which almost reached an epidemic situation. The possible reasons behind the infection could be a cytokine storm (interleukin 6), islet damage, elevated ferritin, and endotheliitis. Increased hyperglycemia and impaired phagocytosis due to uncontrolled use of corticosteroids, a cytokine storm, and ketoacidosis are suspected to contribute significantly to mucormycosis associated with COVID-19 (Acosta-España and Voigt, 2022).

9.2.2 HOST DEFENSE

It has been seen that patients with neutropenia are at higher risk of developing mucormycosis, suggesting the importance of neutrophils (but not necessarily T lymphocytes) for immunity against mucormycosis. Now, it is known that oxidative metabolites and the cationic peptides, defensins, are generated by mononuclear and polymorphonuclear phagocytes of normal hosts which are responsible for killing mucorale (Diamond et al, 1982; Waldorf et al, 1984; Waldorf, 1989; Ibrahim et al, 2012). Further, it has been seen that pulmonary alveolar macrophages harvested from immunocompetent patients have an ability to inhibit the growth of *R. oryzae* sporangiospores, but they had limited ability *in vitro*. Macrophages from immunosuppressed individuals are unable to inhibit germination (Waldorf et al, 1984).

9.3 FORMS OF MUCORMYCOSIS

Based on the anatomic location, mucormycosis can be divided into six forms.

9.3.1 RHINOCEREBRAL MUCORMYCOSIS (RCM)

During the COVID-19 pandemic, it was observed that this form was most common in India. It involves infection of the nose, paranasal sinuses, and brain. It is mainly reported in diabetic patients. Reports show that 77% of RCM infections occurred in the diabetic population (Prakash and Chakrabarti, 2021). Facial pain, headache, nasal congestion, and retro-orbital headache are among the symptoms (Acosta-España and Voigt, 2022) (Figure 9.2A).

9.3.2 PULMONARY MUCORMYCOSIS

This form was observed to be most common in developed countries. Neutropenic patients with hematologic malignancies were at high risk (Prakash and Chakrabarti, 2021; Acosta-España and Voigt, 2022). In India, around 38% of patients with postpulmonary tuberculosis were also at higher risk (about 21%) (Prakash et al, 2019; Prakash and Chakrabarti, 2021). The difficulty in diagnosing pulmonary mucormycosis is primarily because the radiological imaging may not be specific and it can show patient with numerous nodules, pleural effusion (as much as 21%), thickly walled cavities (as much as 37%), hilar or mediastinal lymphadenopathy (about 3.3%), air crescent sign (as much as 8%), and pneumothorax (1%–3%) in addition to lung infiltration and consolidation (as much as 96%). To be noted, the '*reverse halo sign*', was observed in about 10% of the patients (Prakash et al, 2019) (Figure 9.2B).

9.3.3 CUTANEOUS MUCORMYCOSIS

This occurs because of a loss of innate immunity; that is the loss of the skin barrier. It is most commonly seen in patients with trauma and burns (Prakash and Chakrabarti, 2021; Acosta-España and

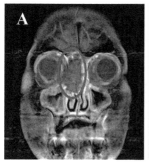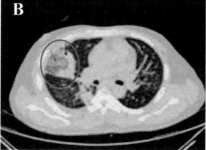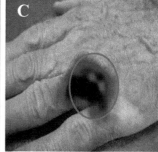

FIGURE 9.2 Select examples of mucormycosis. (A) MRI scan of a patient showing black turbinate, a sign suggestive of rhinocerebral mucormycosis. (B) A reverse halo sign shown in a CT scan, suggesting pulmonary mucormycosis. (C) Image showing cutaneous mucormycosis. (From Wikimedia Commons.)

Voigt, 2022). In India, as much as 79% of cutaneous mucormycosis patients had trauma (Prakash and Chakrabarti, 2021) (Figure 9.2C).

9.3.4 GASTROINTESTINAL MUCORMYCOSIS

This mostly occurs in patient undergoing solid organ transplant and hematological malignancy. In this case stomach, ileum, colon and liver are the effected organs (Acosta-España and Voigt, 2022). It accounts for as much as 8% of cases from India. Pediatric patients have shown higher rates of gastrointestinal mucormycosis, around 60%, out of which 83% were premature neonates. Diabetes mellitus and peritoneal dialysis are the significant risk factors of gastrointestinal mucormycosis in adults. However, in children, broad-spectrum antibiotic use and malnourishment is the major cause (Prakash and Chakrabarti, 2021).

9.3.5 RENAL MUCORMYCOSIS

This form of mucormycosis is responsible for affecting the kidneys. In India, a unique case series has been observed which includes 33%–100% of mucormycosis cases in an immunocompetent host. Imaging findings have shown enlarged kidneys, perinephric stranding, and thickened Gerota's fascia. Symptoms of renal mucormycosis include fever, flank pain, hematuria, kidney injury and white flakes in urine (Prakash and Chakrabarti, 2021).

9.3.6 DISSEMINATED MUCORMYCOSIS

Here, a mortality rate of over 80% has been reported (Cornely et al, 2019).

9.3.7 UNCOMMON FORMS

In these forms, other organs of the body are affected such as the spine, breast, heart, and bone (Prakash and Chakrabarti, 2021; Acosta-España and Voigt, 2022).

9.4 METHODS OF DETECTION

For the proper diagnosis of mucormycosis infection, the first step is the collection of samples which occurs through modification of 1–3 mm biopsy punches, followed by methods discussed below. The morphological features of the fungi which confirms its presence is the broad non-septate hyphae showing coenocytic to irregular branching, with large branching angles almost like rectangles (Acosta-España and Voigt, 2022). In the current scenario, it is hard to confirm mucormycosis just by performing any single test because the same result can be given by other fungi as well. Therefore, combinations of tests are performed to attain maximum accuracy (Acosta-España and Voigt, 2022). Recently, several diagnostic methods have been developed which claim to give accurate and specific results.

9.4.1 RADIOLOGY

Conventional computed tomography (CT) and contrast-enhanced magnetic resonance imaging (MRI) are preferably used in detection of mucormycosis. MRI distinguishes between necrotic and non-necrotic tissue. Radiography can help in distinguishing the involvement of sinus, lung lesion, or cerebral abscesses. However, radiography cannot identify the specific fungi responsible for the infection but can show the severity of the infection. Furthermore, it can help in biopsy sampling and monitoring the condition of the patient and planning the treatment (Honavar, 2021; Ponnaiyan et al, 2022).

CT is a radiological imaging technique that is better to scan bones and can also be used in patients having a pacemaker, metallic fragments, or cardiac monitors since MRI cannot be used in such patients because its powerful magnets can trigger change in the setting of such devices. CTs can be used to scan nasal and paranasal sinus; mucosal thickening with irregular patchy enhancement provides an early sign of mucormycosis, and if the medial rectus is showing thickening, then it means that orbital invasion is taking place (Honavar, 2021; Ponnaiyan et al, 2022). It can identify the infection if the CT scan is showing a '*reverse halo sign*' (Figure 9.2B). This reverse halo sign can be seen even in the first week of infection and gives positive results in almost 94% of cases (Ponnaiyan et al, 2022).

9.4.2 SEROLOGICAL OR ANTIGEN-BASED TESTS

Serological tests can be helpful in the detection of mucormycosis, but it depends on the presence of antigens and antibodies in the blood of the patient. Enzyme-linked immunosorbent assay (ELISA), radio immunosorbent assay (RIA), immunodiffusion, counter-immunoelectrophoresis, and complement fixation (CF) can be used in the identification of specific antibodies in blood and saliva (Ponnaiyan et al, 2022). In comparison to ELISA with lateral flow immunoassay (LFIA), LFIA has been found to be more convenient because blood, saliva, and urine all can be tested easily (Ponnaiyan et al, 2022).

Using serological tests for the diagnosis can be a challenging task. This is primarily because mucorale spores are so abundant in nature that human beings are frequently exposed to these spores. This can generate an antibody titer that can interfere with the results (Gupta et al, 2022). These tests are also not reliable in the case of patients with immunosuppression because they cannot produce antibodies in large numbers, thereby limiting the utility of these tests. Often, these tests can also be time consuming (Ponnaiyan et al, 2022).

9.4.3 MOLECULAR METHODS

Molecular techniques are important and reliable methods that can be used for detection of mucormycosis. These methods can detect infections at an early stage but can only test for the DNA, as shown in Figure 9.3.

Galactomannan and beta-D-glucan cannot be used to detect this group of fungi. Due to the angioinvasion nature of mucorale, its DNA load has been found to be very high in serum as compared to that of aspergillosis. Because of this, blood can be taken as a sample in this case for detection (Lackner et al, 2021).

DNA-based detection mostly involves amplification of genomes using polymerase chain reaction (PCR), which shows high analytic sensitivities. Detection of target DNA can be done during quantitative real-time PCR (qPCR) or after PCR by gel electrophoresis, microarray, and electrospray ionization mass spectrometry (ESI-MS) sequencing. The specificity of assay can be increased by using probes or high-resolution melt (HRM) analysis or use of microarray or ESI-MS. These methods show reduced rates of false positivity (Lackner et al, 2021).

The significance of ESI-MS is that it can help in identifying organisms by comparing the molecular fingerprint with the reference database and it has given good results in the case of *Candida* and *Aspergillus*, but the results can only be described as moderate for mucorale. ESI-MS uses electronspray ionization mass spectroscopy with PCR (multiple pairs of broad-range primers) (Lackner et al, 2021; Ponnaiyan et al, 2022).

The biggest merit of using a molecular method, especially PCR based, for diagnosis of mucormycosis, is that it can give fast results and help in starting treatment without any delay. This decreases the rate of mortality because of mucormycosis. Also, these methods involve identification of fungi based on its DNA sequence irrespective of their morphology (Ponnaiyan et al, 2022).

Pandey et al. have developed a more specific novel qPCR assay for detection and melting curve analysis for differentiation of mucormycosis from three other molds (*Aspergillus* species, Mucorales and *Fusarium*) in a single reaction (using one set of primers), which is clinically relevant in terms

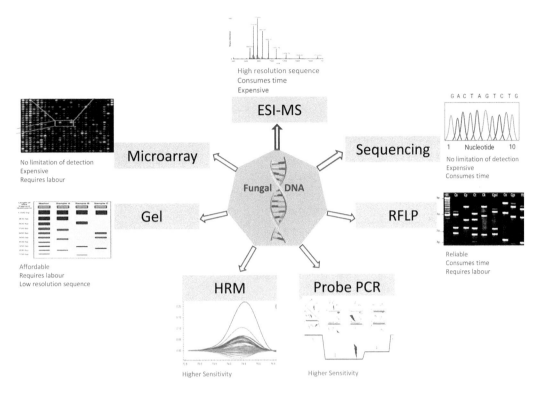

FIGURE 9.3 Select examples of molecular methods of detection of mucormycosis. (From Wikimedia Commons.)

of fungal infections. The primers were designed (using HYDEN and Primer3 software) against the conserved ribosomal region (Clustal X software) of the genome of the mucorale (from NCBI and GenBank databases). The pre-requisite was that these primers become annealed to the target sequence of the mold at 60°C and not with unrelated genera (Pandey et al, 2022). To identify the specificity of the primer for a specific fungal genome, the DNA isolated from the clinical isolate was sequenced using Sanger sequencing of the ITS region. The obtained set of primers were M1D and M1R, which amplify a fragment of gene coding for a small ribosomal subunit (18SrRNA). Finally, BLAST search was used for the selected primers to avoid cross-homology with other unrelated microorganisms (Pandey et al, 2022). This study has proved that PCR using newly designed primers has more specificity and sensitivity in diagnosing mucormycosis and can also distinguish between mucormycosis and aspergillosis.

9.4.4 Biomarkers

Identification of biomarkers related with mucormycosis can help in rapid detection of the infection, leading to early treatment, which will decrease the rate of mortality among patients. The biomarkers in this case are mostly carbohydrates, which are derived from the cell wall of fungi. Examples of theses biomarker can be galactomannan and beta-D glucan. Since gluconic cell wall sugars are not exposed on the surface of the hyphae, galactomannan levels are very low in the case of mucorale. By using matrix-assisted laser desorption/ionization time-of-flight mass spectrometry (MALDI-ToF) (Figure 9.4), about 90% of mucormycosis patients have shown dihexasaccharide (MS-DS) in their serum (Acosta-España and Voigt, 2022).

In mice, another biomarker, a cell wall carbohydrate named fucomannan, has been seen in invasive mucormycosis (Orne et al, 2018; Acosta-España and Voigt, 2022). Orne et al. have developed

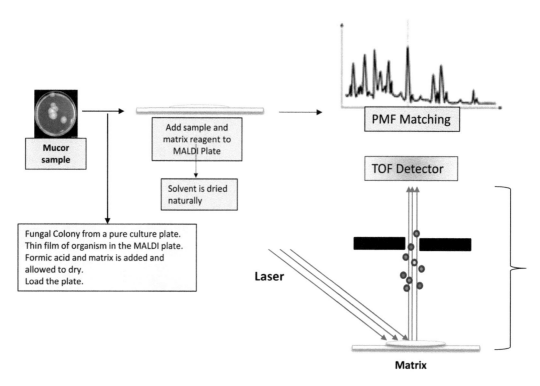

FIGURE 9.4 A simplified layout for detection of mucormycosis using MALDI-ToF. (From Wikimedia Commons.)

lateral flow assay (mAb2DA6), which has shown promising results in the detection of spores of mucorale. This method is based on detection of the cell wall carbohydrate fucomannon. For the detection of infection samples of blood, urine, serum, and bronchoalveolar lavage fluid were collected after 3–4 days of intratracheal infection from immunosuppressed mice. The obtained positive results showed the usefulness of LFA in early diagnosis of mucormycosis (Orne et al, 2018). Moreover, SNPs can help in determining the susceptibility for mucormycosis by determining mutation in TLR. Taken together, detection and study of biomarkers for mucormycosis can help in diagnosing infection at an early stage (Acosta-España and Voigt, 2022).

9.4.5 Histopathological and Microscopic Diagnosis

Microscopy and cell cultures are the fundamental methods for the diagnosis of mucormycosis. Mucorale takes around 3–7 days for growth on most of the fungal culture media, including Sabouraud agar and Potato dextrose agar at 25°C to 30°C (Skiada et al, 2018). Fungal cultures are negative in half of the cases, even when fungal hyphae are seen in histopathologic analysis. The reason behind this phenomenon can be excessive homogenization of tissue which can damage the hyphae (Skiada et al, 2018). Microscopy plays a significant role in diagnosing mucormycosis, but it is mostly useful in an advanced state of mucormycosis. So, this method can confirm infection in invasive conditions. Invasive mucormycosis can cause angioinvasion and prominent infarcts. Patients with neutropenia are more susceptible to the invasion due to mucormycosis (Lackner et al, 2021).

For the detection and confirmation of the fungal infection, the collected samples can be stained for the process of screening under a microscope to find the characteristic feature of the mucorale such as non-septate hyphae with right-angle branching (Gupta et al, 2022; Ponnaiyan et al, 2022).

FIGURE 9.5 Examples of histopathological staining for detection of fungi. (A) Hematoxylin and eosin staining demonstrating *Mucor* hyphae. (B) PAS staining of *Aspergillus*. (C) GMS staining of *Cryptococcus*. (A & B: From Wikipedia Commons. C: From Flickr (CC BY-NC-SA 4.0); Creator: Ed Uthman.)

9.4.5.1 Hematoxylin and Eosin Staining

The sample (excised tissue) is treated with 10% formalin, sectioned, followed by dehydration, thereafter embedded with paraffin. About 3–5 μm sections of paraffin block is stained by Hematoxylin and eosin (H and E), and observed under microscope to find fungal hyphae, necrosis, etc. (Gupta et al, 2022). This method of detection cannot detect fungi if it their number is too low, and also distinguishes fungal hyphae from thin blood vessels in CNS and lung tissue (Gupta et al, 2022) (Figure 9.5A).

9.4.5.2 Periodic Acid-Schiff (PAS) Staining and Grocott-Gomori's Methanamine Silver (GMS) Staining

PAS staining helps in the detection of carbohydrate components present in the cell wall of fungi. It does not stain protein and nucleic acids present in the cell. Periodic acid oxidizes carbohydrate content to aldehyde, which is then applied with Schiff reagent giving pink, red, and magenta colors (Gupta et al, 2022) (Figure 9.5B). GMS staining gives the best contrast for screening of the slides. It can stain fungal organisms irrespective of old or new. GMS staining can also be used to visualize carbohydrate content in dark and black colors (Gupta et al, 2022). The main merit of using this technique during detection is its ability to distinguish between septate and non-septate hyphae of fungi, as mucorale have non-septate hyphae (Lackner et al, 2021) (Figure 9.5C).

9.4.6 ANTIFUNGAL SUSCEPTIBILITY

The API50CH kit developed by bioMérieux – Campus de l'étoile, is a testing kit that is used for the identification of *Mucor* species but it has failed to distinguish between *M. circinelloides* and *M. rouxii*. However, antifungal susceptibility testing studies have allowed identification of probable drugs for treating mucormycosis. For example, isavuconazole is licensed to be used in the United States, as its hepatotoxicity is low, although it results in shortening of the QTc interval. A drug named posaconazole is also now available for first line of treatment in cases of mucormycosis (Cornely et al, 2019).

9.5 FUTURE PERSPECTIVES

Research is still ongoing for improvement in methods of detection and treatment of mucormycosis. Recently, biosensors have been developed for the easy and early diagnosis of fungal infection. Biosensors have the ability to translate biological, chemical, or physical data by using its components, and electrochemical biosensors have shown promising results in the detection of *C. albicans* and *A. fumigatus* (Ponnaiyan et al, 2022). Bioanalysis of skin using microneedles has also been proposed. Newer technologies, such as CRISPR-Cas, are currently not used for the detection of fungal infections, but they can be developed in future as promising methods of detection for such infections (Lackner et al, 2021).

9.6 CONCLUSION

Detection of mucormycosis can be the deciding factor in its diagnosis and successful remediation. Late diagnosis of mucormycosis results in a higher mortality rate from this disease. Several studies have shown that even when detected, the treatment of mucormycosis in some cases, such as in aggressive rhinocerebral mucormycosis, is quite difficult. As predisposing factors are more widely understood, increasing awareness among common people as well as medical professionals to take all precautions against those responsible for causing the disease can help in controlling mucormycosis. Depending upon the type and location, combining different methods might be needed for diagnosis and treatment of mucormycosis. Overall, accurate early diagnoses remain the key to treatment and provide a better quality of life to mucormycosis patients.

ACKNOWLEDGMENTS

AS acknowledges the funding supported by the DBT, Govt. of India (BT/PR38505/MED/29/1513/2020), DST, Govt. of India (CRG/2022/001047; SR/PURSE Phase 2/29(C)), ICMR, Govt. of India (No.52/08/2019-BIO/BMS), UP Higher Education (No. 10/2021/281/-4-Sattar-2021-04(2)/2021) and the University of Lucknow. NB thanks support from ICMR No. 56/2/Hae/BMS and UGC start-up grant No. 30-496/2019. SC, KA, and SAU are grateful for JRF fellowships from DST (CRG/2022/001047), ICMR (Ref. No 3/1/3/JRF2021/HRD(LS)), and UGC (NTA Ref. No. 22161023985), respectively.

REFERENCES

Acosta-España, J. D., and Voigt, K. 2022. Mini review: Risk assessment, clinical manifestation, prediction, and prognosis of mucormycosis: Implications for pathogen- and human-derived biomarkers. Front Microbiol. 13: 895989.

Agarwal, R., Muthu, V., Sehgal, I. S., Dhooria, S., Prasad, K. T., and Aggarwal, A. N. 2022. Allergic bronchopulmonary aspergillosis. Clin Chest Med. 43(1): 99–125.

Chakrabarti, A., Chatterjee, S. S., Das, A., Panda, N., Shivaprakash, M. R., Kaur, A., Varma, S. C., Singhi, S., Bhansali, A., and Sakhuja, V. 2009. Invasive zygomycosis in India: Experience in a tertiary care hospital. Postgrad Med J. 85: 573–581.

Chakrabarti, A., Das, A., Mandal, J., Shivaprakash, M. R., George, V. K., Tarai, B., Rao, P., Panda, N., Verma, S. C., and Sakhuja, V. 2006. The rising trend of invasive zygomycosis in patients with uncontrolled diabetes mellitus. Med Mycol. 44: 335–342.

Cornely, O. A., Alastruey-Izquierdo, A., Arenz, D., Chen, S. C. A., Dannaoui, E., Hochhegger, B., Hoenigl, M., Jensen, H. E., Lagrou, K., Lewis, R. E., et al. 2019. Global guideline for the diagnosis and management of mucormycosis: An initiative of the European Confederation of Medical Mycology in Cooperation with the Mycoses Study Group Education and Research Consortium. Lancet Infect Dis. 19(12): e405–e421.

Diamond, R. D., Haudenschild, C. C., and Erickson, N. F. 1982. Monocyte-mediated damage to *Rhizopus oryzae* hyphae *in vitro*. Infect Immun. 38: 292–297.

Gupta, M. K., Kumar, N., Dhameja, N., Sharma, A., and Tilak, R. 2022. Laboratory diagnosis of mucormycosis: Present perspective. J Family Med Prim Care. 11(5): 1664–1671.

Honavar, S. G. 2021. Code mucor: Guidelines for the diagnosis, staging and management of rhino-orbito-cerebral mucormycosis in the setting of COVID-19. Indian J Ophthalmol. 69(6): 1361–1365.

Ibrahim, A. S., Spellberg, B., Walsh, T. J., and Kontoyiannis, D. P. 2012. Pathogenesis of mucormycosis. Clin Infect Dis. 54(Suppl 1): S16–S22.

Jeong, W., Keighley, C., Wolfe, R., Lee, W. L., Slavin, M. A., Kong, D. C. M., and Chen, S. C. A. 2019. The epidemiology and clinical manifestations of mucormycosis: A systematic review and meta-analysis of case reports. Clin Microbiol Infect. 25: 26–34.

Lackner, N., Posch, W., and Lass-Flörl, C. 2021. Microbiological and molecular diagnosis of mucormycosis: From old to new. Microorganisms. 9(7): 1518.

Leung, A. K. C., Hon, K. L., Leong, K. F., Barankin, B., and Lam, J. M. 2020. *Tinea capitis*: An updated review. Recent Pat Inflamm Allergy Drug Discov. 14(1): 58–68.

Muthu, V., Rudramurthy, S. M., Chakrabarti, A., and Agarwal, R. 2021. Epidemiology and pathophysiology of COVID-19-associated mucormycosis: India versus the rest of the world. Mycopathologia. 186(6): 739–754.

Orne, C., Burnham-Marusich, A., Baldin, C., Gebremariam, T., Ibrahim, A., Kvam, A., and Kozel, T. 2018. Cell wall fucomannan is a biomarker for diagnosis of invasive murine mucormycosis. 28th European Congress of Clinical Microbiology and Infectious Disease (ECCMID) 1.

Pandey, M., Xess, I., Sachdev, J., Yadav, U., Singh, G., Pradhan, D., Xess, A. B., Rana, B., Dar, L., Bakhshi, S., Seth, R., Mahapatra, M., Jyotsna, V. P., Jain, A. K., Kumar, R., Agarwal, R., and Mani, P. 2022. Development of a sensitive and specific novel qPCR assay for simultaneous detection and differentiation of mucormycosis and aspergillosis by melting curve analysis. Front Fungal Biol. 2: 800898.

Ponnaiyan, D., Anitha, C. M., Prakash, P. S. G., Subramanian, S., Rughwani, R. R., Kumar, G., and Nandipati, S. R. 2022. Mucormycosis diagnosis revisited: Current and emerging diagnostic methodologies for the invasive fungal infection (Review). Exp Ther Med. 25(1): 47.

Prakash, H., and Chakrabarti, A. 2021. Epidemiology of mucormycosis in India. Microorganisms. 9(3): 523.

Prakash, H., Ghosh, A. K., Rudramurthy, S. M., Singh, P., Xess, I., Savio, J., Pamidimukkala, U., Jillwin, J., Varma, S., Das, A., et al. 2019. A prospective multicenter study on mucormycosis in India: Epidemiology, diagnosis, and treatment. Med Mycol. 57(4): 395–402.

Ray, A., Aayilliath K, A., Banerjee, S., Chakrabarti, A., and Denning, D. W. 2022. Burden of serious fungal infections in India. Open Forum Infect Dis. 9(12): ofac603.

Roden, M. M., Zaoutis, T. E., Buchanan, W. L., Knudsen, T. A., Sarkisova, T. A., Schaufele, R. L., Sein, M., Sein, T., Chiou, C. C., Chu, J. H., et al. 2005. Epidemiology and outcome of zygomycosis: A review of 929 reported cases. Clin Infect Dis. 41: 634–653.

Skiada, A., Lass-Floerl, C., Klimko, N., Ibrahim, A., Roilides, E., and Petrikkos, G. 2018. Challenges in the diagnosis and treatment of mucormycosis. Med Mycol. 56(suppl_1): 93–101.

Sree Lakshmi, I., Kumari, B. S., Jyothi, C., Devojee, M., Padma Malini, K., Sunethri, P., Bheemrao Somalwar, S., and Kavitha, T. 2023. Histopathological study of mucormycosis in post COVID-19 patients and factors affecting it in a tertiary care hospital. Int J Surg Pathol. 31(1): 56–63.

Sugar, A. M. 2005. Agents of mucormycosis and related species. In: Mandell, G. L., Bennett, J. E., and Dolin, R., editors. Principles and practice of infectious diseases. 6th ed. Philadelphia, PA: Elsevier, p. 2979.

Waldorf, A. R. 1989. Pulmonary defense mechanisms against opportunistic fungal pathogens. Immunol Ser. 47: 243–271.

Waldorf, A. R., Ruderman, N., and Diamond, R. D. 1984. Specific susceptibility to mucormycosis in murine diabetes and bronchoalveolar macrophage defense against *Rhizopus*. J Clin Invest. 74: 150–160.

10 Antifungal Therapies and Drug Resistance

Hafsa Qadri, Manzoor Ahmad Mir*, and Abdul Haseeb Shah*

10.1 INTRODUCTION

The prevalence of fungal infections is rising globally as the number of patients with impaired immune systems expands. Over 1,350,000 people are believed to have perished as a result of deadly fungal infections affecting more than 300,000,000 individuals (1–4). The escalating need to address infections caused by various pathogenic fungi has significantly heightened the utilization of antifungal drugs for both disease prevention and treatment. Unfortunately, there are very few categories of chemicals that are covered by the current antifungal agents, which severely restrict the antifungal-therapy possibilities. Azoles, polyenes, etc., are some of the known categories of antifungal agents that are currently under usage (5, 6). Polyenes (like amphotericin B) act by targeting ergosterol in the fungal plasma membrane, creating large pores that disrupt cellular processes. In contrast, azoles (e.g., fluconazole) and allylamines (e.g., terbinafine) inhibit the synthesis of ergosterol. Antifungal drug Flucytosine (5-fluorocytosine) disrupts DNA synthesis and the pyrimidine metabolism. Echinocandins like caspofungin are cell wall-active agents that hinder the formation of 1,3-β-glucan, a crucial structural element of the fungal cell wall (7).

The emergence of drug resistance is thought to be facilitated by the extensive utilization of antifungal drugs (7). The management of patients is altered by the establishment of acquired resistance to commonly employed drugs. To enhance therapies, diagnosis, and treatment techniques that may eliminate and avert resistance, it is essential to have a better knowledge of the different antifungal therapies and mechanism-specific resistance and the biological processes that lead to its establishment (5). Pharmacological limitations in treatments for *Candida* spp. have raised interest in the topic of antifungal drug resistance and its molecular basis. Consequently, this chapter provides a summary of potential antifungal therapies, mechanisms of action, and their resistance. The chapter also focuses on the progress made in the development of different antifungals.

10.2 CONVENTIONAL ANTIFUNGAL DRUGS

Currently, the arsenal of antifungal drugs effective against fungal diseases is limited, primarily focusing on specific elements of the fungal plasma membrane or its biosynthetic mechanisms. Notably, echinocandins represent a more recent antifungal category that targets elements of the fungal cell wall. The prevailing use of predominantly fungistatic antifungals in clinical settings has often led to the establishment of resistance by fungal species toward these drugs. Given the restricted supply of antifungals and the rise of multidrug resistance (MDR) in fungal diseases, there persists an ongoing imperative for the creation of new, wide-ranging antifungals characterized by enhanced efficiency (8). Currently, available different antifungal drug categories and their sites of action (9) have been illustrated in Figure 10.1 and are described in the next subsections.

---

* Corresponding Author: abdulhaseeb@kashmiruniversity.ac.in

DOI: 10.1201/9781032642864-10

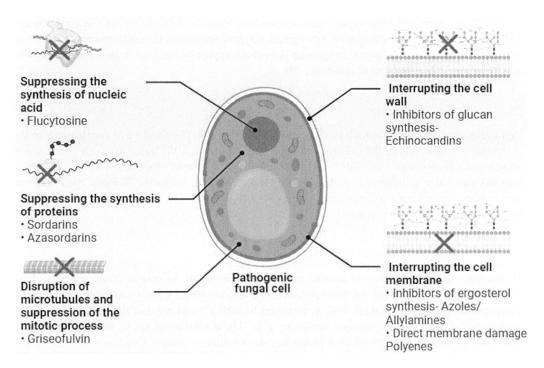

FIGURE 10.1 Commonly used antifungals and their specific target sites. Antifungals target various stages of fungal cell growth and replication. Nucleic acid synthesis inhibitors, such as flucytosine, act on fungal DNA and RNA synthesis enzymes. Cell membrane disruptors, like amphotericin B, target fungal cell membranes. Ergosterol synthesis inhibitors, including azoles interfere with the production of a crucial component found in the cell membrane. Finally, cell wall synthesis inhibitors, such as echinocandins, disrupt the formation of the cell wall, leading to fungal cell destruction.

10.2.1 AZOLES

Azoles stand as the predominant class of antifungal medications, with the initial developments centered on imidazole-based compounds. Notable pioneers in this category, such as Miconazole, clotrimazole, ketoconazole, etc., marked significant strides in antifungal therapy. Luliconazole, a more recent addition from 2013, belongs to the imidazole class, primarily employed topically for dermatophytic infections. Subsequent advancements led to the development of triazole-based azole antifungals like fluconazole and itraconazole, distinguished by a wide range of effectiveness. Triazoles find application in both systemic and mucosal fungal infections, while imidazoles are predominantly utilized for mucosal infections (10).

The process underlying azole action involves the destabilization of the fungal cell membrane. This is elucidated by the *ERG11* gene, responsible for encoding 14-lanosterol demethylase, a pivotal enzyme transforming lanosterol into ergosterol. An iron protoporphyrin unit is present in the active site of the enzyme. Azoles impede the ergosterol biosynthetic pathway by binding to iron, thereby suppressing ergosterol synthesis. Consequently, the formation of 14-methyl sterols alters the membrane stability, permeability, and functionality, as well as the associated enzymes (5).

10.2.2 POLYENES

Polyenes like amphotericin B (AmpB), first discovered in Streptomyces, are amphipathic macrolides (11). They attach to ergosterol, creating pores in the plasma membrane, resulting in a loss of ionic equilibrium and cell death (12). The three primary polyenes are AmpB, natamycin, and nystatin. Natamycin and nystatin are recommended for topical infections because of poor absorption, while AmpB works best against *Aspergillus*, *Cryptococcus*, and *Candida* species (13). Owing to the serious side effects, potent drugs like AmpB are typically used for invasive mycoses only, despite minimal development of resistance (9).

10.2.3 PYRIMIDINE ANALOGS

Pyrimidine analogs, 5-fluorouracil (5-FU), and 5-fluorocytosine (5-FC) serve as counterparts to the nucleotide cytosine. Under the influence of cytidine deaminase, 5-FC transforms into 5-FU and is subsequently incorporated into DNA and RNA during its synthesis processes. This incorporation hampers the entire cellular process by inhibiting DNA/protein synthesis. Notably, these analogs demonstrate activity against *Cryptococcus* and *Candida* spp. (12, 13). The swift absorption of 5-FC contributes to its high bioavailability.

10.2.4 ALLYLAMINE, THIOCARBAMATES, AND MORPHOLINES

Thiocarbamates and Allylamines possess exceptional efficacy in the case of dermatophytes, demonstrating high effectiveness, yet their performance against yeasts is only moderately successful. The presence of the naphthalene moiety, common to both allylamines and thiocarbamates, likely contributes to their enzyme-binding capability (14). These antifungal agents show mild interaction with the synthesis of cholesterol-producing mammalian enzymes. Onychomycosis, a fungal infection of the nails, is treated with amorolfine (morpholine) which inhibits enzymes related to ergosterol synthesis, specifically 14-reductase and 7, 8-isomerase (15, 16). Unlike allylamines and thiocarbamates, which target only the *ERG1* gene, morpholines such as fenpropimorph and amorolfine block the *ERG24* and *ERG2* genes involved in ergosterol synthesis. Terbinafine represents an example of an allylamine, while tolnaftate is an illustration of a thiocarbamate (9).

10.2.5 ECHINOCANDINS

Echinocandins, a recently developed antifungal category, primarily interfere with the formation of cell wall elements. These lipopeptides act as non-competitive inhibitors of 1,3-β-D-glucan synthase, synthesizing β-glucan, a key structural element of fungal cell walls (17). The interference with this synthetic process induces cell wall stress, resulting in various cellular abnormalities upon echinocandin treatment. These abnormalities include decreased sterol levels, enhanced osmotic sensitivity, pseudohyphae development, etc. (18). Importantly, echinocandins are often harmless to mammalian cells because they exert their effects on the formation of fungal cell wall components, absent in mammalian cells (19).

10.3 ANTIFUNGAL DRUG RESISTANCE

Given the increasing public health concern posed by fungal pathogens resistant to drugs, a comprehension of the processes leading to antifungal resistance becomes imperative for effectively managing infections induced by these microorganisms (20). Here we present some of the molecular pathways/processes that drive antifungal resistance in both prevalent and newly identified fungal pathogens (Figure 10.2).

10.3.1 DRUG TARGET ALTERATION

Acquired antifungal resistance commonly arises through alterations in drug targets, particularly evident in the case of azoles. Resistance to azoles is associated with modifications in the drug target 14-α-demethylase encoding genes (*ERG11* in yeast and *cyp51* in molds). In *Candida albicans*

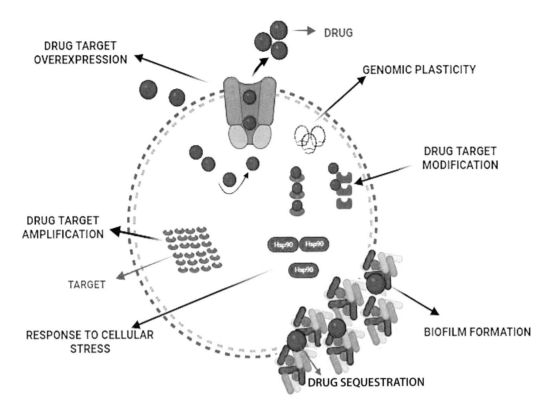

FIGURE 10.2 The mechanisms underlying antifungal drug resistance encompass various processes, including the overexpression of drug efflux pumps, modification of drug targets, and genomic plasticity in different fungi that pose a threat to human health.

clinical strains, more than 140 different amino acid modifications (substitutions) in the *ERG11* gene have been documented, primarily concentrating in three hotspot regions (21, 22). The examination of the *ERG11* gene in *C. albicans* has revealed that certain substitutions affect residues that are present in the active-site chamber, influencing binding with azoles directly, while other substitutions take place on the enzyme's proximal surface, potentially contributing indirectly to resistance to azoles, such as by impacting associations with other enzymes like cytochrome P450 reductase (23).

Moreover, in *C. auris ERG11* mutations have also been found to perform a role in diminished azole susceptibility (24–27). The latest research identified that azole-resistant clade IV clinical strains had F444L substitution in *ERG11* that had not been documented previously (27). Fewer *ERG11* mutations have been associated with clinical strains of *C. neoformans* that are resistant to azoles (28). In contrast to *Candida* or *Cryptococcus*, *Aspergillus fumigatus* expresses a pair of genes, cyp51A and cyp51B, which encode Cyp51 isoenzymes that share 60% of their identity. Although the function of cyp51B mutations in azole drug resistance remains unknown, cyp51A mutations are a common and frequently reported factor for resistance to triazoles in the case of *A. fumigatus* (29, 30).

10.3.2 Drug Target Overexpression

Elevated expression of drug targets represents a common mechanism for developing resistance to azole drugs. Alterations in the zinc cluster transcription factor gene *UPC2* have the potential to cause gain-of-function mutation in *C. albicans* clinical isolates that exhibit azole resistance resulting in an overproduction of *ERG11* (31). Upc2 possesses an N-terminal nuclear localization signal (NLS) and a C-terminal ligand-binding domain (LBD), with the latter capable of sensing the amounts of ergosterol in cells (32). The ligand-dependent conformational switching of *Upc2* LBD plays a crucial role, and mutations in the glycine-rich loop can disrupt the binding of ligands through steric clashes, ultimately causing constitutive activation (33). Interestingly, the latest research has unveiled a broader function of *C. glabrata Upc2A* (the homolog of *C. albicans Upc2*) in regulating the expression of genes beyond those related to ergosterol biosynthesis, extending to those involved in translation and plasma membrane constituents (34). Even though the precise functions of these target genes in conferring resistance to antifungals remain unclear, it is intriguing to directly investigate their influence on susceptibility to azoles. In both *A. fumigatus* and *C. neoformans*, the expression of *cyp51A/ERG11* is governed by the sterol regulatory element-binding protein (SREBP), specifically *SrbA* and *Sre1* in response to azole exposure (35, 36). In *A. fumigatus*, additional contributors to azole resistance involve the species-specific transcription factors AtrR and SltA (37, 38). The regulatory mechanisms orchestrated by these factors shed light on the complex interplay influencing azole susceptibility in these fungal species (20). Furthermore, *AtrR* performs a crucial part in modulating the expression of *cyp51A* in *A. fumigatus* (37, 39).

10.3.3 Increased Drug Efflux

In addition to altering the drug target or increasing its expression, enhanced efflux of drugs represents a prominent resistance mechanism against various antimicrobials. Azole therapy resistance commonly involves the upregulation of efflux pumps in the plasma membrane, which work to reduce the drug accumulation within the cell. Unlike azoles, echinocandins or polyenes do not induce adaptive resistance through efflux-mediated mechanisms, as they are poor substrates for pump proteins (40).

In the case of *C. albicans*, azole resistance is connected with the increased expression of ATP-binding cassette (ABC) drug transporters Cdr1p, Cdr2p, and the major facilitator transporter Mdr1p (41–43). Gain-of-function mutations in the transcriptional activator gene *TAC1* lead to the constitutive overexpression of both *CDR1* and *CDR2* while activating mutations in *MRR1* are responsible for upregulating *MDR1* (40, 44). Similarly, in *C. auris*, gain-of-function mutations in *TAC1B* (homolog of *C. albicans TAC1*) and *MRR1A* (homolog of *C. albicans MRR1*) led to fluconazole resistance (45, 46). Multidrug efflux transporters overexpression is implicated in various pathogenic fungal species in the case of azole resistance, where key contributors include the ABC transporters *Afr1* for *C. neoformans* and *AtrF* for *A. fumigatus* (47, 48). This underscores the significance of targeting drug efflux as a potential strategy to overcome azole resistance in diverse fungal infections (20).

10.3.4 Genomic Plasticity

Apart from particular point mutations that cause elevated expression of drug targets or efflux pumps in fungal pathogens, another mechanism of enhanced drug resistance involves genomic modifications. In *C. albicans*, drug-resistant clinical isolates and laboratory-derived resistant strains exhibit the duplication of the left arm of chromosome 5, forming an isochromosome (i5(L)) (49). This results in elevated gene dosage of the azole target *ERG11* and the transcription factor Tac1, which regulates drug export (50). Similarly, studies on *C. auris* suggest that an additional copy of chromosome 5 (housing various drug-resistance genes), led to fluconazole resistance (51, 52). Furthermore, loss of heterozygosity (LOH) events can occur in some genomic regions containing azole resistance determinants (*ERG11* and *TAC1*), causing mutations in these genes to become homozygous and consequently enhancing drug resistance (53).

In *C. albicans*, a substantial contributor to the swift establishment of resistance to azoles stems from substantial variations in copy number variations (CNVs) within the genome. These intricate CNVs amplify large genomic regions and are marked by unique long inverted repeat sequences. They are distributed throughout the genome and frequently encompass genes linked to drug resistance (54). Fungal pathogens exhibit impressive genomic flexibility, facilitating their adaptation to diverse environments further contributing to the development of resistance to antifungals (20, 55).

10.3.5 Stress Responses

Cellular stress response plays a crucial role in facilitating pathogen survival in the face of various environmental challenges and is essential for alleviating stress induced by antifungal treatments. The central hub governing tolerance and resistance to antifungal drugs involves a key player known as the molecular chaperone Hsp90. Hsp90 plays a crucial role in stabilizing various signal transducers that respond to stress induced by antifungal agents in *Candida*, *Aspergillus*, and *Cryptococcus* (52, 56). In *C. albicans*, Hsp90 interacts with important client proteins, including the Ca^{2+}-calmodulin-activated protein phosphatase calcineurin, as well as several members of the PKC-MAPK cell wall integrity cascade such as Pkc1, Bck1, Mkk2, and Mkc1 (56–59). Inhibition of Hsp90 leads to the blockade of calcineurin-dependent stress responses and PKC signaling, ultimately disrupting azole/echinocandin drug resistance (58, 59). Additionally, the development of resistance to polyenes in *Candida* is also reliant on Hsp90, underlining its conserved function in acquiring resistance to various antifungals (60). The significance of Hsp90 in regulating cellular responses to azoles in *C. neoformans* is further highlighted by recent findings. Researchers showed that HSP90 reduction using a copper-repressible promoter-enhanced vulnerability to various cellular stressors and azoles in this fungal pathogen (61).

Azole resistance, a crucial aspect involving stress responses, is linked to mutations in the Δ-5,6-desaturase gene, *ERG3*. *ERG3* loss-of-function mutations hinder the accumulation of 14-α-methyl-3,6-diol, the toxic sterol that would typically build up in response to azole-induced *ERG11* inhibition (17). In *C. albicans*, the resistance dependent on *ERG3* requires the involvement of Hsp90, calcineurin, and Pkc1 (56, 58). Although the impact of *ERG3* mutations on resistance to azoles is less pronounced in *C. glabrata*, latest studies have shown that exposure to echinocandins *in vitro* can lead to the acquisition of *ERG3* mutations, conferring cross-resistance to fluconazole (62). Additionally, *C. auris* has been found to exhibit *ERG3* mutations after being subjected to echinocandin exposure (52).

Extensive research in the model yeast *S. cerevisiae* has elucidated the crucial role of *Pkc1* in cell wall integrity signaling (63). This kinase is also integral to membrane stress responses in *C. albicans*, where its functionality relies significantly on *Rlm1* transcription factor which plays role as the primary regulator of cell wall integrity in the yeast *S. cerevisiae* (58, 63). Beyond Pkc1, various other kinases have been identified as key players in mediating resistance to antifungal drugs. Notably, the target of rapamycin (*TOR*), a pivotal regulator in nutrient sensing and metabolism across eukaryotes, is among these influential kinases. In *C. albicans*, the hyperactivation of *TOR* emerges as a strategic mechanism to circumvent azole-induced toxicity (64).

10.3.6 Biofilms

Biofilm formation serves as a fascinating mechanism of resistance employed by fungal pathogenic organisms. The role of biofilm development in the pathogenicity of such pathogens has long been acknowledged (65). Biofilms, intricate 3D structures, consist of an elaborate arrangement of yeast/filamentous cells embedded in a polymeric matrix predominantly composed of nucleic acids, sugars, and proteins (5). The resilience of *Candida* biofilms is ascribed to various factors, including high biofilm cell density, the presence of persister cells, and more (66). *C. albicans*, forms well-organized biofilms comprising diverse cell types enclosed in an ECM (67, 68). Biofilms are notorious for thriving on medical equipment like pacemakers, catheters, etc., providing an ideal surface for the generation of biofilms. Infections rooted in biofilms pose a significant clinical challenge, as those formed by *C. albicans* inherently resist conventional antifungal therapies, host immune responses, and environmental disruptions (69).

10.4 ANTIFUNGAL DRUG DISCOVERY/NEED FOR NEW ANTIFUNGALS

Fungal infections currently pose a significant challenge (Table 10.1) in the realm of human diseases (2). The surge in fungal infections is a pressing issue, exacerbated by insufficient understanding of the resistance processes and mechanisms adopted by pathogenic fungi. *Candida* species, in particular,

TABLE 10.1

List of Priority Fungal Pathogens by the World Health Organization (WHO) (72)

Critical-Priority Group	High-Priority Group	Medium-Priority Group
Cryptococcus neoformans	*Candida glabrata*	*Scedosporium* species
Aspergillus fumigatus	*Candida paropsilosis*	*Cryptococcus gattii*
Candida albicans	*Candida tropicalis*	*Lomentospora prolificans*
Candida auris	*Histoplasma* species	*Talaromyces marneffei*
	Eumycetoma causative agents	*Coccidioides* species
	Fusarium species	*Pneumocystis jirovecii*
	Mucorales	*Pichia kudriavzeveii*
		Paracoccidioides

pose a significant global health threat, contributing to escalating mortality and morbidity (55, 70). The limited treatment options for fungal infections are compounded by the scarcity of available antifungals (6). Consequently, addressing the challenge of fungal diseases necessitates the exploration of novel therapeutic approaches (70). There is a need for the development of alternative antifungal agents that surpass the efficacy of traditional ones. The following section outlines potential compounds (with promising antifungal potential) as well as some antifungal-based approaches against different fungal pathogens, representing a crucial step in addressing this medical challenge (see also Figure 10.3). Table 10.2 presents comprehensive information on different investigational antifungal agents, encompassing their mode of action, and prospective benefits (71).

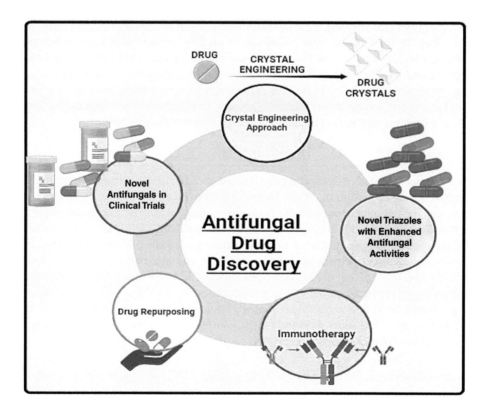

FIGURE 10.3   Schematic outline of the potential compounds with antifungal properties aimed at addressing drug resistance in pathogenic fungi.

TABLE 10.2

Antifungal Agents Currently Undergoing Clinical Trials for Novel Investigations (71)

S. No.	New Agent	Category	Display Activity Against	Mode of Action	Clinical Trials Status	Possible Benefits
1.	Rezafungin	Echinocandins	*Candida* species, *Aspergillus* species, etc.	Blocks the activity of 1,3-β-D-glucan synthase	Finished	A prolonged half-life enables weekly administration. Does not pose a risk of liver toxicity, etc.
2.	Encochleated amphotericin B	Polyenes	*Candida* species, and *Cryptococcus* species, *Aspergillus* species	Form pores in the cell membrane of fungi by binding to ergosterol	Finished	Oral formulation. Reduced toxicity compared to deoxycholate. Lipid-based formulations of amphotericin B.
3.	Ibrexafungerp	Triterpenoids	*Candida* species, *Aspergillus* species, etc.	Blocks the activity of 1,3-β-D-glucan synthase	Finished	Effectively combats resistant *Candida* species. Pioneers oral bioavailability as an inhibitor of (1(3)-β-D-glucan synthase. Enables combination therapy for invasive aspergillosis. Provides oral fungicidal treatment for *Candida* spp., etc.
4.	Fosmanogepix	Glycosylphos-phatidylinositol (GPI) inhibitors	*Candida* species and *Cryptococcus* species, *Aspergillus* species	Hindering the activity of the fungal enzyme Gwt1, thereby disrupting the post-translational modification of proteins through GPI-anchor	In progress	Effective against a wide range of fungi, including those that are highly resistant.
5.	PC945	Triazoles	*Candida* species and *Cryptococcus* species, *Aspergillus* species	Inhibiting the lanosterol 14-alpha-demethylase enzyme to interfere with the synthesis of ergosterol	None	Delivery through inhalation. Efficacy against *A. fumigatus* that is resistant to azoles.
6.	Oteseconazole	Tetrazoles	*Candida* species and *Cryptococcus* species, *Aspergillus* species	Blocking the lanosterol 14-alpha-demethylase enzyme to interfere with the synthesis of ergosterol	Finished	Targeting enzymes specific to fungi reduces the likelihood of drug-drug interactions. Effective against a wide range of yeasts, molds, etc.
7.	Olorofim	Orotomides	*Aspergillus* species, etc.	Blocking the activity of dihydroorotate dehydrogenase, an enzyme involved in pyrimidine biosynthesis	In progress	Effective against molds that are highly resistant.

10.4.1 CELL WALL TARGETING AGENTS

Fosmanogepix, formerly known as APX001 and E1210 and developed by Eisai Company in Japan, represents a groundbreaking antifungal prodrug. It operates by targeting the fungal Gwt1 gene, which codes for a unique acyltransferase crucial in the early stages of the GPI biosynthetic pathway (73). Fosmanogepix exhibits a broad range of activity against various yeasts and molds. It demonstrates strong *in vitro* efficacy against the majority of *Candida* species, although *C. krusei* is an exception to this broad-spectrum potency (73). Ibrexafungerp, formerly known as MK-3118 and SCY-078, created by New Jersey-based Scynexis, functions similarly to echinocandins by disrupting the formation of cell wall by inhibiting the β-D-glucan synthase. It can inhibit the growth of *Candida* species. Despite its structural uniqueness, ibrexafungerp is a semisynthetic derivative of enfumafungin, a naturally occurring hemiacetal triterpene glycoside. This derivative includes a pyridine triazole at position 15 of the core phenanthropyran carboxylic acid and a 2-amino-2,3,3-trimethyl-butyl ether at position 14, enhancing its antifungal potential and pharmacokinetic characteristics. Consequently, ibrexafungerp represents the first compound in the innovative class of triterpenoid antifungals (74). Ibrexafungerp demonstrates strong fungicidal efficacy against *Candida* species, such as *C. glabrata*, and *C. auris* (75–77). Rezafungin, previously known as SP3025 and CD101, is a pioneering member of the echinocandins. It works by preventing (1→3)-β-D-glucan from being synthesized. Rezafungin shares structural similarities with anidulafungin, yet it possesses enhanced stability and solubility (78). Rezafungin demonstrates robust *in vitro* efficacy comparable to other echinocandins against both wild-type and azole-resistant *Aspergillus* and *Candida* spp. (79–81). Earlier research has also evaluated the antibacterial and antifungal properties exhibited by certain β-nitrostyrenes. Moreover, a mechanistic study from our group related to a class of β-nitrostyrene derivatives suggests that these compounds exhibit potent antifungal properties. Their target of action was mostly fungal cell wall wherein we observed loss of cellular integrity by disruption of the cell membrane and cell wall on exposure to β-nitrostyrenes. *Candida CAS5* (regulating cell wall dynamics) transcription factor mutants particularly showed hypersensitivity to these nitrostyrene derivatives (55).

10.4.2 CELL MEMBRANE TARGETING AGENTS

Matinas BioPharma, based in Bedminster, has developed Encochleated AmpB, which introduces an innovative formulation facilitating oral administration while minimizing toxicity. The distinctive cochleate structure involves layers created by phosphatidylserine and a divalent calcium cation, offering protection to AmpB from degradation in the GIT (82). Studies indicate similar *in vitro* effectiveness against *Candida* and *Aspergillus* spp. when comparing Encochleated AmpB to deoxycholate AmpB (83, 84).

Moving on to the next generation of azoles, Mycovia Pharmaceuticals, Inc. in Durham (North Carolina), has introduced Oteseconazole (VT-1161), VT-1598, and VT-1129. These compounds selectively inhibit the fungal enzyme CYP51 by substituting the 1-(1,2,4-triazole) metal-binding group with a tetrazole (85). Oteseconazole, VT-1598, and VT-1129 demonstrate robust *in vitro* efficacy against various strains of *Cryptococcus* and *Candida* (86, 87).

Pulmocide, based in London, has developed PC945, an innovative triazole antifungal designed for inhaled administration to address and prevent IFIs in the sinopulmonary tract. PC945's structure shares similarities with posaconazole but maintains distinct differences. PC945 shows efficacy in the case of different *Aspergillus* and *Candida* spp. (71).

10.4.3 NUCLEIC ACID METABOLISM TARGETING AGENTS

Olorofim, formerly known as F910318 and initially uncovered in Australia by F2G Ltd, belongs to a novel antifungal class called orotomides. Its function involves inhibiting dihydroorotate dehydrogenase enzyme, disrupting the pyrimidine biosynthesis process. This disruption hampers nucleic acid synthesis and results in the cessation of hyphal extension (88). Notably, Olorofim stands out from existing antifungal agents by exhibiting no activity against *Candida* species. Instead, it demonstrates good activity against wild-type and resistant *Aspergillus* spp. and *Coccidioides* spp. (89).

10.4.4 Novel Triazoles for Enhanced Antifungal Defense

Owing to the limited availability and ineffectiveness of existing antifungals, ongoing research is focused on the development of advanced triazoles such as albaconazole, ravuconazole, and isavuconazole. Preliminary analyses indicate that these drugs exhibit improved pharmacokinetic profiles, low toxicity, and *in vitro* efficacy against various species of *Candida*. Albaconazole, in particular, demonstrates a wide range of antifungal efficacy with excellent tolerance and potency against various *Candida* species, in both controlled laboratory environments, and within the living cells. Its efficacy is comparable to the widely utilized antifungal fluconazole, as indicated by research conducted by (90) and (91). Investigations have revealed that a significant majority of isolates from *C. albicans* and *C. glabrata* are responsive to albaconazole, as highlighted in studies by Pasqualotto et al. (92).

Isavuconazole, classified as a second-generation triazole compound, exhibits broad-spectrum antifungal efficacy and promising activity against resistant *Candida* species, comparable to other triazoles (92). Similarly, ravuconazole, structurally identical to isavuconazole, demonstrates enhanced antimycotic action against *Candida* isolates resistant to fluconazole (92).

10.4.5 Crystal Engineering Approach

Crystal engineering (CE) is emerging as a cost-effective and environmentally friendly approach for developing materials with desired physicochemical characteristics (93). It surpasses the complexities of organic synthesis, offering a greener alternative that is both time-efficient and economically viable (94, 95). The application of CE principles holds great promise in pharmaceutical chemistry, particularly in designing drug co-crystals/salts with enhanced physicochemical and pharmacokinetic attributes (96–98). The formation of co-crystals/salts involves the supramolecular complexation of a drug molecule with a biologically safe coformer, making it a straightforward process (99). The selection of the coformer performs a critical part in modulating the behavior of new drug forms, relying on factors such as functional compatibility, hydrogen bonding tendency, and alignment of functionalities to create robust synthons.

Numerous research studies have emphasized the synthesis and potential antifungal properties of various co-crystals (100–102). CE serves as an ecological/greener option to traditional organic production/synthesis, allowing the manipulation of molecular behavior promptly and economically. Research conducted by our group details the synthesis and analysis of a pharmaceutical organic salt named FLC-C, which results from the combination of fluconazole (FLC) and organo-sulfonate (NDSA-2H). This salt is created through the sulfonate-pyridinium supramolecular synthon. Examinations on solubility demonstrate higher water solubility of FLC-C in comparison to pure FLC under physiological pH conditions of 2 and 7. Notably, *in vitro* studies suggest that FLC-C could be a viable substitute for FLC. This organic salt exhibits a wide spectrum of antifungal efficacy and is expected to show better bioavailability and least host toxicity. Importantly, FLC-C demonstrates fluorescence characteristics within *Candida* cells, adding an extra dimension to its potential applications (100).

10.4.6 Drug Repurposing

The integration of existing drugs with potential antifungal capabilities into the pipeline for antifungal development can offer significant time and resource savings, particularly considering the existing knowledge of their pharmacokinetic and pharmacodynamic profiles. Tamoxifen, a medication utilized in the treatment of breast cancer and sertraline, a serotonin reuptake inhibitor prescribed for depression, stand out as promising candidates for inclusion in antifungal regimes. Tamoxifen has demonstrated effectiveness against cryptococcal infections and holds promise as a synergistic agent with fluconazole (103, 104).

Sertraline provides a notable example of drug repurposing, as it is presently in phase III trials to evaluate its potential as an adjuvant therapy in combination with standard treatment for cryptococcal meningitis. Another instance is AR-12, initially designed as a derivative of celecoxib and initially assessed for safety in a phase I oncology clinical trial. However, it demonstrated repeated antifungal efficacy against specific yeasts and molds, leading to its repurposing as a potential adjuvant therapy alongside fluconazole in the management of IFIs (104, 105).

10.4.7 IMMUNOTHERAPY-BASED APPROACH

Enhancing the body's immune system to combat fungal infections holds great potential (106), and immunotherapy offers a range of methods to achieve this goal. These approaches include utilizing recombinant growth factors, cytokines, and antibodies. Among these, antifungal antibodies, a form of passive immunotherapy, are of particular interest (107). At present, clinical development is focused on two antifungal antibodies. One is MAb 2G8, a monoclonal antibody designed to target laminarin, which is primarily composed of β-glucans. This antibody binds to the walls of *C. albicans* and *C. neoformans*, preventing the latter two from growing and forming capsules (108). By selectively targeting β-1,3 glucans, a humanized variant of MAb 2G8 exhibits significant antifungal efficacy against *C. auris in vitro* (104).

Efungumab, commonly referred to as mycograb, is a type of single-chain variable-fragment antibody designed to target HSP90. Clinical trials combining efungumab with amphotericin B have demonstrated reduced mortality and improved survival rates in *Candida*-infected patients. Advancing beyond antifungal antibodies, radio-immunotherapy involves linking these antibodies to radioisotopes to deliver fungicidal radiation specifically to fungal cells. This innovative approach has shown promise in the treatment of *C. neoformans* infections (109).

10.5 CONCLUSION AND FUTURE PERSPECTIVES

In conclusion, the escalating global prevalence of fungal infections, coupled with the restricted range of antifungals and the rise in resistance, underscores the urgent need for advancements in antifungal therapy. This comprehensive overview has delved into the mechanisms of action and resistance of conventional antifungals, including azoles, polyenes, pyrimidine analogs, allylamines, and echinocandins. The exploration of novel antifungals in clinical trials, such as fosmanogepix, ibrexafungerp, and rezafungin, as well as promising compounds like encochleated AmpB, oteseconazole, and olorofim, highlights ongoing efforts to address therapeutic gaps. Additionally, the discussion on drug repurposing, crystal engineering, and immunotherapy-based approaches reflects innovative strategies in the quest for effective antifungal solutions. As we navigate the challenges of drug resistance and limited treatment options, future perspectives emphasize the continuous exploration of novel antifungals, combination therapies, and immunomodulatory techniques, emphasizing a multidisciplinary approach against the evolving landscape of fungal infections.

ABBREVIATIONS

IFIs	Invasive fungal infections	**AmpB**	Amphotericin-B
U.S.	United States	**5-FC**	5-Fluorocytosine
MDR	Multidrug resistance	**Cdr**	*Candida* drug resistance
AMR	Antimicrobial resistance	**HSP90**	Heat shock protein 90
ABC	ATP-binding cassette	**CE**	Crystal engineering
MFS	Major Facilitator Superfamily	**CNVs**	Copy number variations
ECM	Extracellular matrix		

REFERENCES

1. Brown G, Denning D, Gow N, Levitz S, Netea M, White T. Hidden killers: Human fungal infections. Science Translational Medicine, 2012;4:165rv13..
2. Qadri H, Qureshi MF, Mir MA, Shah AH. Glucose—The X factor for the survival of human fungal pathogens and disease progression in the host. Microbiological Research, 2021;247:126725.
3. Qadri H, Shah AH, Mir MA, Qureshi MF, Prasad R. Quinidine drug resistance transporter knockout *Candida* cells modulate glucose transporter expression and accumulate metabolites leading to enhanced azole drug resistance. Fungal Genetics and Biology, 2022;161:103713.
4. Qadri H, Shah AH, Alkhanani M, Almilaibary A, Mir MA. Immunotherapies against human bacterial and fungal infectious diseases: A review. Frontiers in Medicine, 2023;10:1135541.
5. Cowen LE, Sanglard D, Howard SJ, Rogers PD, Perlin DS. Mechanisms of antifungal drug resistance. Cold Spring Harbor Perspectives in Medicine, 2015;5(7):a019752.
6. Kathiravan MK, Salake AB, Chothe AS, Dudhe PB, Watode RP, Mukta MS, et al. The biology and chemistry of antifungal agents: A review. Bioorganic & Medicinal Chemistry, 2012;20(19):5678–98.
7. Cowen LE, Steinbach WJ. Stress, drugs, and evolution: The role of cellular signaling in fungal drug resistance. Eukaryotic Cell, 2008;7(5):747–64.
8. Prasad R, Shah AH, Rawal MK. Antifungals: Mechanism of action and drug resistance. Advances in Experimental Medicine and Biology, 2016;892:327–49.
9. Hossain CM, Ryan LK, Gera M, Choudhuri S, Lyle N, Ali KA, et al. Antifungals and drug resistance. Encyclopedia, 2022;2(4):1722–37.
10. Sanglard D, Coste AT. Activity of isavuconazole and other azoles against *Candida* clinical isolates and yeast model systems with known azole resistance mechanisms. Antimicrobial Agents and Chemotherapy, 2016;60(1):229–38.
11. Vandeputte P, Ferrari S, Coste AT. Antifungal resistance and new strategies to control fungal infections. International Journal of Microbiology, 2012;2012:713687.
12. Sanglard D, Coste A, Ferrari S. Antifungal drug resistance mechanisms in fungal pathogens from the perspective of transcriptional gene regulation. FEMS Yeast Research, 2009;9(7):1029–50.
13. Lemke A, Kiderlen A, Kayser O. Amphotericin B. Applied Microbiology and Biotechnology, 2005;68:151–62.
14. Polak A. Mode of action studies. Chemotherapy of Fungal Diseases: Springer; 1990. p. 153–82.
15. Mercer EI. Morpholine antifungals and their mode of action. Biochemical Society Transactions, 1991;19(3):788–93.
16. Haria M, Bryson HM. Amorolfine: A review of its pharmacological properties and therapeutic potential in the treatment of onychomycosis and other superficial fungal infections. Drugs, 1995;49:103–20.
17. Shapiro RS, Robbins N, Cowen LE. Regulatory circuitry governing fungal development, drug resistance, and disease. Microbiology and Molecular Biology Reviews, 2011;75(2):213–67.
18. Ghannoum MA, Rice LB. Antifungal agents: Mode of action, mechanisms of resistance, and correlation of these mechanisms with bacterial resistance. Clinical Microbiology Reviews, 1999;12(4):501–17.
19. Perlin DS. Current perspectives on echinocandin class drugs. Future microbiology, 2011;6(4):441–57.
20. Lee Y, Robbins N, Cowen LE. Molecular mechanisms governing antifungal drug resistance. NPJ Antimicrobials and Resistance, 2023;1(1):5.
21. Morio F, Loge C, Besse B, Hennequin C, Le Pape P. Screening for amino acid substitutions in the *Candida albicans* Erg11 protein of azole-susceptible and azole-resistant clinical isolates: New substitutions and a review of the literature. Diagnostic Microbiology and Infectious Disease, 2010;66(4):373–84.
22. Marichal P, Koymans L, Willemsens S, Bellens D, Verhasselt P, Luyten W, et al. Contribution of mutations in the cytochrome P450 14α-demethylase (Erg11p, Cyp51p) to azole resistance in *Candida albicans*. Microbiology, 1999;145(10):2701–13.
23. Hargrove TY, Friggeri L, Wawrzak Z, Qi A, Hoekstra WJ, Schotzinger RJ, et al. Structural analyses of *Candida albicans* sterol 14α-demethylase complexed with azole drugs address the molecular basis of azole-mediated inhibition of fungal sterol biosynthesis. Journal of Biological Chemistry, 2017;292(16):6728–43.
24. Chowdhary A, Prakash A, Sharma C, Kordalewska M, Kumar A, Sarma S, et al. A multicentre study of antifungal susceptibility patterns among 350 *Candida auris* isolates (2009–17) in India: Role of the ERG11 and FKS1 genes in azole and echinocandin resistance. Journal of Antimicrobial Chemotherapy, 2018;73(4):891–9.
25. Healey KR, Kordalewska M, Ortigosa CJ, Singh A, Berrío I, Chowdhary A, et al. Limited ERG11 mutations identified in isolates of *Candida auris* directly contribute to reduced azole susceptibility. Antimicrobial Agents and Chemotherapy, 2018;62(10):10.1128/aac.01427-18.

26. Rybak JM, Sharma C, Doorley LA, Barker KS, Palmer GE, Rogers PD. Delineation of the direct contribution of *Candida auris* ERG11 mutations to clinical triazole resistance. Microbiology Spectrum, 2021;9(3):e01585–21.

27. Li J, Coste AT, Liechti M, Bachmann D, Sanglard D, Lamoth F, et al. Novel ERG11 and TAC1b mutations associated with azole resistance in *Candida auris*. Antimicrobial Agents and Chemotherapy, 2021;65(5):10.1128/aac.02663-20.

28. Rodero L, Mellado E, Rodriguez AC, Salve A, Guelfand L, Cahn P, et al. G484S amino acid substitution in lanosterol 14-α demethylase (ERG11) is related to fluconazole resistance in a recurrent *Cryptococcus neoformans* clinical isolate. Antimicrobial Agents and Chemotherapy, 2003;47(11):3653–6.

29. Handelman M, Meir Z, Scott J, Shadkchan Y, Liu W, Ben-Ami R, et al. Point mutation or overexpression of Aspergillus fumigatus cyp51B, encoding lanosterol 14α-sterol demethylase, leads to triazole resistance. Antimicrobial Agents and Chemotherapy, 2021;65(10):10.1128/aac.01252-21.

30. Howard SJ, Arendrup MC. Acquired antifungal drug resistance in Aspergillus fumigatus: Epidemiology and detection. Medical Mycology, 2011;49(Supplement_1):S90–5.

31. Silver PM, Oliver BG, White TC. Role of *Candida albicans* transcription factor Upc2p in drug resistance and sterol metabolism. Eukaryotic Cell, 2004;3(6):1391–7.

32. Yang H, Tong J, Lee CW, Ha S, Eom SH, Im YJ. Structural mechanism of ergosterol regulation by fungal sterol transcription factor Upc2. Nature Communications, 2015;6(1):6129.

33. Tan L, Chen L, Yang H, Jin B, Kim G, Im YJ. Structural basis for activation of fungal sterol receptor Upc2 and azole resistance. Nature Chemical Biology, 2022;18(11):1253–62.

34. Vu BG, Stamnes MA, Li Y, Rogers PD, Moye-Rowley WS. The *Candida glabrata* Upc2A transcription factor is a global regulator of antifungal drug resistance pathways. PLOS Genetics, 2021;17(9):e1009582.

35. Willger SD, Puttikamonkul S, Kim K-H, Burritt JB, Grahl N, Metzler LJ, et al. A sterol-regulatory element binding protein is required for cell polarity, hypoxia adaptation, azole drug resistance, and virulence in *Aspergillus fumigatus*. PLOS Pathogens, 2008;4(11):e1000200.

36. Chang YC, Bien CM, Lee H, Espenshade PJ, Kwon-Chung KJ. Sre1p, a regulator of oxygen sensing and sterol homeostasis, is required for virulence in *Cryptococcus neoformans*. Molecular Microbiology, 2007;64(3):614–29.

37. Paul S, Stamnes M, Thomas GH, Liu H, Hagiwara D, Gomi K, et al. AtrR is an essential determinant of azole resistance in *Aspergillus fumigatus*. mBio, 2019;10(2):10.1128/mbio.02563-18.

38. Du W, Zhai P, Wang T, Bromley MJ, Zhang Y, Lu L, et al. The C_2H_2 transcription factor SltA contributes to azole resistance by coregulating the expression of the drug target Erg11A and the drug efflux pump Mdr1 in *Aspergillus fumigatus*. Antimicrobial Agents and Chemotherapy, 2021;65(4):10.1128/aac.01839-20.

39. Hagiwara D, Miura D, Shimizu K, Paul S, Ohba A, Gonoi T, et al. A novel Zn2-Cys6 transcription factor AtrR plays a key role in an azole resistance mechanism of *Aspergillus fumigatus* by co-regulating cyp51A and cdr1B expressions. PLOS Pathogens, 2017;13(1):e1006096.

40. Cannon RD, Lamping E, Holmes AR, Niimi K, Baret PV, Keniya MV, et al. Efflux-mediated antifungal drug resistance. Clinical Microbiology Reviews, 2009;22(2):291–321.

41. Sanglard D, Kuchler K, Ischer F, Pagani J, Monod M, Bille J, et al. Mechanisms of resistance to azole antifungal agents in *Candida albicans* isolates from AIDS patients involve specific multidrug transporters. Antimicrobial Agents and Chemotherapy, 1995;39(11):2378–86.

42. Sanglard D, Ischer F, Monod M, Bille J. Cloning of *Candida albicans* genes conferring resistance to azole antifungal agents: Characterization of CDR2, a new multidrug ABC transporter gene. Microbiology, 1997;143(2):405–16.

43. Prasad R, Banerjee A, Khandelwal NK, Dhamgaye S. The ABCs of *Candida albicans* multidrug transporter Cdr1. Eukaryotic cell, 2015;14(12):1154–64.

44. Coste A, Turner V, Ischer F, Morschhäuser J, Forche A, Selmecki A, et al. A mutation in Tac1p, a transcription factor regulating CDR1 and CDR2, is coupled with loss of heterozygosity at chromosome 5 to mediate antifungal resistance in *Candida albicans*. Genetics, 2006;172(4):2139–56.

45. Rybak JM, Muñoz JF, Barker KS, Parker JE, Esquivel BD, Berkow EL, et al. Mutations in TAC1B: A novel genetic determinant of clinical fluconazole resistance in *Candida auris*. mBio, 2020;11(3):10.1128/mbio.00365-20.

46. Li J, Coste AT, Bachmann D, Sanglard D, Lamoth F. Deciphering the Mrr1/Mdr1 pathway in azole resistance of *Candida auris*. Antimicrobial Agents and Chemotherapy, 2022;66(4):e00067–22.

47. Slaven JW, Anderson MJ, Sanglard D, Dixon GK, Bille J, Roberts IS, et al. Increased expression of a novel *Aspergillus fumigatus* ABC transporter gene, atrF, in the presence of itraconazole in an itraconazole resistant clinical isolate. Fungal Genetics and Biology, 2002;36(3):199–206.

48. Posteraro B, Sanguinetti M, Sanglard D, La Sorda M, Boccia S, Romano L, et al. Identification and characterization of a *Cryptococcus neoformans* ATP binding cassette (ABC) transporter-encoding gene, CnAFR1, involved in the resistance to fluconazole. Molecular Microbiology, 2003; 47(2):357–71.

49. Selmecki A, Forche A, Berman J. Aneuploidy and isochromosome formation in drug-resistant *Candida albicans*. Science, 2006;313(5785):367–70.

50. Selmecki A, Gerami-Nejad M, Paulson C, Forche A, Berman J. An isochromosome confers drug resistance in vivo by amplification of two genes, ERG11 and TAC1. Molecular Microbiology, 2008;68(3):624–41.

51. Bing J, Hu T, Zheng Q, Muñoz JF, Cuomo CA, Huang G, et al. Experimental evolution identifies adaptive aneuploidy as a mechanism of fluconazole resistance in *Candida auris*. Antimicrobial Agents and Chemotherapy, 2020;65(1):10.1128/aac.01466-20.

52. Carolus H, Pierson S, Muñoz JF, Subotić A, Cruz RB, Cuomo CA, et al. Genome-wide analysis of experimentally evolved *Candida auris* reveals multiple novel mechanisms of multidrug resistance. mBio, 2021;12(2):10.1128/mbio.03333-20.

53. Coste A, Selmecki A, Forche A, Diogo D, Bougnoux M-E, d'Enfert C, et al. Genotypic evolution of azole resistance mechanisms in sequential *Candida albicans* isolates. Eukaryotic Cell, 2007;6(10): 1889–904.

54. Todd RT, Selmecki A. Expandable and reversible copy number amplification drives rapid adaptation to antifungal drugs. eLife, 2020;9:e58349.

55. Ramzan A, Padder SA, Masoodi KZ, Shafi S, Tahir I, Rehman RU, et al. β-Nitrostyrene derivatives as broad range potential antifungal agents targeting fungal cell wall. European Journal of Medicinal Chemistry, 2022;240:114609.

56. Cowen LE, Lindquist S. Hsp90 potentiates the rapid evolution of new traits: Drug resistance in diverse fungi. Science, 2005;309(5744):2185–9.

57. Caplan T, Polvi EJ, Xie JL, Buckhalter S, Leach MD, Robbins N, et al. Functional genomic screening reveals core modulators of echinocandin stress responses in *Candida albicans*. Cell Reports, 2018;23(8):2292–8.

58. LaFayette SL, Collins C, Zaas AK, Schell WA, Betancourt-Quiroz M, Gunatilaka AL, et al. PKC signaling regulates drug resistance of the fungal pathogen *Candida albicans* via circuitry comprised of Mkc1, calcineurin, and Hsp90. PLOS Pathogens, 2010;6(8):e1001069.

59. Singh SD, Robbins N, Zaas AK, Schell WA, Perfect JR, Cowen LE. Hsp90 governs echinocandin resistance in the pathogenic yeast *Candida albicans* via calcineurin. PLOS Pathogens, 2009; 5(7):e1000532.

60. Vincent BM, Lancaster AK, Scherz-Shouval R, Whitesell L, Lindquist S. Fitness trade-offs restrict the evolution of resistance to amphotericin B. PLOS Biology, 2013;11(10):e1001692.

61. Fu C, Beattie SR, Jezewski AJ, Robbins N, Whitesell L, Krysan DJ, et al. Genetic analysis of Hsp90 function in *Cryptococcus neoformans* highlights key roles in stress tolerance and virulence. Genetics, 2022;220(1):iyab164.

62. Ksiezopolska E, Schikora-Tamarit MÀ, Beyer R, Nunez-Rodriguez JC, Schüller C, Gabaldón T. Narrow mutational signatures drive acquisition of multidrug resistance in the fungal pathogen *Candida glabrata*. Current Biology, 2021;31(23):5314–26.e10.

63. Levin DE. Cell wall integrity signaling in *Saccharomyces cerevisiae*. Microbiology and Molecular Biology Reviews, 2005;69(2):262–91.

64. Khandelwal NK, Chauhan N, Sarkar P, Esquivel BD, Coccetti P, Singh A, et al. Azole resistance in a *Candida albicans* mutant lacking the ABC transporter CDR6/ROA1 depends on TOR signaling. Journal of Biological Chemistry, 2018;293(2):412–32.

65. Martinez LR, Fries BC. Fungal biofilms: Relevance in the setting of human disease. Current Fungal Infection Reports, 2010;4:266–75.

66. Silva S, Rodrigues CF, Araújo D, Rodrigues ME, Henriques M. *Candida* species biofilms' antifungal resistance. Journal of Fungi, 2017;3(1):8.

67. Ramage G, Mowat E, Jones B, Williams C, Lopez-Ribot J. Our current understanding of fungal biofilms. Critical Reviews in Microbiology, 2009;35(4):340–55.

68. Ramage G, Saville SP, Thomas DP, Lopez-Ribot JL. *Candida* biofilms: An update. Eukaryotic Cell, 2005;4(4):633–8.

69. Nobile CJ, Johnson AD. *Candida albicans* biofilms and human disease. Annual Review of Microbiology, 2015;69:71–92.

70. de Andrade Monteiro C, dos Santos JRA. Phytochemicals and their antifungal potential against pathogenic yeasts. Phytochemicals in Human Health: IntechOpen Limited; 2019. p. 1–31.

71. Jacobs SE, Zagaliotis P, Walsh TJ. Novel antifungal agents in clinical trials. F1000Research, 2021; 10:507.

72. Coordination G, Alastruey-Izquierdo A, World Health Organization. WHO fungal priority pathogens list to guide research, development and public health action. Organización Mundial de la Salud (OMS); 2022. Report No.: 9240060251.

73. Miyazaki M, Horii T, Hata K, Watanabe N-a, Nakamoto K, Tanaka K, et al. In vitro activity of E1210, a novel antifungal, against clinically important yeasts and molds. Antimicrobial Agents and Chemotherapy, 2011;55(10):4652–8.

74. Wring SA, Randolph R, Park S, Abruzzo G, Chen Q, Flattery A, et al. Preclinical pharmacokinetics and pharmacodynamic target of SCY-078, a first-in-class orally active antifungal glucan synthesis inhibitor, in murine models of disseminated candidiasis. Antimicrobial Agents and Chemotherapy, 2017;61(4):10.1128/aac.02068-16.

75. Ghannoum M, Long L, Isham N, Hager C, Wilson R, Borroto-Esoda K, et al. Activity of a novel 1, 3-beta-D-glucan synthase inhibitor, ibrexafungerp (formerly SCY-078), against *Candida glabrata*. Antimicrobial Agents and Chemotherapy, 2019;63(12):10.1128/aac.01510-19.

76. Berkow EL, Angulo D, Lockhart SR. In vitro activity of a novel glucan synthase inhibitor, SCY-078, against clinical isolates of *Candida auris*. Antimicrobial Agents and Chemotherapy, 2017;61(7):e00435-17.

77. Scorneaux B, Angulo D, Borroto-Esoda K, Ghannoum M, Peel M, Wring S, et al. SCY-078 is fungicidal against *Candida* species in time-kill studies. Antimicrobial Agents and Chemotherapy, 2017;61(3):10.1128/aac.01961-16.

78. Sofjan AK, Mitchell A, Shah DN, Nguyen T, Sim M, Trojcak A, et al. Rezafungin (CD101), a next-generation echinocandin: A systematic literature review and assessment of possible place in therapy. Journal of Global Antimicrobial Resistance, 2018;14:58–64.

79. Pfaller MA, Carvalhaes C, Messer SA, Rhomberg PR, Castanheira M. Activity of a long-acting echinocandin, rezafungin, and comparator antifungal agents tested against contemporary invasive fungal isolates (SENTRY Program, 2016 to 2018). Antimicrobial Agents and Chemotherapy, 2020;64(4):10.1128/aac.00099-20.

80. Tóth Z, Forgács L, Locke JB, Kardos G, Nagy F, Kovács R, et al. In vitro activity of rezafungin against common and rare *Candida* species and *Saccharomyces cerevisiae*. Journal of Antimicrobial Chemotherapy, 2019;74(12):3505–10.

81. Wiederhold NP, Locke JB, Daruwala P, Bartizal K. Rezafungin (CD101) demonstrates potent in vitro activity against *Aspergillus*, including azole-resistant *Aspergillus fumigatus* isolates and cryptic species. Journal of Antimicrobial Chemotherapy, 2018;73(11):3063–7.

82. Aigner M, Lass-Flörl C. Encochleated amphotericin B: Is the oral availability of amphotericin B finally reached? Journal of Fungi, 2020;6(2):66.

83. Delmas G, Park S, Chen Z, Tan F, Kashiwazaki R, Zarif L, et al. Efficacy of orally delivered cochleates containing amphotericin B in a murine model of aspergillosis. Antimicrobial Agents and Chemotherapy, 2002;46(8):2704–7.

84. Zarif L, Graybill JR, Perlin D, Najvar L, Bocanegra R, Mannino RJ. Antifungal activity of amphotericin B cochleates against *Candida albicans* infection in a mouse model. Antimicrobial Agents and Chemotherapy, 2000;44(6):1463–9.

85. Hoekstra WJ, Garvey EP, Moore WR, Rafferty SW, Yates CM, Schotzinger RJ. Design and optimization of highly-selective fungal CYP51 inhibitors. Bioorganic & Medicinal Chemistry Letters, 2014;24(15):3455–8.

86. Schell W, Jones A, Garvey E, Hoekstra W, Schotzinger R, Alexander BD. Fungal CYP51 inhibitors VT-1161 and VT-1129 exhibit strong in vitro activity against *Candida glabrata* and *C. krusei* isolates clinically resistant to azole and echinocandin antifungal compounds. Antimicrobial Agents and Chemotherapy, 2017;61(3):10.1128/aac.01817-16.

87. Nishimoto AT, Wiederhold NP, Flowers SA, Zhang Q, Kelly SL, Morschhäuser J, et al. In vitro activities of the novel investigational tetrazoles VT-1161 and VT-1598 compared to the triazole antifungals against azole-resistant strains and clinical isolates of *Candida albicans*. Antimicrobial Agents and Chemotherapy, 2019;63(6):10.1128/aac.00341-19.

88. Oliver JD, Sibley GE, Beckmann N, Dobb KS, Slater MJ, McEntee L, et al. F901318 represents a novel class of antifungal drug that inhibits dihydroorotate dehydrogenase. Proceedings of the National Academy of Sciences of the United States of America, 2016;113(45):12809–14.

89. Wiederhold NP, Najvar LK, Jaramillo R, Olivo M, Birch M, Law D, et al. The orotomide olorofim is efficacious in an experimental model of central nervous system coccidioidomycosis. Antimicrobial Agents and Chemotherapy, 2018;62(9):10.1128/aac.00999-18.

90. Bartroli J, Merlos M, Sisniega H. Overview of albaconazole. European Infectious Disease, 2011;5(2):88–91.

91. de Oliveira Santos GC, Vasconcelos CC, Lopes AJ, de Sousa Cartágenes MdS, Filho AK, do Nascimento FR, et al. *Candida* infections and therapeutic strategies: Mechanisms of action for traditional and alternative agents. Frontiers in Microbiology, 2018;9:1351.

92. Pasqualotto AC, Thiele KO, Goldani LZ. Novel triazole antifungal drugs: Focus on isavuconazole, ravuconazole and albaconazole. Current Opinion in Investigational Drugs, 2010;11(2):165–74.

93. Grothe E, Meekes H, Vlieg E, Ter Horst J, de Gelder R. Solvates, salts, and cocrystals: A proposal for a feasible classification system. Crystal Growth & Design, 2016;16(6):3237–43.

94. Duggirala NK, Perry ML, Almarsson Ö, Zaworotko MJ. Pharmaceutical cocrystals: Along the path to improved medicines. Chemical Communications, 2016;52(4):640–55.

95. Dar AA, Rashid SJC. Organic co-crystal semiconductors: A crystal engineering perspective. CrystEngComm, 2021;23(46):8007–26.

96. Bolla G, Sarma B, Nangia AK. Crystal engineering of pharmaceutical cocrystals in the discovery and development of improved drugs. Chemical Reviews, 2022;122(13):11514–603.

97. Guo M, Sun X, Chen J, Cai T. Pharmaceutical cocrystals: A review of preparations, physicochemical properties and applications. Acta Pharmaceutica Sinica B, 2021;11(8):2537–64.

98. Bolla G, Nangia A. Pharmaceutical cocrystals: Walking the talk. Chemical Communications, 2016;52(54):8342–60.

99. Ganie AA, Vishnoi P, Dar AA. Utility of bis-4-pyridines as supramolecular linkers for 5-sulfosalicylic acid centers: Structural and optical investigations. Crystal Growth & Design, 2019;19(4):2289–97.

100. Ahangar AA, Qadri H, Malik AA, Mir MA, Shah AH, Dar AA. Physicochemical and anti-fungal studies of the pharmaceutical co-crystal/salt of fluconazole. Molecular Pharmaceutics, 2023;20(7):3471–83.

101. Benito M, Frontera A, Molins E. Cocrystallization of antifungal compounds mediated by halogen bonding. Crystal Growth & Design, 2023;23(4):2932–40.

102. Vemuri VD, Lankalapalli S, Guntaka PCR. Posaconazole-amino acid cocrystals for improving solubility and oral bioavailability while maintaining antifungal activity and low in vivo toxicity. Journal of Drug Delivery Science and Technology, 2022;74:103491.

103. Dolan K, Montgomery S, Buchheit B, DiDone L, Wellington M, Krysan DJ. Antifungal activity of tamoxifen: In vitro and in vivo activities and mechanistic characterization. Antimicrobial Agents and Chemotherapy, 2009;53(8):3337–46.

104. Bouz G, Doležal MJP. Advances in antifungal drug development: An up-to-date mini review. Pharmaceuticals, 2021;14(12):1312.

105. Koselny K, Green J, DiDone L, Halterman JP, Fothergill AW, Wiederhold NP, et al. The celecoxib derivative AR-12 has broad-spectrum antifungal activity in vitro and improves the activity of fluconazole in a murine model of cryptococcosis. Antimicrobial Agents and Chemotherapy, 2016;60(12):7115–27.

106. Qadri H, Shah AH, Mir MA. Role of immunogenetics polymorphisms in infectious diseases. A Molecular Approach to Immunogenetics: Elsevier; 2022. p. 169–91.

107. Posch W, Wilflingseder D, Lass-Flörl C. Immunotherapy as an antifungal strategy in immune compromised hosts. Current Clinical Microbiology Reports, 2020;7:57–66.

108. Rachini A, Pietrella D, Lupo P, Torosantucci A, Chiani P, Bromuro C, et al. An anti-β-glucan monoclonal antibody inhibits growth and capsule formation of *Cryptococcus neoformans* in vitro and exerts therapeutic, anticryptococcal activity in vivo. Infection and Immunity, 2007;75(11):5085–94.

109. Nosanchuk JD, Dadachova E. Radioimmunotherapy of fungal diseases: The therapeutic potential of cytocidal radiation delivered by antibody targeting fungal cell surface antigens. Frontiers in Microbiology, 2012;2:283.

11 Antifungal Prevention and Control Strategies

Sushil Kumar and Dibyendu Banerjee

11.1 INTRODUCTION

In this chapter, we discuss the prevention and control strategies for human fungal diseases. Fungal diseases can affect various parts of our body, from our skin to our lungs, and even our internal organs. Fungi are tiny organisms that can sometimes make us sick, like athlete's foot or thrush [1]. Fungal pathogens can cause a wide range of infections in humans and other animals, from mildly irritating to life-threatening infections [2]. The prevalence of fungal diseases is global; they may become severe in immunocompromised individuals such as those with HIV/AIDS, cancer, or organ transplant recipients [3]. In recent years, the incidence of fungal infections has been on the rise, driven by factors like increased use of immunosuppressive medications, changing climate patterns, and the globalization of travel and trade [4, 5]. Furthermore, fungi exhibit a remarkable ability to adapt to various environments and can develop resistance to antifungal medications over time [6, 7]. Therefore, establishing effective ways to properly combat these illnesses is necessary [8]. The urgency of addressing fungal diseases as a global health priority to treat individuals, especially those with compromised immune systems. Prevention and control strategies for human fungal diseases involve a combination of public health measures, individual hygiene practices, and medical interventions [9–11]. Fungal diseases can range from mild and self-limiting to severe and life-threatening, so it's important to take a multifaceted approach to reduce the risk of infection [12]. Control strategies refer to the plans or methods used to manage and deal with something (Figure 11.1). In this case, it's about how to handle fungal diseases if they do happen. This might involve medical treatments, quarantine, or other actions to control the spread of the disease.

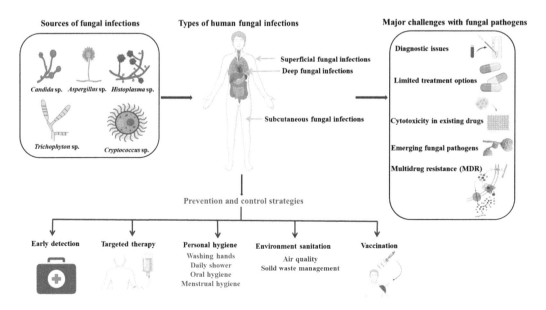

FIGURE 11.1 Types of human fungal infections and their control strategies.

DOI: 10.1201/9781032642864-11

11.2   PUBLIC HEALTH MEASURES

1. Establish surveillance systems to monitor the prevalence and distribution of fungal diseases. This allows for early detection and response to outbreaks [13].
2. Raise public awareness about fungal diseases, their causes, and prevention strategies. Target high-risk populations, such as those with compromised immune systems [14].
3. The fungal diseases transmitted by vectors (e.g., mosquitoes for fungal diseases like histoplasmosis and coccidioidomycosis), implement vector control measures to reduce transmission [15, 16].
4. Implement measures to reduce fungal contamination in public spaces, including healthcare facilities, and in agricultural settings where fungal spores can be a source of infection [17].
5. Eat a balanced diet, exercise regularly, and get enough sleep to keep your immune system in good shape [18].

11.3   PERSONAL HYGIENE

1. Encourage regular hand washing with soap and water, especially before eating and after using the restroom. Fungi love warm and moist environments (areas prone to moisture like between toes and under breasts), helps in preventing fungal growth. Keeping yourself dry can go a long way [19].
2. Promote covering the mouth and nose when coughing or sneezing to prevent the spread of respiratory fungal infections [20].
3. Maintain good personal hygiene, including keeping skin clean and dry and avoiding sharing personal items like towels and combs or razors as they can spread infections [21].
4. Keep your feet clean and dry, especially between the toes. Fungal infections like athlete's foot often start here. Wear clean, dry socks and well-ventilated shoes, and try not to go barefoot in public places [22].
5. Wearing loose-fitting, breathable clothing made of natural fibers like cotton can help in reducing the risk of fungal infections. These fabrics allow air circulation and absorb moisture, preventing fungal growth on the skin [23].

11.4   MEDICAL INTERVENTIONS

1. For some fungal diseases like histoplasmosis and coccidioidomycosis, vaccine research and development are ongoing. Promote the use of vaccines which is available for fungi [24].
2. When necessary, provide antifungal medications for the treatment of fungal infections. Timely diagnosis and appropriate treatment are crucial [25].
3. Strengthening the immune system, especially in individuals with weakened immune systems (e.g., due to HIV/AIDS or immunosuppressive medications), can help prevent fungal infections [26].
4. Implement rigorous infection control practices in healthcare settings to prevent healthcare-associated fungal infections [27].
5. Antifungal creams can help treat and prevent fungal infections on the skin [28].

11.5   ENVIRONMENTAL CONTROL

1. Minimize exposure to environments where fungal pathogens thrive, such as construction sites or areas with high humidity and bird or bat droppings [29].
2. Ensure proper ventilation in indoor spaces to reduce the accumulation of airborne fungal spores [30].
3. Prevent mold growth in buildings by addressing water leaks, controlling humidity levels, and promptly removing and cleaning affected materials [31].

4. Properly handle and store food to prevent contamination by mycotoxins produced by molds, which can cause foodborne fungal infections [32].
5. Individuals traveling to regions where certain fungal diseases are endemic should take precautions, such as wearing appropriate clothing and using insect repellent [33].
6. Monitor and manage antifungal drug resistance to ensure that treatment options remain effective [34].
7. Ensuring proper ventilation and keeping living spaces clean and dry is essential to reduce the risk of fungal infections [35].

11.6 NOVEL STRATEGIES TO COMBAT MDR

Multidrug-resistant (MDR) fungal infections cause a significant threat to human health, particularly in immunocompromised individuals, and the emergence of MDR strains has made these infections even more challenging to treat. In this context, novel strategies to combat MDR in human fungal pathogens are essential [36]. Here, we discuss some of these innovative approaches:

a. *Nanotechnology*: This provides a promising approach to combating these resistant pathogens due to its ability to enhance the delivery and efficacy of antifungal agents. Nanotechnology can address MDR in human fungal pathogens through advanced drug delivery systems, enhanced antifungal activity, combination therapies, innovative treatment options like photodynamic therapy and photothermal therapy, and improved diagnostic tools.

b. *Phytochemicals and natural compounds*: Compounds like essential oils, alkaloids, flavonoids, and terpenoids exhibit strong antifungal properties by disrupting fungal cell membranes and inhibiting vital cellular processes. These natural agents often work synergistically with existing drugs, enhancing their efficacy, and reducing resistance development. Additionally, their broad-spectrum activity and minimal side effects make them promising candidates for MDR fungal infection treatments.

c. *RNAi*: RNA interference (RNAi) provides a targeted approach to combat MDR in human fungal pathogens by silencing specific genes crucial for fungal survival and drug resistance. Using small interfering RNAs (siRNAs) or short hairpin RNAs (shRNAs), RNAi can precisely inhibit the expression of resistance-related genes, effectively reducing fungal virulence and enhancing drug susceptibility.

d. *Phage therapy*: Engineered mycoviruses or Mycophages that carry antifungal genes can specifically target and kill resistant fungi. This targeted approach minimizes damage to beneficial microbiota and reduces the likelihood of resistance development.

e. *Anti-biofilm agents*: These agents combat MDR fungal pathogens by disrupting biofilms, which are protective layers that enhance fungal resistance to drugs. These agents, including enzymes, peptides, and small molecules, can degrade the biofilm matrix, inhibit its formation, or enhance drug penetration. By breaking down biofilms, these treatments improve the efficacy of antifungal drugs and reduce resistance.

f. *Combination therapy*: Similar to antibacterial therapy, combining multiple antifungal drugs with different mechanisms of action can be effective in treating MDR fungal infections [37]. This approach reduces the likelihood of resistance development and enhances the overall antifungal activity.

g. *Repurposing existing drugs*: Researchers are exploring the repurposing of existing drugs approved for other purposes as antifungal agents [38]. Drug libraries are screened to identify compounds that exhibit antifungal activity, potentially offering new treatment options.

h. *Pharmacokinetic enhancements*: Enhancing the delivery of antifungal drugs to the site of infection through improved drug formulations or drug delivery systems can improve treatment outcomes [39]. For example, lipid-based formulations of existing antifungals can enhance drug penetration and efficacy.

i. *Targeting virulence factors*: Rather than killing the fungi directly, targeting their virulence factors or regulatory mechanisms is an emerging strategy [40]. This approach weakens the pathogen's ability to cause disease and may reduce the selective pressure for resistance development.

j. *Host-pathogen interactions*: Understanding the host-pathogen interactions at the molecular level can lead to the development of therapies that modulate the host's immune response to better control fungal infections [41]. Immunomodulatory drugs may be used in conjunction with antifungal agents.

k. *Antifungal peptides*: Natural and synthetic antifungal peptides are being explored for their potential to combat MDR fungi [42]. These peptides can target specific fungal cell components, disrupting their integrity and function.

l. *CRISPR/Cas9-based approaches*: Gene editing techniques such as CRISPR/Cas9 can be employed to create genetically modified fungi that are more susceptible to existing antifungal drugs [43]. This approach aims to reverse resistance or enhance drug susceptibility.

m. *Antifungal vaccines*: Developing vaccines against fungal pathogens is an emerging strategy [44]. Vaccines can stimulate the host's immune system to recognize and target specific fungal species, reducing the risk of infection and MDR development.

n. *Diagnostic advances*: Rapid and accurate diagnostics are crucial for timely and targeted treatment. Advances in diagnostic technologies, such as molecular techniques and next-generation sequencing, can help identify MDR fungal strains more quickly [45, 46].

o. *Surveillance and infection control*: Strengthening surveillance systems and infection control practices in healthcare settings can help prevent the spread of MDR fungi [47]. Early detection and isolation of infected patients are vital components of this strategy.

p. *Precision medicine*: These combats MDR fungal pathogens by customizing treatments based on the genetic profile of the infection and the patient's unique characteristics. This approach utilizes genomic sequencing to identify specific fungal strains and their resistance mechanisms, allowing for the selection of the most effective antifungal drugs. Personalized therapies enhance treatment efficacy, minimize side effects, and reduce the likelihood of resistance development.

q. *Antifungal management*: The development of resistance can be slowed down by encouraging careful and conscientious use of antifungal medications in healthcare settings [48]. This includes maximizing dose regimes, reducing needless use, and making sure that the treatment course is sufficient in duration.

11.7 NON-TRADITIONAL APPROACHES

Preventing and controlling human fungal diseases typically involves traditional approaches such as antifungal medications and hygiene practices. However, non-traditional approaches are being explored as potential complements or alternatives to existing strategies [49]. Many non-traditional approaches are still in the experimental or early research stages and may not yet be widely available for clinical use. Here are some non-traditional approaches for the prevention and control of human fungal diseases:

a. *Probiotics and microbiome manipulation*
   i. Probiotics, which are beneficial bacteria, can help maintain a healthy balance of microbiome in the human body. Some research suggests that certain probiotics may be effective in preventing fungal infections, especially in the gastrointestinal and vaginal tracts [50, 51].

    ii. Bacteriophages are viruses that can infect and kill bacteria. They are being investigated as a potential treatment for bacterial infections that can predispose individuals to fungal infections [52]. By eliminating the bacterial source, fungal infections may be reduced.

   iii. Microbiota transplantation, such as fecal microbiota transplantation (FMT), is a non-traditional approach to combating MDR fungal pathogens by restoring healthy microbial balance. This procedure introduces beneficial microorganisms to outcompete and inhibit pathogenic fungi, enhancing the host's natural defenses.

  b. *Immunomodulation*: Enhancing the immune system's response to fungal pathogens is a non-traditional approach. Immunomodulatory therapies, such as cytokine therapy or monoclonal antibodies, can be used to boost the body's ability to combat fungal infections [53].

  c. *Nanotechnology*: Nanoparticles and nanomaterial's can be designed to deliver antifungal agents more effectively [54]. These materials can improve the stability and bioavailability of antifungal drugs, making them more potent against fungal infections.

  d. *Plant-based compounds*: Some natural compounds derived from plants, such as essential oils, have shown antifungal properties. These compounds can be used as topical treatments or in conjunction with traditional antifungal medications [55].

  e. *Photodynamic therapy*: Photodynamic therapy involves using light-activated compounds to kill fungal cells [56]. This approach can be used in localized infections, such as fungal nail infections or oral thrush.

  f. *Vaccination*: While vaccines for fungal diseases are limited compared to viral and bacterial diseases, research is ongoing to develop vaccines for certain fungal infections like *Candida* and *Aspergillus* [57].

  g. *Genetic modification of fungi*: Genetic engineering techniques can be employed to modify fungal pathogens, making them less virulent or more susceptible to the host's immune response [58]. This approach is still in its experimental stages.

  h. *Environmental control*: Managing the environment to reduce exposure to fungal spores can be an effective preventive measure. This may involve improving indoor air quality, reducing humidity in susceptible areas, or using air filtration systems [59].

  i. *Antifungal surfaces*: Creating highly efficient antifungal surfaces involves altering their nanostructural and chemical characteristics, along with their wettability, to prevent fungal cell adhesion, growth, and reproduction.

  j. *Education and awareness*: Raising public awareness about the risks of fungal infections, especially among vulnerable populations, and promoting proper hygiene practices can help prevent the spread of these diseases [60].

  k. *International collaborations*: Engage with international collaborations and initiatives to control fungal MDR as this is a global problem requiring a coordinated response.

11.8 ANTIFUNGAL PROPHYLAXIS

Antifungal prophylaxis is a medical strategy used to prevent fungal infections in individuals who are at high risk of developing these infections. It involves the administration of antifungal medications to prevent the growth and spread of fungal pathogens before an infection occurs [61]. Here are some key aspects of antifungal prophylaxis:

High-risk patient populations: Antifungal prophylaxis is typically considered for individuals who are at an increased risk of developing fungal infections. This includes patients with compromised immune systems, such as those undergoing chemotherapy, organ transplant recipients, and individuals with advanced HIV/AIDS [62]. It may also be considered for patients with certain medical conditions, like severe neutropenia or prolonged use of broad-spectrum antibiotics.

Types of antifungal prophylaxis: There are several antifungal medications that can be used for prophylaxis, including azoles (e.g., fluconazole, itraconazole, voriconazole), echinocandins (e.g., caspofungin), and amphotericin B formulations [63]. The choice of antifungal agent depends on the patient's specific risk factors, underlying medical conditions, and the types of fungal infections commonly encountered in their clinical setting.

Duration of prophylaxis: The duration of antifungal prophylaxis varies depending on the patient population and clinical circumstances. It may be administered for a limited period, such as during the neutropenic phase following chemotherapy, or it may be continued for a longer duration in individuals with ongoing immune suppression [64].

Risk assessment: Before initiating antifungal prophylaxis, healthcare providers perform a risk assessment to determine if the benefits exceed the potential risks and side effects of the medication [65]. Factors considered include the patient's overall health, immune status, and the likelihood of encountering fungal pathogens in their clinical environment.

Monitoring: Patients receiving antifungal prophylaxis are typically monitored closely for signs of fungal infection, as well as for any adverse effects of the medication. Laboratory tests, imaging, and clinical assessments may be used to detect early signs of infection or medication toxicity [66].

Adjustments and discontinuation: Antifungal prophylaxis may be adjusted or discontinued based on the patient's clinical progress and risk factors. If a patient's immune status improves or they are no longer at high risk for fungal infections, the prophylactic treatment may be stopped [67].

Resistance and breakthrough infections: There is a risk of developing antifungal resistance during prolonged prophylaxis [68]. In some cases, breakthrough fungal infections can occur despite prophylactic treatment. Monitoring for resistance and adjusting antifungal therapy accordingly is essential.

11.9 CONCLUSION

Fungal infections can be serious and sometimes life-threatening. Prevention and control strategies are crucial to avoid the spread of these diseases. Prevention and control strategies for human fungal diseases are essential for protecting the public's health and lowering the incidence of these frequently impaired infections. One key strategy is to avoid close contact with infected individuals or contaminated materials. In conclusion, a comprehensive strategy that combines education, hygiene, vaccination, effective treatment, and research efforts is necessary to prevent and control human fungal diseases. Public health initiatives and international cooperation are essential in the fight against these often overlooked but significant health threats.

REFERENCES

1. H. Sakaguchi, Treatment and prevention of oral candidiasis in elderly patients, Med. Mycol. J. 58 (2017) J43–J49. https://doi.org/10.3314/mmj.17.004.
2. J.B. Konopka, A. Casadevall, J.W. Taylor, J. Heitman, L. Cowen, One health: Fungal pathogens of humans, animals, and plants, Am. Acad. Microbiol. (2019) 1–40. https://www.ncbi.nlm.nih.gov/books/NBK549988/.
3. F. Bongomin, S. Gago, R.O. Oladele, D.W. Denning, Global and multi-national prevalence of fungal diseases—Estimate precision, J. Fungi. 3 (2017). https://doi.org/10.3390/jof3040057.
4. R.E. Baker, A.S. Mahmud, I.F. Miller, M. Rajeev, F. Rasambainarivo, B.L. Rice, S. Takahashi, A.J. Tatem, C.E. Wagner, L.F. Wang, A. Wesolowski, C.J.E. Metcalf, Infectious disease in an era of global change, Nat. Rev. Microbiol. 20 (2022) 193–205. https://doi.org/10.1038/s41579-021-00639-z.
5. WHO releases first-ever list of health-threatening fungi, Saudi Med. J. 43 (2022) 1284–1285.

6. A. Arastehfar, T. Gabaldón, R. Garcia-Rubio, J.D. Jenks, M. Hoenigl, H.J.F. Salzer, M. Ilkit, C. Lass-Flörl, D.S. Perlin, Drug-resistant fungi: An emerging challenge threatening our limited antifungal armamentarium, Antibiotics. 9 (2020) 1–29. https://doi.org/10.3390/antibiotics9120877.

7. A. Vitiello, F. Ferrara, M. Boccellino, A. Ponzo, C. Cimmino, E. Comberiati, A. Zovi, S. Clemente, M. Sabbatucci, Antifungal drug resistance: An emergent health threat, Biomedicines. 11 (2023). https://doi.org/10.3390/biomedicines11041063.

8. Jain T, Muktapuram PR, Sharma K, Ravi O, Pant G, Mitra K, Bathula SR, Banerjee D. Biofilm inhibition and anti-Candida activity of a cationic lipo-benzamide molecule with twin-nonyl chain. Bioorg Med Chem Lett. 2018 Jun 1;28(10):1776–1780. doi: 10.1016/j.bmcl.2018.04.024. Epub 2018 Apr 12. PMID: 29678464.

9. M. Haque, J. McKimm, M. Sartelli, S. Dhingra, F.M. Labricciosa, S. Islam, D. Jahan, T. Nusrat, T.S. Chowdhury, F. Coccolini, K. Iskandar, F. Catena, J. Charan, Strategies to prevent healthcare-associated infections: A narrative overview, Risk Manag. Healthc. Policy. 13 (2020) 1765–1780. https://doi.org/10.2147/RMHP.S269315.

10. D.J. Smith, J.A.W. Gold, K. Benedict, K. Wu, M. Lyman, A. Jordan, N. Medina, S.R. Lockhart, D.J. Sexton, N.A. Chow, B.R. Jackson, A.P. Litvintseva, M. Toda, T. Chiller, Public health research priorities for fungal diseases: A multidisciplinary approach to save lives, J. Fungi. 9 (2023). https://doi.org/10.3390/jof9080820.

11. Kumar, S., Jain, T., Banerjee, D. (2019). Fungal Diseases and Their Treatment: A Holistic Approach. In: Hameed, S., Fatima, Z. (eds) Pathogenicity and Drug Resistance of Human Pathogens. Springer, Singapore. https://doi.org/10.1007/978-981-32-9449-3_6.

12. G.K.K. Reddy, A.R. Padmavathi, Y. V. Nancharaiah, Fungal infections: Pathogenesis, antifungals and alternate treatment approaches, Curr. Res. Microb. Sci. 3 (2022). https://doi.org/10.1016/j.crmicr.2022.100137.

13. T.R. Kozel, B. Wickes, Fungal diagnostics, Cold Spring Harb. Perspect. Med. 4 (2014). https://doi.org/10.1101/cshperspect.a019299.

14. C. Ibe, A.A. Otu, Recent advances and challenges in the early diagnosis and management of invasive fungal infections in Africa, FEMS Yeast Res. 22 (2022). https://doi.org/10.1093/femsyr/foac048.

15. M.T. Rahman, M.A. Sobur, M.S. Islam, S. Ievy, M.J. Hossain, M.E.E. Zowalaty, A.M.M.T. Rahman, H.M. Ashour, Zoonotic diseases: Etiology, impact, and control, Microorganisms. 8 (2020) 1–34. https://doi.org/10.3390/microorganisms8091405.

16. Y. Qin, X. Liu, G. Peng, Y. Xia, Y. Cao, Recent advancements in pathogenic mechanisms, applications and strategies for entomopathogenic fungi in mosquito biocontrol, J. Fungi. 9 (2023). https://doi.org/10.3390/jof9070746.

17. A.A. Haleem Khan, S. Mohan Karuppayil, Fungal pollution of indoor environments and its management, Saudi J. Biol. Sci. 19 (2012) 405–426. https://doi.org/10.1016/j.sjbs.2012.06.002.

18. M. Yadav, Diet, sleep and exercise: The keystones of healthy lifestyle for medical students, J. Nepal Med. Assoc. 60 (2022) 841–843. https://doi.org/10.31729/jnma.7355.

19. R. Hay, Superficial fungal infections, Med. (United Kingdom). 49 (2021) 706–709. https://doi.org/10.1016/j.mpmed.2021.08.006.

20. T. O'Connor, Preventing health care-associated infections, Nurs. N. Z. 15 (2009) 28.

21. A. Ghiffari, R. Asmalia, R. Pamudji, H. Nurdita, Health education to promote and prevent tinea cruris at Darul Fadhli Elementary School, Palembang City, Indonesia. Berdaya. 3 (2022) 303–308. https://doi.org/10.47679/ib.2022222.

22. N.F. Makola, N.M. Magongwa, B. Matsaung, G. Schellack, N. Schellack, Managing athlete's foot, South African Fam. Pract. 60 (2018) 37–41. https://doi.org/10.4102/safp.v60i5.4911.

23. A.J. Stratigos, A.D. Katsambas, Medical and cutaneous disorders associated with homelessness. Skinmed. 2 (2003) 168–172.

24. E. Santos, S.M. Levitz, Fungal vaccines and immunotherapeutics, Cold Spring Harb. Perspect. Med. 4 (2014). https://doi.org/10.1101/cshperspect.a019711.

25. M. Gupte, P. Kulkarni, B.N. Ganguli, Antifungal antibiotics, Appl. Microbiol. Biotechnol. 58 (2002) 46–57. https://doi.org/10.1007/s002530100822.

26. C.Y. Low, C. Rotstein, Emerging fungal infections in immunocompromised patients, F1000 Med. Rep. 3 (2011). https://doi.org/10.3410/M3-14.

27. H. Kanamori, W.A. Rutala, E.E. Sickbert-Bennett, D.J. Weber, Review of fungal outbreaks and infection prevention in healthcare settings during construction and renovation, Clin. Infect. Dis. 61 (2015) 433–444. https://doi.org/10.1093/cid/civ297.

28. R.C. Raval, K. Singh, A. V. Gandhi, Topical therapy in neonates, J. Neonatol. 22 (2008) 60–64. https://doi.org/10.1177/097321790802200114.
29. K. Benedict, M. Richardson, S. Vallabhaneni, B.R. Jackson, T. Chiller, Emerging issues, challenges, and changing epidemiology of fungal disease outbreaks, Lancet Infect. Dis. 17 (2017) e403–e411. https://doi.org/10.1016/S1473-3099(17)30443-7.
30. M.J. Mendell, T. Brennan, L. Hathon, J.D. Odom, F.J. Offerman, B.H. Turk, K.M. Wallingford, R.C. Diamond, W.J. Fisk, Causes and prevention of symptom complaints in office buildings: Distilling the experience of indoor environmental quality investigators, Facilities. 24 (2006) 436–444. https://doi.org/10.1108/02632770610701549.
31. EPA, Mold Course Chapter 2, United States Environ. Prot. Agency. (2021).
32. R.A. El-Sayed, A.B. Jebur, W. Kang, F.M. El-Demerdash, An overview on the major mycotoxins in food products: Characteristics, toxicity, and analysis, J. Futur. Foods. 2 (2022) 91–102. https://doi.org/10.1016/j.jfutfo.2022.03.002.
33. M.A. Tolle, Evaluating a sick child after travel to developing countries, J. Am. Board Fam. Med. 23 (2010) 704–713. https://doi.org/10.3122/jabfm.2010.06.090271.
34. M.C. Fisher, A. Alastruey-Izquierdo, J. Berman, T. Bicanic, E.M. Bignell, P. Bowyer, M. Bromley, R. Brüggemann, G. Garber, O.A. Cornely, S.J. Gurr, T.S. Harrison, E. Kuijper, J. Rhodes, D.C. Sheppard, A. Warris, P.L. White, J. Xu, B. Zwaan, P.E. Verweij, Tackling the emerging threat of antifungal resistance to human health, Nat. Rev. Microbiol. 20 (2022) 557–571. https://doi.org/10.1038/s41579-022-00720-1.
35. H.S. Maji, R. Chatterjee, D. Das, S. Maji, Fungal infection: An unrecognized threat, in: Viral, Parasitic, Bacterial, and Fungal Infections: Antimicrobial, Host Defense, and Therapeutic Strategies, (2022) Academic Press, pp. 625–644. https://doi.org/10.1016/B978-0-323-85730-7.00059-X.
36. J. Tanwar, S. Das, Z. Fatima, S. Hameed, Multidrug resistance: An emerging crisis, Interdiscip. Perspect. Infect. Dis. 2014 (2014). https://doi.org/10.1155/2014/541340.
37. D. Armstrong-James, G.D. Brown, M.G. Netea, T. Zelante, M.S. Gresnigt, F.L. Van De Veerdonk, S.M. Levitz, Series Fungal infections 6: Immunotherapeutic approaches to treatment of fungal diseases, Lancet Infect. Dis. 17 (2017) e393–e402.
38. G. Wall, J.L. Lopez-Ribot, Screening repurposing libraries for identification of drugs with novel antifungal activity, Antimicrob. Agents Chemother. 64 (2020). https://doi.org/10.1128/AAC.00924-20.
39. R. Mali, J. Patil, Nanoparticles: A novel antifungal drug delivery system, Mater. Proc. 14 (2023) 61. https://doi.org/10.3390/iocn2023-14513.
40. M. Ivanov, A. Ćirić, D. Stojković, Emerging antifungal targets and strategies, Int. J. Mol. Sci. 23 (2022). https://doi.org/10.3390/ijms23052756.
41. A. Rana, M. Ahmed, A. Rub, Y. Akhter, A tug-of-war between the host and the pathogen generates strategic hotspots for the development of novel therapeutic interventions against infectious diseases, Virulence. 6 (2015) 566–580. https://doi.org/10.1080/21505594.2015.1062211.
42. T. Ciociola, L. Giovati, S. Conti, W. Magliani, C. Santinoli, L. Polonelli, Natural and synthetic peptides with antifungal activity, Future Med. Chem. 8 (2016) 1413–1433. https://doi.org/10.4155/fmc-2016-0035.
43. D. Uthayakumar, J. Sharma, L. Wensing, R.S. Shapiro, CRISPR-based genetic manipulation of candida species: Historical perspectives and current approaches, Front. Genome Ed. 2 (2020). https://doi.org/10.3389/fgeed.2020.606281.
44. N.P. Medici, M. Del Poeta, New insights on the development of fungal vaccines: From immunity to recent challenges, Mem. Inst. Oswaldo Cruz. 110 (2015) 966–973. https://doi.org/10.1590/0074-02760150335.
45. S.E. Kidd, S.C.A. Chen, W. Meyer, C.L. Halliday, A new age in molecular diagnostics for invasive fungal disease: Are we ready? Front. Microbiol. 10 (2020). https://doi.org/10.3389/fmicb.2019.02903.
46. S. Jiang, Y. Chen, S. Han, L. Lv, L. Li, Next-generation sequencing applications for the study of fungal pathogens, Microorganisms. 10 (2022). https://doi.org/10.3390/microorganisms10101882.
47. G.J. Buckley, G.H. Palmer, Combating antimicrobial resistance and protecting the miracle of modern medicine (2022). https://doi.org/10.17226/26350.
48. A. Srinivasan, J.L. Lopez-Ribot, A.K. Ramasubramanian, Overcoming antifungal resistance, Drug Discov. Today Technol. 11 (2014) 65–71. https://doi.org/10.1016/j.ddtec.2014.02.005.
49. T. Mehra, M. Köberle, C. Braunsdorf, D. Mailänder-Sanchez, C. Borelli, M. Schaller, Alternative approaches to antifungal therapies, Exp. Dermatol. 21 (2012) 778–782. https://doi.org/10.1111/exd.12004.
50. T.K. Das, S. Pradhan, S. Chakrabarti, K.C. Mondal, K. Ghosh, Current status of probiotic and related health benefits, Appl. Food Res. 2 (2022). https://doi.org/10.1016/j.afres.2022.100185.

51. L. Lehtoranta, R. Ala-Jaakkola, A. Laitila, J. Maukonen, Healthy vaginal microbiota and influence of probiotics across the female life span, Front. Microbiol. 13 (2022). https://doi.org/10.3389/fmicb.2022.819958.

52. A. Górski, P.L. Bollyky, M. Przybylski, J. Borysowski, R. Międzybrodzki, E. Jończyk-Matysiak, B. Weber-Dabrowska, Perspectives of phage therapy in non-bacterial infections, Front. Microbiol. 10 (2019). https://doi.org/10.3389/fmicb.2018.03306.

53. Q.H. Sam, W.S. Yew, C.J. Seneviratne, M.W. Chang, L.Y.A. Chai, Immunomodulation as therapy for fungal infection: Are we closer? Front. Microbiol. 9 (2018). https://doi.org/10.3389/fmicb.2018.01612.

54. S. Nami, A. Aghebati-Maleki, L. Aghebati-Maleki, Current applications and prospects of nanoparticles for antifungal drug delivery, EXCLI J. 20 (2021) 562–584. https://doi.org/10.17179/excli2020-3068.

55. A.A. Rashed, D.N.G. Rathi, N.A.H.A. Nasir, A.Z.A. Rahman, Antifungal properties of essential oils and their compounds for application in skin fungal infections: Conventional and nonconventional approaches, Molecules. 26 (2021). https://doi.org/10.3390/molecules26041093.

56. J.J. Shen, G.B.E. Jemec, M.C. Arendrup, D.M.L. Saunte, Photodynamic therapy treatment of superficial fungal infections: A systematic review, Photodiagnosis Photodyn. Ther. 31 (2020). https://doi.org/10.1016/j.pdpdt.2020.101774.

57. R. Kumar, V. Srivastava, Application of anti-fungal vaccines as a tool against emerging anti-fungal resistance, Front. Fungal Biol. 4 (2023). https://doi.org/10.3389/ffunb.2023.1241539.

58. D. Malavia, N.A.R. Gow, J. Usher, Advances in molecular tools and *in vivo* models for the study of human fungal pathogenesis, Microorganisms. 8 (2020). https://doi.org/10.3390/microorganisms8060803.

59. R. Araujo, J.P. Cabral, Fungal air quality in medical protected environments, in: Air Quality, (2010) Sciyo. https://doi.org/10.5772/9766.

60. K. Benedict, N.A.M. Molinari, B.R. Jackson, Public awareness of invasive fungal diseases—United States, 2019, MMWR. Morb. Mortal. Wkly. Rep. 69 (2020) 1343–1346. https://doi.org/10.15585/mmwr.mm6938a2.

61. S. Ascioglu, B.E. De Pauw, J.F.G.M. Meis, Prophylaxis and treatment of fungal infections associated with haematological malignancies, Int. J. Antimicrob. Agents. 15 (2000) 159–168. https://doi.org/10.1016/S0924-8579(00)00159-X.

62. C.D. Hornik, D.S. Bondi, N.M. Greene, M.P. Cober, B. John, Review of fluconazole treatment and prophylaxis for invasive candidiasis in neonates, J. Pediatr. Pharmacol. Ther. 26 (2021) 115–122. https://doi.org/10.5863/1551-6776-26.2.115.

63. J.R. Wingard, H. Leather, A new era of antifungal therapy, Biol. Blood Marrow Transplant. 10 (2004) 73–90. https://doi.org/10.1016/j.bbmt.2003.09.014.

64. O. Lortholary, B. Dupont, Antifungal prophylaxis during neutropenia and immunodeficiency, Clin. Microbiol. Rev. 10 (1997) 477–504. https://doi.org/10.1128/cmr.10.3.477.

65. T. Lehrnbecher, B.T. Fisher, B. Phillips, M. Beauchemin, F. Carlesse, E. Castagnola, N. Duong, L.L. Dupuis, V. Fioravantti, A.H. Groll, G.M. Haeusler, E. Roilides, M. Science, W.J. Steinbach, W. Tissing, A. Warris, P. Patel, P.D. Robinson, L. Sung, Clinical practice guideline for systemic antifungal prophylaxis in pediatric patients with cancer and hematopoietic stem-cell transplantation recipients, J. Clin. Oncol. 38 (2020) 3205–3216. https://doi.org/10.1200/JCO.20.00158.

66. M.M. Riwes, J.R. Wingard, Diagnostic methods for invasive fungal diseases in patients with hematologic malignancies, Expert Rev. Hematol. 5 (2012) 661–669. https://doi.org/10.1586/ehm.12.53.

67. A.H. Limper, K.S. Knox, G.A. Sarosi, N.M. Ampel, J.E. Bennett, A. Catanzaro, S.F. Davies, W.E. Dismukes, C.A. Hage, K.A. Marr, C.H. Mody, J.R. Perfect, D.A. Stevens, An official American Thoracic Society statement: Treatment of fungal infections in adult pulmonary and critical care patients, Am. J. Respir. Crit. Care Med. 183 (2011) 96–128. https://doi.org/10.1164/rccm.2008-740ST.

68. A.A. Rabaan, T. Sulaiman, S.H. Al-Ahmed, Z.A. Buhaliqah, A.A. Buhaliqah, B. AlYuosof, M. Alfaresi, M.A. Al Fares, S. Alwarthan, M.S. Alkathlan, R.S. Almaghrabi, A.A. Abuzaid, J.A. Altowaileb, M. Al Ibrahim, E.M. AlSalman, F. Alsalman, M. Alghounaim, A.S. Bueid, A. Al-Omari, R.K. Mohapatra, Potential strategies to control the risk of antifungal resistance in humans: A comprehensive review, Antibiotics. 12 (2023). https://doi.org/10.3390/antibiotics12030608.

12 Nanotechnology
An Advanced Approach toward the Treatment of Fungal Infections

Hemlata Kumari, Minakshi, Shaurya Prakash, Ritu Pasrija, and Antresh Kumar*

12.1 INTRODUCTION

Fungal infection is a severe health issue encountered worldwide with significantly higher morbidity and mortality in the human population, especially in immunocompromised and debilitated patients [1]. The global morbidity rate of invasive fungal infections accounts for over a billion and a recorded mortality of 1.6 million; of these, 80% of deaths are caused by *Candida, Aspergillus*, and *Cryptococcus* spp. [2]. Based on the site of infection or the nature of the pathogen and host tissue interactions, fungal infections are categorized as cutaneous, sub-cutaneous, mucosal, superficial, invasive, and systemic infections. The most prevalent type of fungal infection is superficial infection and the most severe one is systemic infection. Fungal infections are relatively prevalent in women and more than 135 million women underwent vulvovaginal fungal infections caused by different *Candida* spp. [3]. Medical conditions like cancer, HIV, burns, organ transplantation, and medications like corticosteroids and tumor necrosis factor (TNF) inhibitors enhance the chances of fungal infection [4]. Recently, there were mucormycosis fungal outbreaks as secondary infections in different continents among COVID-19-infected patients with high morbidity and mortality. Over a million people have undergone surgery to excise infected limbs for fungal treatment. It was also reported that the rate of fungal infection has dramatic during the COVID-19 pandemic [5]. Candidiasis is the most prominent type of infection incurred in the population and contributes to 40% mortality. *Candida* spp. are commensal microbes found in the human body that cause both superficial and invasive acute and chronic infections. Fungal infections are spread by inhalation of fungal spores or by disturbance in commensal microbes of the host [6]. The etiology of fungal infection is associated with poor health hygiene, insanitation, and indiscriminate use of antifungals that liberate systemic infections such as thrush (affects the mouth), invasive (affects the eye, lungs, brain, heart, etc.), and vaginal candidiasis [4]. Among 15 different pathogenic *Candida* spp., *C. albicans, C. tropicalis, C. glabrata, C. parapsilosis*, and *C. krusei* are responsible for most candidiasis [3]. The second is aspergillosis, caused by *Aspergillus* sp., *A. nidulans, A. flavus, A. niger, A. fumigatus, A. terrus*, and *A. clavatus* are the pathogenic *Aspergillus* spp. with the first being the most common. The next is cryptococcosis caused by *C. neoformans* and *C. gatti* with a 20%–70% mortality rate [5].

Azoles, polyenes, and echinocandins are classes of antifungals mainly used for the prevention of fungal infections [7]. Azoles (itraconazole, voriconazole, and fluconazole) primarily target to 14α-demethylase enzyme involved in ergosterol biosynthesis and block its activity. Polyenes also target ergosterol present in the fungal cell membrane and inhibit its production. However, echinocandins target differently to β-1,3-glucan synthase; an enzyme responsible for the β-1,3 glucan synthesis in fungal cell walls [8]. The excessive and improper use of antifungals with our doctor's prescription has augmented drug resistance that emerges as a major challenge in the treatment of fungal infections, especially in developing countries. Various contributory factors of drug resistance in fungi

---

* Corresponding Author: antreshkumar@gmail.com

DOI: 10.1201/9781032642864-12

155

have been reported [9–13]. Horizontal gene transfer, adaptive phenotypic plasticity, mutation, over-expressing of drug efflux pumps, and biofilm formation are some examples [14]. The most common mechanism of azole resistance is associated with GOF (gain of function) mutation in transcription factors that regulate the expression of MFS and ABC transports, like MRR1, and TAC1 in *C. albicans*, PDR1 in *C. glabrata* [8]. Similarly, polyene resistance mainly occurs due to the loss of the ERG3 gene and the upregulation of genes ERG5, ERG6, and ERG25 involved in ergosterol biosynthesis [15]. To overcome and fight that type of challenge there is an urgent need to find new antifungal drugs or treatments to treat fungal infections. Nanotechnology is an effective approach to overcoming multidrug resistance and enhancing the drug delivery system by improving drug efficacy, increasing penetration of drugs, improving aqueous solubility, etc. [16]. Antifungal drugs are also fabricated with NPs to enhance their antifungal effect. For example, biogenic AgNPs conjugated with AmB more actively inhibit the growth of *C. tropicalis and albicans*, as compared to AmB alone [17], SLN formed by using AmB enhances the bioavailability and decreases the toxicity of the drug [18]. Nanotechnology shows very significant results in drug delivery systems by using various types of NPs like nanocarriers, polymeric NPs, liposomes, solid lipid NPs, nanostructure lipid carriers, etc. They help in targeting the particular drug at the site of infection, which helps in proper and maximum utilization of antifungals at the particular site [19]. NPs like nitric oxide-releasing NPs, metal and metal oxide NPs, and chitosan NPs help in fighting against antifungal drug resistance by various mechanisms like ROS production, DNA, RNA, protein damage, and cell membrane damage. [20].

12.2   CURRENT ANTIFUNGAL TREATMENTS

Polyenes, azoles, allyl amines, pyrimidine, and echinocandins are the drugs used against fungal infection (Table 12.1). Polyenes like Amphotericin B a broad-spectrum drug widely used against fungal infection mainly targets the ergosterol present in fungal cells and disturbs the cell homeostasis leading to death. Azoles like voriconazole, itraconazole, and fluconazole inhibit the sterol

TABLE 12.1
Antifungal Classes of Drugs Used against Fungal Infections, Their Target, Biological Effect, and Recommended Dose

S. No.	Class	Common Name	Target(s)	Affected spp.	Therapeutic Effect	Administration	Recommended Dose
1.	Polyene	Amphotericin B	Lanosterol 14-α-demethylase	*Candida* *Aspergillus* *Cryptococcus*	Fungicidal	Intravenous Oral	3–6 mg/kg of body weight
		Nystatin					200–400 KIU four times a day
2.	Allylamines	Terbinafine	Squalene epoxidase	*Microsporum*	Fungicidal and Fungistatic	Oral	250 mg once a day
		Naftifine		*Trichophyton*		Oral	100 mg/kg /day
		Butenafine		*Epidermophyton* *Aspergillus*		Oral/Topically	25 mg/kg/day or 1% cream
3.	Echinocandins	Caspofungin	β-1,3-glucan synthase	*Aspergillus*	Fungicidal and fungistatic	Intravenous	70 mg/day
		Micafungin		*Candida*		Intravenous	150 mg/day
		Anidulafungin				Intravenous	200 mg/per day
4.	Azoles	Miconazole	Lanosterol 14-α-demethylase	candida	Fungistatic	Topical	2% cream
		Econazole		aspergillus		Topical	1% cream
		Ketoconazole		mucor		Oral	200 mg/day
		Clotrimazole		rhizopus		Oral/Topical	10 mg/day, 1% cream
		Itraconazole				Oral	200 mg/day
		Fluconazole				Oral/Intravenous	12 mg/kg, 200 mg/day
		Voriconazole				Intravenous	6 mg/kg

14α-demethylase enzyme, which helps in the sterol biosynthesis change of lanosterol into ergosterol and is essential for maintaining the integrity of fungal cell membranes [1]. Echinocandins like anidulafungin, micafungin, and caspofungin are semisynthetic antifungal drugs used against *candida* and some *Aspergillus* species, inhibit the β-1,3-glucan-synthase which help in the synthesis of β-1,3 glucan, an essential component of fungal cell [8]. Allyl amines like naftifine and terbinafine are mainly used in superficial fungal infections and inhibit fungal growth by targeting the squalene epoxidase which changes squalene to lanosterol during ergosterol biosynthesis [21]. Pyrimidine analogs like 5-fluorocytosine (5-FC) enter into fungal cells through cytosine permeases and inhibit nucleic acid and protein synthesis by being deaminated to 5-fluorouracil [22]. Despite the availability of these topical antifungals, the incidences of fungal infections are continuously increasing due to rapidly emerging drug resistance which made them redundant to work effectively [23]. To overcome the problem of drug resistance in fungal pathogens to the existing drugs, the nanoparticle (NP)-based drug delivery system is a potential substitute. Nanoparticles are characteristically small 1–100 nm size particles having different shapes like conical, spiral, spherical, cylindrical, etc. [24]. Due to their antimicrobial, diagnostic, and therapeutical properties, NPs have gained much attention in the research communities. NPs have also been explored in the drug delivery and diagnosis of disease. Based on size, shape, and preparation method nanoparticles are divided into different types like nanospheres, nanocapsules, polymersomes, nanotubes, liposomes, etc. [16] and all have separate self-life and loading capacities. These are also classified in organic and inorganic, polymeric or macromolecular, natural or synthetic, as discussed in this chapter. Nanomaterials are arranged in two types of composition for drug delivery systems: A drug can be conjugated with a nanoparticle or encapsulated in a nanocarrier [25,26]. Nanotechnology is an effective approach to overcoming multidrug resistance and enhancing the drug delivery system by improving drug efficacy, increasing penetration of drugs, and improving aqueous solubility [16].

12.3 CHALLENGES WITH CURRENT DRUGS

Drug resistance and toxicity are the main challenges that limit the treatment of fungal infections. Antifungal drugs used for the treatment of fungal infection are azoles, polyenes, and echinocandins but due to continuous, excessive, and improper use of these drugs, microbes start attaining resistance against them. Fungal cells can overcome the effect of antifungal drugs by attaining resistance mechanisms like activating the stress response signaling, altering the cellular pathway, and drug targets, overexpression of efflux pumps, biofilm formation, target incompatibility, and so on (Figure 12.1) [27]. Sometimes biofilms include populations of both bacteria and fungus, which makes the treatment challenging. Infection caused by multidrug resistance *Candida* species like *C. auris* and azole-resistant *Aspergillus* species is associated with a high mortality rate. According to a study, the drug-resistant rate in *Candida* species increased from 4.2% to 7.8% between 2008 and 2014 [28]. Polyenes basically target the ergosterol present in the plasma membrane of microbes, create pores, and also cause oxidative damage in the membrane. To resist the effect of polyenes like AmB microbes start altering the content of β-1,3 glucan in the microbial cell wall, AmB resistance in *A. flavus* is an example of glucan alteration [29]. Azoles target and inhibit the lanosterol demethylase enzyme responsible for ergosterol biosynthesis which is an important component of the fungal cell membrane. Nitrogen atoms of the azole ring directly bind with the iron group of the heme domain present in the active site of lanosterol demethylase and prevent its demethylation [30]. A study by Vuffray et al. suggests that microbial resistance occurs due to longer exposure to a single dose of fluconazole in HIV-infected patients [31], apart from that, a fluconazole-resistant *C. neoformans* was isolated from an immunocompetent patient who was not exposed to fluconazole previously [32]. Alteration in the drug efflux system is also associated with microbial resistance. *Candida* species contains approximately five candida drug-resistant (CDR1 to CDR5) genes for azole resistance, out of them CDR1 and CDR2 belong to the ABC transporters family. According to recent studies, gain of-function mutation in genes TAC1B, PDR1 leads to azole resistance in *C. auris* and *C. glabrata*

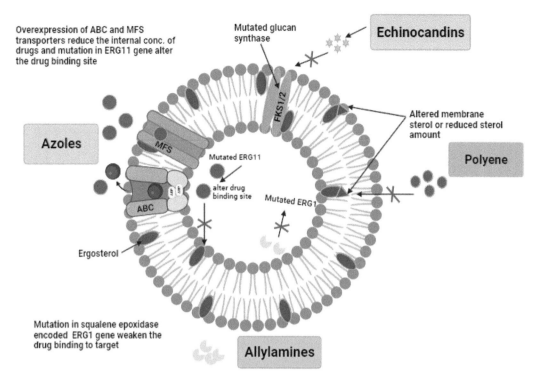

FIGURE 12.1   Development of resistance against the conventional antifungal drugs.

[33, 34]. In the case of echinocandin resistance, S645P and S639P/F/Y are the most common amino acid substitutions found in HS1 (hotspot 1) of FKS1 in clinical isolates of *C. tropicalis* and *C. auris* [35]. These problems demonstrate the urgent need for a new antifungal treatment or drugs to fight the widely spread fungal infections.

12.4   NANOPARTICLES (NPs), A PROMISING ANTIFUNGAL

NPs are prepared using three methods including chemical, physical, and biosynthesis methods. The stability of NPs is enhanced by adding a stabilizer during their preparation. The limitations like toxicity effect and biocompatibility are also overcome by using an eco-friendly approach of nanotechnology which is a green synthesis of NPs, here NPs are synthesized using natural resources and they come under the chemical reduction method of preparation and use biological products like plant extracts, bacteria or fungal filtrates, and animal sources as a stabilizer. AgNPs, CuO NPs, ZnO NPs, and Au NPs are some examples of NPs fabricated using the green synthesis method [27]. A number of previous studies have revealed antifungal properties of different metal NPs which can inhibit the growth of yeast and various fungal pathogens. The antifungal nature of different NPs makes them amenable to use at the industry scale for the protection of food and beverage, etc. The metal NPs have been successfully used as nanogels, and nano solutions, are used in metal-coated devices for different applications [36]. Table 12.2 provides examples of some nanoparticles that have been effectively used against different fungal pathogens. The antifungal activity of NPs has been studied against different *Candida* species including *C. albicans*, *C. tropicalis*, *C. glabrata*, and *C. krusei* [26]. Nanotechnology also helps in improving drug stability and drug efficacy, enhancing the aqueous solubility of drugs (Figure 12.2).

Metal nanoparticles show antimicrobial activity using various mechanisms that are prepared using metals like Ag, Bi, Cu, Mg, ZnO, Au, etc., and their synthesis is affected by many factors

TABLE 12.2

Nanoparticles Effectively Used against Different Fungal Infections

Fungal Infection	Causing Agent	Nanoparticles Used	References
Aspergillosis	*Aspergillus fumigatus*	Marketed AgNPs	[38]
Coccidioidomycosis	*Coccidioides immitis* and *Coccidioides posadasii*	Lipid complexed Amphotericin B (ABLC)	[39]
Mucormycosis	*Rhizopus, Mucor*	Zirconium oxide nanoparticles (ZrO2NPs)	[40]
Candidiasis	*Candida auris*	Trimetallic nanoparticles (Ag-Cu-Co NPs)	[41]

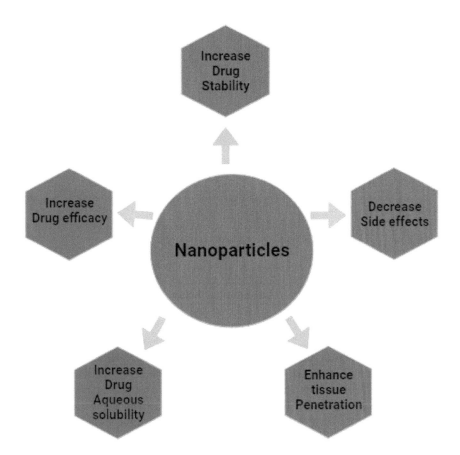

FIGURE 12.2 Nanomaterials resolve many drug delivery problems.

including biomass concentration, pH, culture media, temperature, method of preparation and time of incubation. Metal NPs already have intrinsic antimicrobial properties which are well-marked in silver, copper, and zinc [7]. In order to prevent microbial resistance, nanoparticles use various strategies. Nanoparticles like chitosan NPs, NO NPs, and metal-containing NPs (Ag NPs, Cu NPs, ZnO NPs) limit the emergence of resistance by employing numerous mechanisms like increasing uptake and reducing the release of drugs from the microbial cell [36,37] (see Figure 12.3). Drug targeting using nanotechnology can increase the dose of the drug at the site of infection, and taking multiple antimicrobial drugs in one nanoparticle also help in the fight with drug resistance [20].

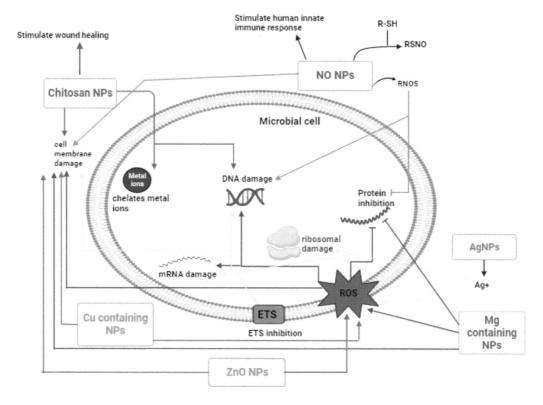

FIGURE 12.3   Different pathways followed by nanoparticles to fight multidrug resistance.

12.4.1   Antifungal Effect of Silver Nanoparticles

Silver nanoparticles characteristically elicit a strong therapeutic activity against a wide range of microbial pathogens. The antimicrobial effect of silver nanoparticles has also been extensively studied against multiresistant bacterial strains and considered for the prevention of fungal infections. The biological activity of AgNPs is regulated by their physiochemical properties that expand the wide applications of AgNPs including antifungals. It is well-accepted that smaller AgNPs are more toxic due to the high feasibility of entering into the cell to cause oxidation dissolution [42]. Different groups of studies reveal that AgNPs were able to show high antifungal activity against yeast, pathogenic *Candida* spp. [43]. Silver NPs have gained attention due to their antimicrobial activity against various microbes like *A. fumigates, A. niger, A. flavus, T. rubrum, C. albicans,* and *Penicillum* species. Their large number of applications in the fields of agriculture and healthcare make it an effective tool for the treatment of fungal infection, drug targeting, and diagnosis purposes [44]. The antimicrobial activity of silver NPs is due to the Ag+ ion which is released when silver is dissolved in an aqueous medium. Ag+ reacts with phosphors or sulfur-containing moieties present in the cell membranes and cell wall proteins of microbes, this binding of positively charged Ag and negatively charged elements creates pores in the membrane, inhibits cytochrome of ETC, inhibits DNA replication, DNA and RNA damage, inhibit protein translation by denaturing 30S ribosomal subunit, ROS formation, and so on. [20]. Silver NPs show biofilm inhibition activity against *C. albicans* and it is also reported that AgNPs show antifungal activity against MDR fungal stains *C. auris* in both preformed and mature biofilms [45].

12.4.2   NO NPs (Nitric Oxide-Releasing NPs)

Nitric oxide is an essential element of host defense mechanisms against disease-causing microbes. It acts as a potent immunostimulatory signaling molecule and it's inherent to preventing the growth and eradication of infections, making it a strong antimicrobial compound [46]. Nitric oxide reacts with superoxide and forms reactive nitrogen oxide intermediates (RNOS) including peroxynitrite, nitrogen dioxide, and dinitrogen trioxide, which are responsible for antimicrobial activity. When the concentration of NO exceeds 1 m, the synthesis of these RNOS becomes adequate to have antimicrobial activity through several methods [37], including damaging DNA (by deamination of G, A, and C bases), generating H_2O_2, inhibiting DNA alkyl transferase, lipid peroxidation, inactivating zinc metalloproteins, etc. [37, 47]. According to a recent study, NO NPs show an anti-biofilm effect against *T. rubrum* [48] and also show antifungal effects against *C. albicans* and *T. mentagrophyte* [37], and antibacterials against multidrug-resistant microbes like *E. faecalis*, *K. pneumoniae*, *E. coli*, and *P. aeruginosa* [49]. NO NPs have the potential to kill hyphae and yeast forms in the mature biofilm [50]. BiNPs are also reported to have anti-fungal activity against *C. albicans* and *C. auris* and show moderate activity against *C. auris* biofilm including alteration in biofilm structures and morphology of biofilm cells [51]. Copper-containing NPs help in the formation of reactive oxygen species which leads to the inhibition of amino acid biosynthesis and DNA replication. CuO NPs show antimicrobial effects against *S. cerevisiae*, and PCL-CuO-NPs (polycaprolactone-copper fibers) show anticandidal activity against *C. albicans*, *C. tropicalis*, and *C. glabrata* [16]. ZnO NPs target microbial membranes and are responsible for the formation of reactive oxygen species. Their antimicrobial activity is reported in *C. albicans*, *C. tropicalis*, *C. krusei*, *C. parapsilosis*, and *S. aureus* (MRSA) [3, 20].

12.4.3 CHITOSAN NPS (CHITOSAN-CONTAINING NANOPARTICLES)

Chitosan is obtained from chitin which is a polymer made up of randomly arranged N-acetyl glucos-amine and glucosamine. These are positively charged at a pH of less than 6.5 and this positive charge helps in providing an antimicrobial effect by interacting with negatively charged microbial cell walls or membranes by disturbing cell permeability, disturbing cell homeostasis, and leading to the death of microbe [52]. The other mechanism followed by chitosan NPs is inhibiting DNA leads to mRNA and protein synthesis inhibition, showing wound healing properties by hindering the release of inflammatory cytokines and enhancing the release of fibroblasts and collagen III, etc. [53]. Chitosan is conjugated with NPs to improve its poor solubility (in vivo) and enhance its antimicrobial effect against fungi, bacteria as well, and viruses. They show an anti-biofilm effect against *C. albicans C. tropicalis*, and *C. krusei* [54].

12.5 NANOTECHNOLOGY USED IN ANTIFUNGAL DRUG DELIVERY SYSTEMS

Delivery of antifungal drugs at the site of fungal infection is a big challenge in the current scenario and drug delivery is one of the main goals of nanotechnology to increase drug localization at the particular site of infection. Nanotechnology is also an effective alternate option that helps to fight fungal infection, drug resistance developed for in-used antifungal drugs and also in drug delivery systems (Table 12.3). According to recent studies, silver nanoparticles used against *C. auris* show both growth and biofilm inhibitory effects [55] and Ag-Cu-Co a trimetallic nanoparticle shows strong fungicidal activity [41].

12.5.1 NANOCARRIERS

Nanocarriers are used for drug delivery, having a size of between 10 and 1000 nm scale. Its small size helps in the drug delivery system. Nanocarriers include nanostructure lipid carriers, polymeric nanoparticles, phospholipid-based vesicles, nanoemulsions, dendrimers, nanocapsules, and solid lipid nanoparticles (see Figure 12.4).

12.5.1.1 Solid Lipid NPs (SLNs)

These are lipid-based nanocarriers with a size ranging between 40 and 1000 nm, formed by emul-sification of a mixture of solid oils. These oils are solid at room temperature. They contain 0.1 to

TABLE 12.3

Different Nanoparticles Used in Antifungal Drug Delivery Systems

Antifungals	Nanomaterial Used	Fungal Agent	Administration	Reference
Amphotericin B (AmB)	PEGylated Polylactic-polyglycolic Acid Copolymer (PLGA-PEG)	*Candida albicans*	Oral	[72]
Nystatin (NYT) and fluconazole (FLU)	AgNPs	*Candida albicans* and *Aspergillus brasiliensis*	*In vivo* studies required	[73]
Ketoconazole (KTZ)	Poly(lactide-coglycolide) (NPs)	*Candida albicans*	ODD (Ocular Drug Delivery)	[74]
Ketoconazole (KTZ)	Silver NPs	*Malassezia furfur*	Topical	[75]
Fluconazole (FLU)	ZnO NPs	*C. albicans*	*In vivo* studies required	[76]
Itraconazole (ITZ)	Poly (ethylene glycol)/Polylactic Acid (PEG/PLA)	*A. fumigatus*	*In vivo* studies required	[77]
Miconazole nitrate (MN)	Solid Lipid Nanoparticles (SLN)	*Candida* species	Oral	[78]
Ketoconazole (KTZ)	Gelatin NPs	*Aspergillus flavus*	ODD (Ocular Drug Delivery)	[79]
Amphotericin B	Sodium Deoxycholate Sulfate (SDCS) Nanomicelles	*C. neoformans*, *C. albicans*	*In vivo* studies required	[80]
Amphotericin B and Nystatin (NYS)	Lipid Complex; Colloidal Dispersion and Liposomal	*C. immitis*	*In vivo* studies required	[81]
Dodecyl gallate (DOD)	Nanostructured Lipid System (NLS)	*P. brasiliensis* and *P. lutzii*	*In vivo* studies required	[82]
Terbinafine hydrochloride (TH)	Solid Lipid Nanoparticles (SLN)	*C. albicans*	Topical	[83]
AmB	Lipid Nanoparticles	*A. fumigatus* and *C. albicans*	Intravenous	[84]
Voriconazole	Polymeric Nanoparticles	*C. albicans*	Oral	[85]
Miconazole	SLNs	*C. albicans*	Topical	[86]
Itraconazole (ITZ)	Lipid Nanoparticles	*A. fumigatus*	Oral	[87]

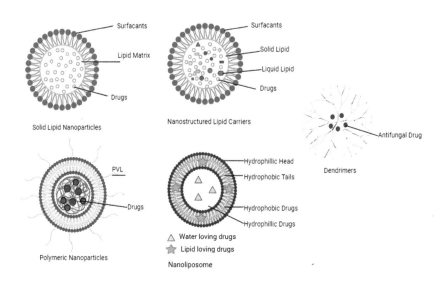

FIGURE 12.4   Various types of nanoparticles used in drug delivery systems.

30% solid fat and up to 5% surfactant. Their lipid part is used for the delivery of antifungal drugs at the targeted site [16]. Because of their tiny particle size and lipophilic nature, they are effective at overcoming physiological barriers. They provide several benefits, including high drug loading capacity, high stability with less drug leakage, low physiological toxicity, controlled or prolonged drug release, and drug targeting [19]. The delivery system of solid lipid NPs is affected by some factors like nanomaterials' interaction with the body, type of administration, and lipid used [56]. Lipase present in the human body is responsible for the degradation of these lipid NPs, the rate of degradation depends on the lipid chain of NPs, the longer the lipid chain lowers the degradation rate. The use of an emulsifier during NP formation helps in lowering the degradation rate [16]. A study by Kamesh et al. reveals the effect of natamycin SLNPs (natamycin is an antifungal drug used to treat eye infections) against *A. fumigatus* increased the zone of inhibition by 8 mm in comparison to natamycin alone [19]. In another study on Amphotericin B conducted in 2016 by Butani et al., it was found that SLNs are useful for topical AmB delivery. He reported notable benefits of utilizing AmB-loaded SLNs, including improved antifungal efficacy, fewer skin irritations, and higher skin deposition. The authors also claimed that SLN5 contains a lipid ratio of 1:10 and Pluronic F 127 1.25% as a surfactant exhibits the best formulation out of all SLN formulations of Amphotericin B utilized in trials [57].

12.5.1.2 Nanostructure Lipid Carriers (NLCs)

This is also a type of lipid-based nanocarrier with a size range of 10–1000 nm. These are biodegradable as well as biocompatible nanomaterials made up of natural lipids, surfactants, and co-surfactants [16]. To overcome the drawbacks of first-generation lipid nanoparticles (SLNs) such as low drug loading capacity, polymorphic transitions, and drug leakage during storage, nanostructure lipid carriers (NLCs) have emerged as second-generation lipid nanoparticles [58]. Since the majority of drugs are lipophilic in nature, NLCs have superior properties as a result of the insertion of a liquid lipid fraction that enables greater drug retention capacity and long-term stability. This sort of system is therefore more effective for drug administration. Additionally, the particle size and charge have an impact on the biodistribution of nanocarriers. While positive NPs are found in the kidney and negative NPs in the liver, large particles are caught by the lungs and small particles by the liver and bone marrow [7, 59]. In a recent study by Fu et al. (2017), it was reported that AmB-CH-NLC nanoparticles can efficiently penetrate the cornea of the rabbit eye as compared to unbound drugs and help in the treatment of fungal keratitis [60].

12.5.1.3 Polymeric Nanoparticles

Polymeric nanoparticles are made up of a chain of similar structures known as monomers. Alginate, chitosan, and PLGA are the three main polymers used for the formation of NPs for the treatment of fungal infections [7, 61, 62]. Alginate is a water-soluble naturally occurring polymer that is mostly found in the organisms of phylum Phaeophyta. α-L-guluronic acid and β-D-mannuronic acid are two monomers present in this polymer. The rigidity of the polymer is decided by monomer concentration and difference in their arrangement. These NPs are not toxic to cells and are biocompatible [63]. Chitin, which is found widely in nature, especially in animals and fungi, is converted into polymer chitosan by the process of deacetylation. Chitosan is the best material for making NPs and delivering various types of compounds through mucous membranes because of its hydrophobic nature and positive charge. Chitosan nanoparticles transport medicines, proteins, peptides, and DNA/RNA efficiently and have good qualities like non-cytotoxicity, biodegradability, and biocompatibility [64]. PLGA polymers are synthetic copolymers made up of lactic acid and glycolic acid monomers in 50–50% concentration. It is one of the best biodegradable polymers, having appealing properties like protecting drugs from degradation, being biocompatible, having potential for sustained release, etc. [65].

12.5.1.4 Liposomes

These are lipid particles made up of natural or synthetic phospholipids arranged in a bilayer with both hydrophilicor and hydrophobic sides like a plasma membrane. These unique properties help it to deliver both hydrophilic and hydrophobic drugs [66]. The best example of this group of nanoparticles is Amphotericin B liposomes, an easily accessible drug [67], and oleic acid liposomes, that easily penetrate the membrane of the MRSA and improve the killing as compared to with the drug alone [68].

12.5.1.5 Dendrimers

These are branched polymer nanoparticles and have several branches with an end group and a multifunctional central core [69]. The components used for the preparation of the central core are poly amidoamine, poly propylene imine, and poly ether hydroxylamine. They are prepared in two ways; divergently which move from center to branches, as well as convergently from branches to center [7, 70]. Dendrimer NPs are mainly used in the transport of drugs for cancer, inflammation, and fungal infections and also for the transport of DNA, plasmids, and RNA [71].

12.6 CONCLUSION

To compete with the increase in fungal infections and rapidly emerging resistance against known drugs, it is necessary to find new alternative ways and effective treatments for mitigating fungal infections. The development of nanotechnology is one of the best approaches employed in the treatment of fungal infections in various ways such as increasing the efficacy of already known drugs, targeting the particular drug at the site of infection, decreasing the side effects of the known drugs, overcoming the resistance, etc. NPs, like solid lipid nanoparticles, nanostructure lipid carriers, polymeric nanoparticles, liposomes, and dendrimers are effectively used for drug delivery systems and increase the amount of the drug at the site of infection. On the other hand, nanotechnology also helps in fighting against antimicrobial drug resistance with various mechanisms used by nanoparticles such as nitric oxide-releasing nanoparticles, chitosan-containing nanoparticles, metal-containing nanoparticles, including silver NPs, copper NPs, zinc oxide NPs, magnesium NPs, and so on. The valuable properties shown by NPs make them the best approach to the treatment of fungal infections and the development of novel drug delivery systems.

REFERENCES

1. Reddy GKK, Padmavathi AR, Nancharaiah YV (2022) Fungal infections: Pathogenesis, antifungals and alternate treatment approaches. Curr Res Microb Sci 3:100137. doi: 10.1016/j.crmicr.2022
2. Janbon G, Quintin J, Lanternier F, d'Enfert C (2019) Studying fungal pathogens of humans and fungal infections: Fungal diversity and diversity of approaches. Microbes and Infect 21(5–6):237–245
3. Du W, Gao Y, Liu L, Sai S, Ding C (2021) Striking back against fungal infections: The utilization of nanosystems for antifungal strategies. Int J Mol Sci 22(18):10104. doi: 10.3390/ijms221810104
4. Brown GD, Denning DW, Gow NA, Levitz SM, Netea MG, White TC (2012) Hidden killers: Human fungal infections. Sci Transl Med 4(165):165rv13–165rv13
5. Shaurya Prakash, Antresh Kumar (2022) Mucormycosis threats: A systemic review. J Basic Microbiol 63(2):119–127
6. Pathakumari B, Liang G, Liu W (2020) Immune defence to invasive fungal infections: A comprehensive review. Biomed Pharmacother 130:110550. doi: 10.1016/j.biopha.2020
7. Kischkel B, Rossi SA, Santos SR, Nosanchuk JD, Travassos LR, Taborda CP (2020) Therapies and vaccines based on nanoparticles for the treatment of systemic fungal infections. Front Cell Infect Microbiol 10:463. doi.org/10.3389/fcimb.2020.00463
8. Berman J, Krysan DJ (2020). Drug resistance and tolerance in fungi. Nature Rev Microbiol 18(6):319–331

9. Banerjee A, Vishwakarma P, Kumar A, Lynn A, Prasad R (2019) Information theoretic measures and muta-genesis identify a novel linchpin residue involved in substrate selection within nucleotide-binding domain of an ABCG family exporter Cdr1p. Arch Biochem Biophys 2019:663:143–150. doi: 10.1016/j.abb.2019.01.013
10. Kumar A, Jaiswal V, Kumar V, Ray A, Kumar A (2018) Functional redundancy in Echinocandin B in-cluster transcription factor ecdB of *Emericella rugulosa* NRRL 11440. Biotech Rep 19:e00264; 1–8
11. Bera K, Rani P, Kishor G, Agarwal S, Kumar A, Singh DV (2017) Structural elucidation of TMD0 Domain of EcdL: An MRP1 family of ABC transporter protein revealed by atomistic simulation. J Biomol Stuc Dyna. doi: 10.1080/07391102.2017.1372311
12. Mandal A, Kumar A, Singh A, Lynn AM, Kapoor K, Prasad R (2012) A key structural domain of the *Candida albicans* Mdr1 protein. Biochem J 445:313–322
13. Kumar A, Shukla S, Mandal A, Shukla S, Ambudkar S, Prasad R (2010) Divergent signature motifs of nucleotide binding domains of ABC multidrug transporter, CaCdr1p of pathogenic *Candida albicans*, are functionally asymmetric and non-interchangeable. BBA Biomemb 1798: 1757–1766
14. Rai V, Gaur M, Kumar A, Shukla S, Komath SS, Prasad R (2008) Novel catalytic mechanism for ATP hydrolysis employed by the N-terminal nucleotide-binding domain of Cdr1p, a multidrug ABC transporter of *C. albicans*. BBA Biomemb 1778:2143–2153
15. Hokken MW, Zwaan BJ, Melchers WJG, Verweij PE (2019) Facilitators of adaptation and antifun-gal resistance mechanisms in clinically relevant fungi. Fungal Genet Biol 132:103254
16. Nami S, Aghebati-Maleki A, Aghebati-Maleki L (2021) Current applications and prospects of nanoparticles for antifungal drug delivery. EXCLI J 20:562
17. Al Aboody MS (2019) Silver/silver chloride (Ag/AgCl) nanoparticles synthesized from *Azadirachta indica* lalex and its antibiofilm activity against fluconazole resistant *Candida tropicalis*. Artif Cells Nanomed Biotechnol 47(1):2107–2113
18. Chaudhari MB, Desai PP, Patel PA, Patravale VB (2016) Solid lipid nanoparticles of amphotericin B (AmbiOnp): *In vitro* and *in vivo* assessment towards safe and effective oral treatment module. Drug Deliv Transl Res 6:354–364
19. Khames A, Khaleel MA, El-Badawy MF, El-Nezhawy AO (2019) Natamycin solid lipid nanopar-ticles–sustained ocular delivery system of higher corneal penetration against deep fungal keratitis: Preparation and optimization. Int J Nanomedicine 14:2515–2531
20. Pelgrift RY, Friedman AJ (2013) Nanotechnology as a therapeutic tool to combat microbial resis-tance. Adv Drug Deliv Rev 65:1803–1815
21. Newland JG, Abdel-Rahman SM (2009) Update on terbinafine with a focus on dermatophytoses. Clin Cosmet Investig Dermatology 2:49–63
22. Scorzoni L, de Paula e Silva AC, Marcos CM, Assato PA, de Melo WC, de Oliveira HC, Fusco-Almeida AM (2017) Antifungal therapy: New advances in the understanding and treatment of mycosis. Front Microbiol 8:36. doi: 10.3389/fmicb.2017.00036
23. León Buitimea A, Garza Cervantes JA, Gallegos Alvarado DY, Osorio Concepción M, Morones Ramírez JR (2021) Nanomaterial-based antifungal therapies to combat fungal diseases *Aspergillosis, Coccidioidomycosis, Mucormycosis*, and candidiasis. Pathogens 10(10):1303. doi: 10.3390/pathogens 10101303
24. Ealia SAM, Saravanakumar MP (2017) A review on the classification, characterization, syn-thesis of nanoparticles and their application. IOP Conf Ser: Mater Sci Eng 263: 032019. doi:10.1088/1757-899X/263/3/032019
25. Bugno J, Hsu HJ, Hong S (2015) Recent advances in targeted drug delivery approaches using dendritic polymers. Biomater Sci 3(7):1025–1034
26. Mohammed Fayaz A, Ao Z, Girilal M, Chen L, Xiao X, Kalaichelvan P T, Yao X (2012) Inactivation of microbial infectiousness by silver nanoparticles-coated condom: A new approach to inhibit HIV-and HSV-transmitted infection. Int J Nanomedicine 7:5007–5018
27. Huang T, Li X, Maier M, O'Brien Simpson NM, Heath DE, O'Connor AJ (2023) Using inorganic nanoparticles to fight fungal infections in the antimicrobial resistant era. Acta Biomater 158:56–79
28. Vallabhaneni S, Cleveland AA, Farley MM, Harrison LH, Schaffner W, Beldavs ZG, Smith RM (2015) Epidemiology and risk factors for echinocandin nonsusceptible *Candida glabrata* bloodstream infec-tions: Data from a large multisite population-based candidemia surveillance program, 2008–2014. Open Forum Infect Dis 2(4):ofv163. doi: 10.1093/ofid/ofv163
29. Loeffler J, Stevens DA (2003) Antifungal drug resistance. Clin Infect Dis 36:S31–S4.
30. Joseph Horne T, Hollomon DW (1997) Molecular mechanisms of azole resistance in fungi. FEMS Microbiol Lett 149(2):141–149

31. Vuffray A, Durussel C, Boerlin P, Boerlin Petzold F, Bille I, Glauser MP, Chave JP (1994) Oropharyngeal candidiasis resistant to single-dose therapy with fluconazole in HIV-infected patients. AIDS 8(5):708–709

32. Orni Wasserlauf R, Izkhakov E, Siegman Igra Y, Bash E, Polacheck I, Giladi M (1999) Fluconazole-resistant *Cryptococcus neoformans* isolated from an immunocompetent patient without prior exposure to fluconazole. Clin Infect Dis 29(6):1592–1593

33. Vale-Silva L, Ischer F, Leibundgut-Landmann S, Sanglard D (2013) Gain-of-function mutations in PDR1, a regulator of antifungal drug resistance in *Candida glabrata*, control adherence to host cells. Infect Immun 81(5):1709–1720

34. Rybak JM, Muñoz JF, Barker KS, Parker JE, Esquivel BD, Berkow EL, Rogers PD (2020) Mutations in TAC1B: A novel genetic determinant of clinical fluconazole resistance in *Candida auris*. Mbio 11(3):e00365–20

35. Arastehfar A, Gabaldón T, Garcia-Rubio R, Jenks JD, Hoenigl M, Salzer HJ, Perlin DS (2020) Drug-resistant fungi: An emerging challenge threatening our limited antifungal armamentarium. Antibiotics 9(12):877. doi: 10.3390/antibiotics9120877

36. Rai M, Yadav A, Gade A (2009) Silver nanoparticles as a new generation of antimicrobials. Biotechnol Adv 27(1):76–83

37. Peretyazhko TS, Zhang Q, Colvin VL (2014) Size-controlled dissolution of silver nanoparticles at neutral and acidic pH conditions: Kinetics and size changes. Environ Sci Technol 48(20):11954–11961

38. Xu Y, Gao C, Li X, He Y, Zhou L, Pang G, Sun S (2013) In vitro antifungal activity of silver nanoparticles against ocular pathogenic filamentous fungi. J Ocul Pharmacol Ther 29(2):270–274

39. Clemons KV, Capilla J, Sobel RA, Martinez M, Tong AJ, Stevens DA (2009) Comparative efficacies of lipid-complexed amphotericin B and liposomal amphotericin B against coccidioidal meningitis in rabbits. Antimicrob Agents Chemother 53(5):1858–1862

40. Mohamed DY (2018) Detection the antifungal effect of zirconium oxide nanoparticles on mold which isolated from domestic's bathroom. Al-Mustansiriyah J Sci 29(1):15–22

41. Kamli MR, Srivastava V, Hajrah NH, Sabir JS, Hakeem KR, Ahmad A, Malik MA (2021) Facile bio-fabrication of Ag-Cu-Co trimetallic nanoparticles and its fungicidal activity against *Candida auris*. J Fungi 7(1):62. doi: 10.3390/jof7010062

42. Panáček A, Kolář M, Večeřová R, Prucek R, Soukupová J, Kryštof V, Kvítek L (2009) Antifungal activity of silver nanoparticles against *Candida* spp. Biomaterials 30(31):6333–6340

43. Mansoor S, Zahoor I, Baba TR, Padder SA, Bhat ZA, Koul AM, Jiang L (2021) Fabrication of silver nanoparticles against fungal pathogens. Front Nanotechnol 3:679358. doi: 10.3389/fnano.2021.679358

44. Hetta HF, Ramadan YN, Al Kadmy IM, Ellah NHA, Shbibe L, Battah B (2023) Nanotechnology-based strategies to combat multidrug-resistant *Candida auris* infections. Pathogens 12(8):1033

45. Schairer DO, Martinez LR, Blecher K, Chouake JS, Nacharaju P, Gialanella P, Friedman AJ (2012) Nitric oxide nanoparticles: Pre-clinical utility as a therapeutic for intramuscular abscesses. Virulence 3(1):62–67

46. Schairer DO, Chouake JS, Nosanchuk JD, Friedman AJ (2012) The potential of nitric oxide releasing therapies as antimicrobial agents. Virulence 3(3):271–279

47. Blecher K, Nasir A, Friedman A (2011) The growing role of nanotechnology in combating infectious disease. Virulence 2(5):395–401

48. Costa Orlandi CB, Martinez LR, Bila NM, Friedman JM, Friedman AJ, Mendes Giannini MJS, Nosanchuk JD (2021) Nitric oxide-releasing nanoparticles are similar to efinaconazole in their capacity to eradicate *Trichophyton rubrum* biofilms. Front Cell Infect Microbiol 11: 684150. doi: 10.3389/fcimb.2021.684150

49. Hajipour MJ, Fromm KM, Ashkarran AA, de Aberasturi DJ, de Larramendi IR, Rojo T, Mahmoudi M (2012) Antibacterial properties of nanoparticles. Trends Biotechnol 30(10):499–451

50. Ahmadi MS, Lee HH, Sanchez DA, Friedman AJ, Tar MT, Davies KP, Martinez LR (2016) Sustained nitric oxide-releasing nanoparticles induce cell death in *Candida albicans* yeast and hyphal cells, preventing biofilm formation in vitro and in a rodent central venous catheter model. Antimicrob Agents Chemother 60(4):2185–2194

51. Vazquez Munoz R, Lopez FD, Lopez Ribot JL (2020) Bismuth nanoantibiotics display anticandidal activity and disrupt the biofilm and cell morphology of the emergent pathogenic yeast *Candida auris*. Antibiotics 9(8):461. doi: 10.3390/antibiotics9080461

52. Huh AJ, Kwon YJ (2011) "Nanoantibiotics": A new paradigm for treating infectious diseases using nanomaterials in the antibiotics resistant era. J Control Release 156(2):128–145

53. Friedman AJ, Phan J, Schairer DO, Champer J, Qin M, Pirouz A, Kim J (2013) Antimicrobial and anti-inflammatory activity of chitosan–alginate nanoparticles: A targeted therapy for cutaneous pathogens. J Invest Dermatol 133(5):1231–1239

54. Gondim BLC, Castellano LRC, de Castro RD, Machado G, Carlo HL, Valença AMG, de Carvalho FG (2018) Effect of chitosan nanoparticles on the inhibition of *Candida* spp. biofilm on denture base surface. Arch Oral Biol 94:99–107

55. Lara HH, Ixtepan Turrent L, Jose Yacaman M, Lopez Ribot J (2020) Inhibition of *Candida auris* biofilm formation on medical and environmental surfaces by silver nanoparticles. ACS Appl Mater Interfaces 12(19):21183–21191

56. Qu J, Zhang L, Chen Z, Mao G, Gao Z, Lai X, Zhu J (2016) Nanostructured lipid carriers, solid lipid nanoparticles, and polymeric nanoparticles: Which kind of drug delivery system is better for glioblastoma chemotherapy? Drug Delivery 23(9):3408–3416

57. Butani D, Yewale C, Misra A (2016) Topical Amphotericin B solid lipid nanoparticles: Design and development. Colloids Surf B 139:17–24

58. Chauhan I, Yasir M, Verma M, Singh AP (2020) Nanostructured lipid carriers: A groundbreaking approach for transdermal drug delivery. Adv Pharm Bull 10(2):150–165

59. Salvi VR, Pawar P (2019) Nanostructured lipid carriers (NLC) system: A novel drug targeting carrier. J Drug Deliv Sci Technol 51:255–267

60. Fu T, Yi J, Lv S, Zhang B (2017) Ocular amphotericin B delivery by chitosan-modified nanostructured lipid carriers for fungal keratitis-targeted therapy. J Liposome Res 27(3):228–233

61. Spadari CDC, Lopes LB, Ishida K (2017) Potential use of alginate-based carriers as antifungal delivery system. Front Microbiol 8:97. doi: 10.3389/fmicb.2017.00097

62. Fernandes Costa A, Evangelista Araujo D, Santos Cabral M, Teles Brito I, Borges de Menezes Leite L, Pereira M, Correa Amaral A (2019) Development, characterization, and *in vitro–in vivo* evaluation of polymeric nanoparticles containing miconazole and farnesol for treatment of vulvovaginal candidiasis. Med Mycol 57(1):52–62

63. Yehia SA, El Gazayerly ON, Basalious EB (2009) Fluconazole mucoadhesive buccal films: In vitro/in vivo performance. Curr Drug Deliv 6(1):17–27

64. Illum L (2003) Nasal drug delivery—Possibilities, problems and solutions. J Control Release 87(1–3):187–198

65. Danhier F, Ansorena E, Silva JM, Coco R, Le Breton A, Préat V (2012) PLGA-based nanoparticles: An overview of biomedical applications. J Control Release 161(2):505–522

66. Stiufiuc R, Iacovita C, Stiufiuc G, Florea A, Achim M, Lucaciu CM (2015) A new class of pegylated plasmonic liposomes: Synthesis and characterization. J Colloid Interface Sci 437:17–23

67. Reis J (2015) Liposomal formulations of amphotericin B: Differences according to the scientific evidence. Rev Esp Quimioter 28:275–281

68. Huang CM, Chen CH, Pornpattananangkul D, Zhang L, Chan M, Hsieh MF, Zhang L (2011) Eradication of drug resistant *Staphylococcus aureus* by liposomal oleic acids. Biomaterials 32(1):214–221

69. Sherje AP, Jadhav M, Dravyakar BR, Kadam D (2018) Dendrimers: A versatile nanocarrier for drug delivery and targeting. Int J Pharm 548(1):707–720

70. Voltan AR, Quindós G, Alarcón KPM, Fusco Almeida AM, Mendes Giannini MJS, Chorilli M (2016) Fungal diseases: Could nanostructured drug delivery systems be a novel paradigm for therapy? Int J Nanomedicine 11:3715–3730

71. Palmerston Mendes L, Pan J, Torchilin VP (2017) Dendrimers as nanocarriers for nucleic acid and drug delivery in cancer therapy. Molecules 22(9):1401. doi: 10.3390/molecules22091401

72. Radwan MA, AlQuadeib BT, Šiller L, Wright MC, Horrocks B (2017) Oral administration of amphotericin B nanoparticles: Antifungal activity, bioavailability and toxicity in rats. Drug Deliv 24(1):40–50

73. Hussain MA, Ahmed D, Anwar A, Perveen S, Ahmed S, Anis I, Khan NA (2019) Combination therapy of clinically approved antifungal drugs is enhanced by conjugation with silver nanoparticles. Int Microbiol 22:239–246

74. Ahmed TA, Aljaeid BM (2017) A potential in situ gel formulation loaded with novel fabricated poly (lactide-co-glycolide) nanoparticles for enhancing and sustaining the ophthalmic delivery of ketoconazole. Int J Nanomedicine 12:1863–1875

75. Mussin JE, Roldán MV, Rojas F, Sosa MDLÁ, Pellegri N, Giusiano G (2019) Antifungal activity of silver nanoparticles in combination with ketoconazole against *Malassezia furfur*. Amb Express 9(1):1–9

76. Hajar AA, Dakhili M, Saghazadeh M, Aghaei S, Nazari R (2020) Synergistic antifungal effect of fluconazole combined with ZnO nanoparticles against *Candida albicans* strains from vaginal candidiasis. Med Lab J 14(3):26–32

77. Essa S, Louhichi F, Raymond M, Hildgen P (2013) Improved antifungal activity of itraconazole-loaded PEG/PLA nanoparticles. J Microencapsul 30(3):205–217

78. Kenechukwu FC, Attama AA, Ibezim EC (2017). Novel solidified reverse micellar solution-based mucoadhesive nano lipid gels encapsulating miconazole nitrate-loaded nanoparticles for improved treatment of oropharyngeal candidiasis. J Microencapsul 34(6):592–609

79. Ahsan SM, Rao CM (2017) Condition responsive nanoparticles for managing infection and inflammation in keratitis. Nanoscale 9(28):9946–9959

80. Usman F, Khalil R, Ul-Haq Z, Nakpheng T, Srichana T (2018) Bioactivity, safety, and efficacy of amphotericin B nanomicellar aerosols using sodium deoxycholate sulfate as the lipid carrier. AAPS PharmSciTech 19:2077–2086

81. González GM, Tijerina R, Sutton DA, Graybill JR, Rinaldi MG (2002) In vitro activities of free and lipid formulations of amphotericin B and nystatin against clinical isolates of *Coccidioides immitis* at various saprobic stages. Antimicrob Agents Chemother 46(5):1583–1585

82. de Lacorte Singulani, J, Scorzoni L, Lourencetti NMS, Oliveira LR, Conçolaro RS, da Silva PB, Giannini MJSM (2018) Potential of the association of dodecyl gallate with nanostructured lipid system as a treatment for paracoccidioidomycosis: *In vitro* and *in vivo* efficacy and toxicity. Int J Pharm 547(1–2):630–636

83. Vaghasiya H, Kumar A, Sawant K (2013) Development of solid lipid nanoparticles based controlled release system for topical delivery of terbinafine hydrochloride. Eur J Pharm Sci 49(2):311–322

84. Jung SH, Lim DH, Jung SH, Lee JE, Jeong KS, Seong H, Shin BC (2009) Amphotericin B-entrapping lipid nanoparticles and their in vitro and in vivo characteristics. Eur J Pharm Sci 37(3–4):313–320

85. Peng HS, Liu XJ, Lv GX, Sun B, Kong QF, Zhai DX, Kostulas N (2008) Voriconazole into PLGA nanoparticles: Improving agglomeration and antifungal efficacy. Int J Pharm 352:29–35

86. Ahmad J (2021) Lipid nanoparticles-based cosmetics with potential application in alleviating skin disorders. Cosmetics 8(3):84. doi: 10.3390/cosmetics8030084

87. Pardeike J, Weber S, Zarfl HP, Pagitz M, Zimmer A (2016) Itraconazole-loaded nanostructured lipid carriers (NLC) for pulmonary treatment of aspergillosis in falcons. Eur J Pharm Biopharm 108:269–276

13 Unraveling the Intricacies of Carbon Catabolic Repression and Glucose in Fungal Pathogenesis

Shaurya Prakash, Hemlata Kumari, Minakshi, Ritu Pasrija, and Antresh Kumar

13.1 INTRODUCTION

Currently, the occurrence of fungal infections is a serious health threat globally. Human fungal infections vary from mild superficial infections to serious life-threatening invasive infections. The pathogens involved in the advent of infections are mostly opportunistic and evade the immunity of an immune-compromised host. The immune-compromised conditions may comprise such as HIV-AIDS, chemotherapy, transplants, pregnancy, and treatment with immune suppressants. Among the various fungal pathogens, the *Candida, Aspergillus,* and *Cryptococcus* genera most abundantly contribute to fungal infections. The occurrence of invasive fungal diseases leads to significant mortality. According to a recent study, almost 4.1% of the total Indian population, i.e. 57,251,328 out of 1,393,400,000 people suffer from serious fungal infections (Ray et al., 2022). Globally, around 1.9 million individuals become infected with acute invasive fungal infections while more than 3 million suffer from chronic fungal infections (Fang et al., 2023). Previous data suggests that fungal pathogens account for more human deaths compared to malaria and tuberculosis (Meyer et al., 2016). Not only among humans, fungal infections are also comparably prevalent among plants as well. Pathogens such as *M. oryzae, A. brassicicola,* and *C. carborum* are among those causing severe plant damage. As per estimates, approximately more than one-third of total food crops are annually destroyed by fungal pathogens (Bhunjun et al., 2021). To establish infection in the host tissue, it is essential to utilize the available nutrients that are crucial for the growth and survival of the pathogen. For successful inhabitation in host niches especially the bloodstream, gastrointestinal tract (GIT), and mucosal surfaces, pathogens tend to adapt to surrounding nutrient availability. Pathogens show metabolic elasticity for nutrient uptake and modulate their metabolism accordingly. These events help the pathogens to gain fitness and thereby enhance their pathogenic potential. Pathogenicity refers to a broad range of traits, including those that contribute to the survival and enhancement of fitness, as well as the pathways that determine the virulence of a pathogen. Among these, nutrient uptake and subsequent metabolism is a significant virulence determinant, as it regulates fungal fitness and survival. Carbon and nitrogen are the prime nutrient sources, and iron and zinc are required in trace amounts, as they are involved in biosynthetic processes. Of all the required sources, the most significant determinant for energy is the carbon source, i.e. sugar. Fungi have a selective preference for carbon sources that can be easily metabolized with less of an energy expenditure and a greater yield of energy. The most favorable carbon source utilized by the organism is glucose, which serves as a preferable carbon source and is most readily catabolized among most of the microbial species. However, other carbohydrates are also utilized as carbon sources by the fungi. Considering the supremacy of glucose as a favorable carbon source for the growth, survival, and pathogenicity of microbes, this chapter

DOI: 10.1201/9781032642864-13

delineates different glucose-centered processes for virulent aspects of several fungal pathogens. In the presence of a favored energy source, i.e. glucose, there is a repression for the use of alternative energy sources, via a process termed carbon catabolite repression (CCR). Glucose and other highly advantageous sugars are assimilated by microbes before they move on to less advantageous carbon sources. The CCR system is considered to be important for appropriate screening and control of energy-producing and readily accessible glucose as a carbon source instead of the more limited or less energetic supplying one. This system helps organisms get the most benefit possible out of the available options. CCR system supports the organisms to maintain an energy-conserving state by switching off the enzymes required for utilization of less favorable carbon sources, thus maintaining the optimal metabolic rate (Ries et al., 2016; Adnan et al., 2018). Other than nutrient uptake, the CCR system aids the microbes in physiological adaptation for their survival. The CCR system has been well studied in *S. cerevisiae* and has revealed that utilization machinery of alternative carbon sources shut down in the presence of glucose and are utilized as a prime carbon source (Kayikci and Nielsen, 2015). Similar findings have been observed in the case of pathogenic *C. albicans* that adopt a glucose repression pathway wherein the presence of glucose represses the expression of other carbon sources utilizing genes (Lagree et al., 2020).

13.2   CARBON CATABOLITE REPRESSION AMONG FILAMENTOUS FUNGI

The pathogenic fungal flora has evolved its genetic machinery to adapt to the surrounding conditions and effectively cause pathogenesis in the host (Matar et al., 2017; Adnan et al., 2018). Effective regulation of carbon metabolism is one of the crucial aspects of disease development and progression. Carbon catabolite repression (CCR) is the molecular process that aids the microbes to adapt specifically to the surrounding conditions. Under the influence of CCR, the microbes switch to a preferable carbon source, i.e. glucose, and halt the machinery involved in sensing, uptake, and utilization of other complex carbon sources (Figure 13.1).

It has been well-studied that *Cre* (catabolite-responsive element) genes (*Cre A-D*) are involved in the regulation of glucose catabolism via the CCR pathway (Ichinose et al., 2014; Chen et al., 2021).

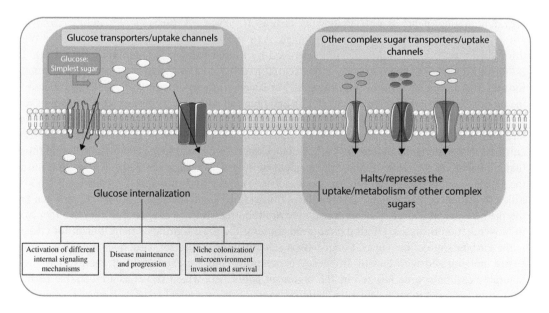

FIGURE 13.1   Schematic representation of carbon catabolite repression (CCR) system among pathogenic fungi.

The C2H2 zinc finger motif which is a DNA binding domain of the CreA transcription factor senses the glucose concentration in the vicinity and represses the expression of genes involved in the utilization of alternative carbon sources (Amore et al., 2013; Ries et al., 2016). However, under normal circumstances, CCR modulates different cellular activities, regulated by CreA and other counterparts. There are three prime signaling pathways intertwined for glucose metabolism in yeast (Kim et al., 2013). The first signal for triggering the regulatory process is glucose itself (New et al., 2014; Weinhandl et al., 2014). Secondly, signals from the surroundings are sensed by G protein-coupled receptors (GPCRs) leading to the regulation of downstream pathways. The glucose signaling pathways such as cAMP/PKA, rgt2/Snf3-Rgt1, and Mig1-Hxk2 involved the glucose metabolism (Kim et al., 2013; New et al., 2014; Peeters and Thevelein, 2014; Shashkova et al., 2015; Tarinejad, 2015). Activation of Ras proteins modulates the action of cAMP levels from low to high glucose. The sensing of glucose by the GPCR, Gpr1 is still unclear; however, the G alpha homolog Gpa2 interacts with the adenylate cyclase. The adenylate cyclase is more strongly influenced by the Ras protein, which is further regulated by Cdc25 and Ira protein (adenylate cyclase controlling elements) (Pentland et al., 2018; Cazzanelli et al., 2018). This pathway is operational under environmental conditions where PKA activation doesn't require adenylate cyclase or cAMP; instead it includes G beta-like proteins krh1, krh2, and Sch9 protein kinase (Peeters et al., 2007; Soulard et al., 2010). Hence, in the discussed pathway, Gpr1 and Gpa2 activate PKA through Krh1, Krh2, and Sch9 proteins in the cAMP-independent pathway (Figure 13.2).

The entrance of glucose inside the yeast cells is facilitated by diffusion through hexose transporters (*HXT1-HXT17, GAL2, SNF3,* and *RGT2*). The function of transcriptional repressor (*Rgt1*) is under the influence of glucose sensors Snf3 and Rgt2. Rgt2 (low-affinity glucose sensor) senses glucose concentration and is induced at higher concentrations, while Snf3 is a high-affinity glucose sensor capable of inducing high-affinity transporters when low levels of glucose are sensed (Kim and Rodriguez, 2021). Rgt1 acts as a repressor of HXT genes in response to Snf3 and Rgt2 signaling. Moreover, levels of PKA affect the phosphorylation of Rgt1 and regulate its binding and blocking of HXT promoters (Palomino et al., 2006; Yao et al., 2015). Consequently, a high glucose level will cause complete activation and an increased level of PKA which would lead to phosphorylation

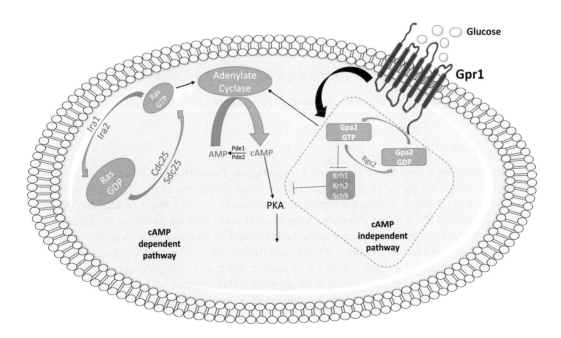

FIGURE 13.2   The PKA/Ras-cAMP pathway in yeast.

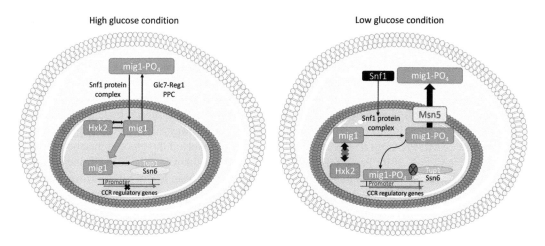

FIGURE 13.3 CCR regulation in yeast during high and low glucose conditions.

of Rgt1, therefore low-affinity HXTs (HXT1, HXT3) will be induced (Kim and Johnston, 2006). On the contrary, low levels of glucose would lead to weak activation of PKA and lead to activation of high-affinity HXTs (HXT2, HXT4) (Roy and Kim, 2016).

The major components of CCR in yeast are Glc7 (protein phosphatase), Snf1 codes for a protein kinase complex, and Mig1, a transcriptional repressor complex (Kim et al., 2013; Tarinejad, 2015). Mig1, a DNA binding protein forms a complex with the co-repressor protein Tup1 and Sen6 to carry out repression of diverse gene families and related transcriptional regulators such as MalR (maltose utilization), Gal4 (galactose utilization), Cat8 (gluconeogenic genes) and Hap4 (respiratory genes). The subcellular localization of Mig1 controls the proper repression activity. Mig1 readily migrates toward the nucleus to repress glucose levels and binds to promoters of glucose-repressible genes (Weinhandl et al., 2014; Welkenhuysen et al., 2017). However, in glucose-starved conditions, the Mig1 becomes phosphorylated and migrates back to the cytoplasm, which is catalyzed by Snf1 (Welkenhuysen et al., 2017), as illustrated in Figure 13.3.

13.3 COMPREHENSIVE ROLE OF CCR REGULATORS AMONG FUNGI

There are several key regulators of CCR among fungi, such as CreA, CreB, CreC, CreD, and Snf1.

13.3.1 CreA

Cre A encodes for a zinc finger family of a transcription factor that has a direct role in glucose metabolism and it serves as its global regulator in fungi (Cupertino et al., 2015; Khosravi et al., 2015; Ries et al., 2016). CreA-encoded protein comprises of the zinc finger of C2H2 (Cys2-His2) class, common S (T)PXX motif, alanine-rich region, and acidic domain (Brown et al., 2014; Benocci et al., 2017). The CreA protein carries a conserved region of 42 amino acids with proline, threonine, and serine rich region residues in fungi (Suzuki and Hikita, 1989). This conserved region was found to be an essential part of growth regulation on carbon, lipid, and, nitrogen sources in *Aspergillus* species and also involved in the utilization of arabinan, xylan, ethanol, and proline (Kubicek et al., 2009; Ries et al., 2016). CreA is a deficient mutant of *A. nidulans* showed alteration in the enzymatic activity, primary metabolism, and metabolic profile shift (David et al., 2005). The expression of CreA downregulates the pathway of xylanase, cellulose, and arabinose utilization (Ruijter and Visser, 1997; Ries et al., 2016; Benocci et al., 2017).

13.3.2 CᴿᴇB ᴀɴᴅ CᴿᴇC

One of the major physiological functions during CCR is ubiquitination which aids the alteration of protein function during macromolecular assembly. Ubiquitin serves as a marker either by catalyzing the modification of any protein or tagging it for proteasomal degradation (Lockington and Kelly, 2002). Similarly, deubiquitination enzymes also aid the regulation and activation of some specific transcription factors targeting their domains. The process of ubiquitination and deubiquitination walk hand in hand to regulate the transcription factors involved in the CCR mechanism (Kubicek et al., 2009; Ries et al., 2016; Alam et al., 2017).

CreB in *A. nidulans* codes for a protein responsible for the deubiquitination process and modulates the function of CreA during CCR (Lockington and Kelly, 2001; Alam et al., 2017). It comprises 6 DUB coiled regions for substrate identification and 4PEST domains that signal for proteolysis (Adnan et al., 2018). Since CreA requires post-translational modification to attain a functional conformation, it can be understood that the process of ubiquitination/deubiquitination catalyzed by CreB is an essential contribution (Alam et al., 2017). CreC however, is also a regulatory gene that encodes a 630aa proline-rich polypeptide comprising 5 WD40 motifs at C terminal, which participates in forming the deubiquitinating complex along with CreB and helps CreA attain functional stability (Matar et al., 2017). In the pathogenic fungus, *Magnaporthe oryzae*, it was found that CreC plays a significant role in vegetative growth, conidiation, and appressorium formation. CreC mutation hindered the penetration ability and reduced the infection which ultimately resulted in attenuated virulence. The mutants also became susceptible to allyl alcohol in the presence of glucose and utilization of a secondary carbon source was not fully repressed. Also, upregulation has been noted in the levels of cell wall degradation enzymes such as feruloyl esterase, beta-glucosidase, and exoglucanase in the mutants (Matar et al., 2017).

The association of CreB and CreC found that CreB acted downstream to the CreC protein. However, overexpression of CreB was able to partially compensate for the CreC deficiency but the same is not true with CreC (Ries et al., 2016; Matar et al., 2017).

13.3.3 CᴿᴇD

CreD holds the responsibility of CreA ubiquitination which results in the beginning of CreB-CreC DUB complex's action (Adnan et al., 2018). HulA is a homolog of Rsp5 (ubiquitin ligase of yeast) in *A. nidulans* which leads to CreA ubiquitination by CreD-HulA ubiquitination ligase complex which may alter the conformation of CreA to be targeted by proteasomes (Alam et al., 2017). Another gene, ApyA was reported to be similar to CreD but displayed a much stronger interaction with HulA than CreD. Hence, it could be thought that ApyA is a product of gene duplication during evolution and it may be responsible for ubiquitination.

13.3.4 Sɴꜰ1

It is a protein kinase that was first studied in *S. cerevisiae*, which is homologous to mammalian cyclic Adenosine Mono Phosphate (cAMP)-dependent protein kinase AMPK (Hardie, 2007). Its major role is studied for its sensing energy status. Snf1 regulates CCR by its protein kinase function in the glucose depression pathway. Moreover, its role is also found in the regulation and repression of cell wall-degrading enzymes (García-Salcedo et al., 2014; Kayikci and Nielsen, 2015; Shashkova et al., 2015). The action of Snf1 causes phosphorylation of its downstream repressor Mig1 and aids the derepression of glucose-repressed genes during low glucose levels (García-Salcedo et al., 2014).

A brief explanation of the interaction of the above-stated regulators in *Aspergillus nidulans* is depicted in Figure 13.4.

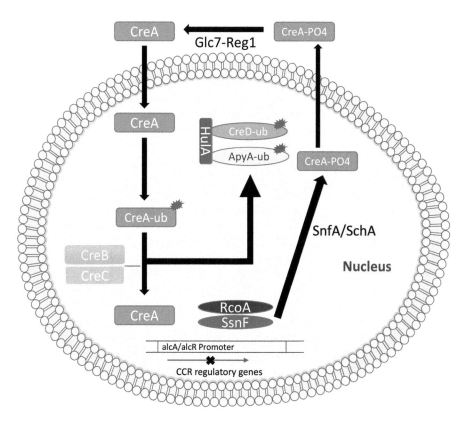

FIGURE 13.4 CCR regulation in *Aspergillus nidualns*. Representation of the interaction of CreA, CreB, CreC, CreD, HulA, ApyA, RcoA, Ssnf, SnfA, and Glc7-Reg1.

13.4 CCR AND FUNGAL VIRULENCE

The CCR regulatory factors either directly or indirectly contribute to pathogenicity in insects, plants, and animals. For the cause of pathogenesis, the pathogens penetrate the host cell, aided by the release of chitinases, esterases, and, proteinases that disturb the target's cellular integrity. An earlier report on *Metarhizium anisopliae* revealed that degradation of cell wall cuticle regulated by CRR1, equivalent to CreA of *A. nidulans* (Screen et al., 1997; Mondal et al., 2016). CreA from *Beauveria bassiana* (BbCre1) was reported to be involved in the carbon source uptake, conidiation, and regulation of the virulence mechanism (Mohamed et al., 2021). The CCR overexpression in *Alternaria citri* led to the eruption of severe symptoms of black rot disease in citrus fruits (Tannous et al., 2018). The role of CreA in the regulation of pathogenicity has also been studied in *F. oxysporum*. The CreA association with F-box protein (Frp1) can regulate carbon utilization and fungal pathogenesis in *F. oxysporum*. (Brown et al., 2014; Hu et al., 2015). Deletion of Frp1 in *F. oxysporum* can compensate for its function by Cre1 overexpression to cause pathogenesis in tomatoes. The Frp1 deletion made the pathogen unable to express genes responsible for the production of cell wall-degrading products and also ICL1 gene which is responsible for aiding the assimilation of C2 carbon sources (Jonkers and Rep, 2009; Gámez-Arjona et al., 2022). The finding suggested that both Frp1 and Cre2 are significant contributors to CCR genes. In *M. oryzae*, three inhibitor proteins, Nmr1-3, sugar sensor Tps1, and multidrug and toxin extrusion MATE pump mdt1 have a crucial role in glucose metabolism for the establishment and progression of infection. The glucose/sugar sensor, Tsp1 senses glucose-6-phosphate that triggers a signaling cascade to inactivate Nmr1 leading to CCR initiation. The assimilation of glucose is sensed by Mdt1 which also contributes to pathogenicity, sporulation, and nutrient utilization. The mutational analysis of all three genes revealed early expression of cell wall-degrading enzymes (CWDEs), suggesting their significant role in the pathogenicity of *M. oryzae* (Wilson et al., 2010; Fernandez et al., 2012).

Mutational analysis in *Alternaria brassicicola* showed no effects on its virulence. Overall, 21 genes were deleted except CreA, XlnR, and Snf1 kinase homologs. The *Alternaria brassicicola* fungus is responsible for causing black spot disease of Brassica. The pathogen expresses CWDEs in a different way where the deletion of only XlnR but not Cre1, Snf1, or Ste12 negatively affected the glucose utilization ability (Cho et al., 2009; Gu et al., 2015; Mohamed et al., 2021; Lengyel et al., 2022). Another pathogen, *Verticilium dahliae* shows pathogenesis regulation by CCR. The VdSnf1 mutant was able to produce the CWDEs but had significantly reduced growth on galactose or pectin as compared to xylose, sucrose, or glucose medium. In *U. virens*, reduced pathogenic potential was observed along with irregular hyphae and spore production due to Snf1 deletion. Additionally, the expression levels of CWDEs were also found to be diminished due to Snf1 deletion (Wen et al., 2023). Similar mutations in *M. oryzae* also displayed reduced pathogenicity and abnormally shaped conidia (Xu et al., 2022).

13.5 GLUCOSE AND ITS ROLE IN FUNGAL PATHOGENESIS

As discussed in the previous section of the chapter, it is well understood that CCR is of great significance in the growth and pathogenesis of many fungal pathogens. The attribute that guides the functioning and regulation of the CCR system is glucose. Glucose being the simplest sugar is most readily utilized by the widest array of microbes, whether pathogenic or non-pathogenic. Its significance has also been well illustrated as a regulator of growth and pathogenesis among a variety of microbial flora. In this section, our focus is on the virulence and pathogenic regulatory attributes of glucose on some serious and concerning human fungal pathogens.

13.5.1 VIRULENCE AND PATHOGENICITY REGULATORY ROLE OF GLUCOSE IN *CANDIDA*

Glucose serves as a primary source of carbon and energy among the majority of the life forms, and hence the sensing and response to glucose is highly developed and closely regulated among them. A significant number of studies have reported the involvement of glucose in the growth and virulence processes in different yeast species, including *Candida albicans* (Brown et al., 2006; Rodaki et al., 2009; Dos Reis et al., 2013; Harpf et al., 2022). The trafficking of glucose through the membrane bi-layer is the initial and limiting step of glucose metabolism. A variety of glucose transporters and their variable expression in response to different glucose levels have been reported in *C. albicans* (Fan et al., 2002; Garbe et al., 2022). The findings suggest that sugar-sensing phenomena in *S. cerevisiae* and *C. albicans* have significance in the regulation of cellular morphology and colonization in the host and attain optimal virulent state (Rolland et al., 2002; Sabina and Brown, 2009; Larcombe et al., 2023). The deletion of drug-resistant genes PDR3/QDR3 resulted in an increase in intracellular levels of glucose and glycerol in yeast. Moreover, a spike in the organic phosphate concentration also occurs to stabilize the glucose and glycerol concentration (Dikicioglu et al., 2014). The encounter of *C. albicans* with glucose induces diverse stress resistance genes involved in regulating the cationic and oxidative stress resistance and also enhancing the drug resistance mechanism (Rodaki et al., 2009; Larcombe et al., 2023). Studies revealed that glucose behaves as a morphogen that aids the non-pathogenic to pathogenic (yeast to hyphal transition) in *C. albicans* (Van Ende et al., 2019). The increased glucose concentration in the culture medium also compromised the activity of Voriconazole and Amphotericin B in *C. albicans* and *C. tropicalis* (Mandal et al., 2014). In contrast, lower glucose concentrations were revealed to induce hyphae formation and expression activity of a variety of high-affinity glucose transporters such as HGT9, HGT12, and HGT17 in *C. albicans* (Van Ende et al., 2019; Harpf et al., 2022). Moreover, many HGTs reported to be involved in hyphal formation, drug resistance, and virulence in *C. albicans* (Van Ende et al., 2019; Harpf et al., 2022). The high-affinity glucose transporters HGT1 and HGT4 predominantly sense the glucose and other sugars like galactose, maltose, and, fructose (Brown et al., 2006; Varma et al., 2000). The significance of glucose can be established from the fact that the deletion of glucose transporters and sensors influences the hyphae formation and virulence-related aspects of the pathogen *Candida*. For example, *Hgt4* deletion mutant displayed significantly lower virulence traits and abnormality in the hyphae formation, compared to the wild-type. These studies suggested that *Hgt4* acts as a modulator of sugar trafficking and metabolism through *Rgt1* (transcriptional repressor of hexose transporter encoding genes), affecting filamentation (Sexton et al., 2007; Broxton et al., 2018). The modulatory effect of higher glucose concentration has also been demonstrated. Glucose accumulation aids the cells to counterbalance stress and maintain cellular pathogenicity and virulence (Qadri et al., 2021). Adaptation of cells with different carbon sources has been shown to impact different virulent traits such as integrity of the cell wall, biofilm formation adhesion properties, hyphae formation, and host cellular damage (Ene et al., 2012; Lagree et al., 2020).

During systemic *C. albicans* infections, an immunometabolic shift occurs due to a complex cross-talk between *Candida* and immune cells, which results in a nutrient-deficient state for the activated immune cells. *Candida* cells exploit this opportunity to the fullest and lead to macrophage destruction (Figure 13.5) (Tucey et al., 2018).

In *C. glabrata*, glucose effectively promotes planktonic growth and biofilm formation. A switch toward an alternative carbon source triggers structural change in the cell wall modulating the oxidative stress resistance and antifungal drug resistance mechanisms (Chew et al., 2019). Different glucose concentrations in *C. glabrata* differently affect both growth and virulence. Higher glucose concentration positively regulates the growth and oxidative stress resistance, however, lower glucose concentration elicits biofilm formation, and amphotericin B resistance. However, higher glucose and lipid levels positively regulate the biofilm formation in *C. parapsilosis*. In contrast, lower glucose concentrations significantly induce the expression of the high-affinity glucose sensor gene *CgSNF3* which affects cell growth, stress tolerance, and virulence factors (Ng et al., 2016).

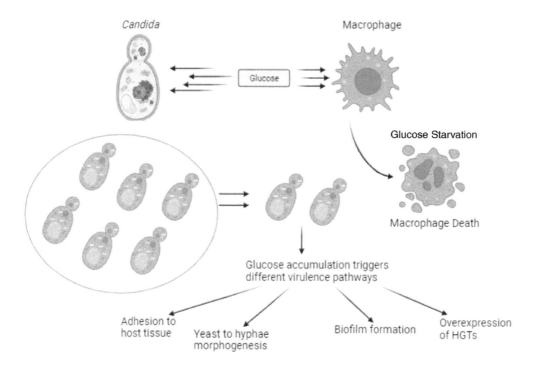

FIGURE 13.5   Glucose competition between a host-invading pathogen and macrophage host. (Created in BioRender.com.)

13.5.2 VIRULENCE AND PATHOGENICITY MODULATORY ROLE OF GLUCOSE IN *ASPERGILLUS*

Carbon catabolite repression is the pathway by which carbon is metabolized in *Aspergillus* for growth promotion and niche colonization (Adnan et al., 2018). The CCR process regulates attributes related to survival, virulence, adaptation, communication, and cell death. Among filamentous fungi, catabolite responsive element A (CreA), yeast Mig1 homolog acts as transcription factors that regulate the CCR system (Adnan et al., 2018). The uptake of glucose from the surroundings is facilitated by various high and low-affinity uptake systems among fungal organisms like *A. nidulans*. The sugar transporters of *A. nidulans* display a high preference for glucose followed by arabinose, galactose, and xylose (Dos Reis et al., 2016). The metabolic plasticity of *A. fumigatus* has a significant role in its survival fitness within the host which aids its competency for disease maintenance and progression (Beattie et al., 2017). Beattie et al. (2017) showed that CCR is dispensable in *A. fumigatus* for the initiation of pulmonary infection. The study also suggested that a CCR-mediated regulatory network is an essential requirement for the fungal thriving and invasion of the microenvironment. More understanding of the molecular network involved in response to high and low glucose environments was provided by Do Reis et al. (2017). They reported novel factors and proteins involved in glucose sensing and signaling systems. The study identified several differentially regulated transcription factors and transporter genes such as *HtxB* (a low-affinity glucose transporter) to participate in the conidiation-related processes contributing further to glucose signaling processes. The deletion GDP (guanosine diphosphate) mannose transporter (GMTs) encoding gene; *gmtA* displayed lower cell surface hydrophobicity, less adherent spores, and reduced biofilm-forming ability in *A. nidulans* (Kadry et al., 2018). The influence of glucose in the establishment of aspergillosis infection is also an interesting finding. The report suggested that the individuals who lacked significant risk factors for aspergillosis such as immune suppression, corticosteroid usage, acute granulocytopenia, etc., were also prone to pulmonary aspergillosis, in conditions where the individuals were diabetic (Shahnazi et al., 2010; Ghanaat and Tayek., 2017).

13.5.3 VIRULENCE AND PATHOGENICITY MODULATORY ROLE OF GLUCOSE IN *CRYPTOCOCCUS*

There are two hexose transport proteins (*Hxs1* and *Hxs2*) from *Cryptococcus* that display very significant sequence homology with *S. cerevisiae* glucose sensors *Snf1* and *Rgt2* and also with *C. albicans* glucose sensor *Hgt4*, all of which have been shown to play a significant role in the glucose-regulated virulence and growth. The study reported about significant contribution of *Hxs1* in the virulence process of *C. neoformans* by contributing toward resistance against oxidative stress and progression of fungal virulence (Liu et al., 2013; Freitas et al., 2022). The macrophages serve as the first line of host defense against cryptococcal infection against which, the pathogen synthesizes an antiphagocytic protein (*App1*) to escape the macrophage-directed phagocytosis.

It was shown that under low glucose conditions or when in contact with body fluids such as bronchoalveolar lavage (BAL) fluid, serum, or cerebrospinal fluid (CSF), the pathogen significantly upregulates *App1* to regulate virulence factor and control macrophage-directed fungal phagocytosis (Luberto et al., 2003; Williams and Del Poeta, 2011; Bhattacharya et al., 2021; Wang et al., 2022). To assess the role of glucose in cryptococcosis, mutational analysis was conducted using a murine model. Findings suggested that *Cryptococcus* mutants were nullified for their ability to utilize glucose, resulting in less virulence in the murine inhalation model of cryptococcosis and deteriorated persistence in rabbit cerebrospinal fluid, which collectively indicated the significance of glucose utilization for virulence and pathogenic persistence (Price et al., 2011; Berguson et al., 2022).

CONCLUSION

The complex interplay between carbon catabolite repression (CCR) and glucose metabolism play a pivotal role in fungal pathogenesis that significantly influence the virulence and adaptability of fungal pathogens. In this chapter, carbon catabolic repression (CCR) linked with the fungal pathogenesis has been described. CCR enables fungi to efficiently utilize glucose for their survival and favors pathogenic potential in various host environments. The regulation of CCR through key factors such as CreA, CreB, CreC, and Snf1 underscores the sophisticated genetic machinery evolved by pathogenic fungi to thrive under diverse and often hostile conditions. These regulatory mechanisms not only optimize metabolic processes but also modulate the expression of virulence factors, essential for infection and fungal colonization. It has been well established that the mechanism involved in glucose uptake directly regulate different physiological alterations in the fungi including morphological switching, biofilm formation, and drug resistance. However, fungal pathogens tend to adapt under varying glucose conditions via the regulation of high-affinity glucose transporters, which exemplifies their metabolic flexibility. Similarly, CCR influences growth, virulence, and stress responses, facilitating effective colonization and infection. The intricate regulation of CCR and its associated pathways indicates the central role of glucose in fungal metabolic networks and pathogenic strategies. Understanding the molecular basis of CCR and its impact on fungal virulence opens avenues for developing targeted therapeutic strategies aimed to disrupt such metabolic pathways. By inhibiting key components of CCR, it may be possible to attenuate fungal virulence and enhance the efficacy of antifungal treatments. As global fungal infections continue to pose a significant health threat, advancing our knowledge in this area is crucial for devising novel interventions to combat fungal diseases.

ABBREVIATIONS

CCR   Carbon catabolite repression
GIT   Gastrointestinal tract
DNA   Deoxyribonucleic acid
GPCRs   G-protein-coupled receptors
cAMP   Cyclic adenosine monophosphate
AMPK   Adenonsine monophosphate kinase
CWDEs   Cell wall-degrading enzymes
GDP   Guanosine diphosphate
GMT   GDP transporter
BAL   Bronchoalveolar liquid
CSF   Cerebrospinal fluid

REFERENCES

Adnan M, Zheng W, Islam W, Arif M, Abubakar YS, Wang Z, Lu G (2018) Carbon catabolite repression in filamentous fungi. Int. J. Mol. Sci. 19: 48.

Alam MA, Kamlangdee N, Kelly JM (2017) The CreB deubiquitinating enzyme does not directly target the CreA repressor protein in *Aspergillus nidulans*. Curr. Genet. 63: 647–667.

Amore A, Giacobbe S, Faraco V (2013) Regulation of cellulase and hemicellulase gene expression in fungi. Curr. Genom. 14: 230–249.

Beattie SR, Mark KM, Thammahong A, Ries LNA, Dhingra S, Caffrey-Carr AK, et al. (2017) Filamentous fungal carbon catabolite repression supports metabolic plasticity and stress responses essential for disease progression. PLOS Pathog. 13: e1006340

Benocci T, Aguilar-Pontes MV, Zhou M, Seiboth B, Vries RP (2017) Regulators of plant biomass degradation in ascomycetous fungi. Biotechnol. Biofuels. 10: 152.

Berguson HP, Caulfield LW, Price MS (2022) Influence of pathogen carbon metabolism on interactions with host immunity. Front. Cell. Infect. Microbiol. 12: 861405.

Bhattacharya S, Oliveira NK, Savitt AG, Silva VKA, Krausert RB, Ghebrehiwet B, et al. (2021) Low glucose mediated fluconazole tolerance in *Cryptococcus neoformans*. J Fungi (Basel). 7(6): 489.

Bhunjun CS, Phillips AJL, Jayawardena RS, Promputtha I, Hyde KD (2021) Importance of molecular data to identify fungal plant pathogens and guidelines for pathogenicity testing based on Koch's postulates. Pathogens 10(9): 1096.

Brown NA, Ries LNA, Goldman GH (2014) How nutritional status signalling coordinates metabolism and lignocellulolytic enzyme secretion. Fungal Genet. Biol. 72: 48–63.

Brown V, Sexton JA, Johnston M (2006) A glucose sensor in *Candida albicans*. Eukaryotic Cell. 5: 1726–1737.

Broxton CN, He B, Bruno VM, Culotta VC (2018) A role for *Candida albicans* superoxide dismutase enzymes in glucose signaling. Biochem Biophys Res Commun 495(1): 814–820.

Cazzanelli G, Pereira F, Alves S, Francisco R, Azevedo L, Dias Carvalho P, et al. (2018) The yeast *Saccharomyces cerevisiae* as a model for understanding RAS proteins and their role in human tumorigenesis. Cells. 7(2): 14

Chen Y, Dong L, Alam MA, Pardeshi L, Miao Z, Wang F, et al., (2021) Carbon catabolite repression governs diverse physiological processes and development in *Aspergillus nidulans*. mBio, 13(1): e0373421.

Chew SY, Ho KL, Cheah YK, Sandai D, Brown AJ, Than LTL (2019) Physiologically relevant alternative carbon sources modulate biofilm formation, cell wall architecture, and the stress and antifungal resistance of *Candida glabrata*. Int. J. Mol. Sci. 20: 3172.

Cho Y, Kim KH, La Rota M, Scott D, Santopietro G, Callihan M, Mitchell TK, Lawrence CB (2009) Identification of novel virulence factors associated with signal transduction pathways in *Alternaria brassicicola*. Mol. Microbiol. 72: 1316–1333.

Cupertino FB, Virgilio S, Freitas FZ, de Souza Candido T, Bertolini MC (2015) Regulation of glycogen metabolism by the CRE-1, RCO-1 and RCM-1 proteins in *Neurospora crassa*. The role of CRE-1 as the central transcriptional regulator. Fungal Genet. Biol. 77: 82–94.

David H, Krogh AM, Roca C, Åkesson M, Nielsen J (2005) CreA influences the metabolic fluxes of *Aspergillus nidulans* during growth on glucose and xylose. Microbiology. 151: 2209–2221.

Dikicioglu D, Oc S, Rash BM, Dunn WB, Pir P, Kell DB, et al., (2014). Yeast cells with impaired drug resistance accumulate glycerol and glucose. Mol. Biosyst. 10: 93–102

Dos Reis TF, Menino JF, Bom VLP, Brown NA, Colabardini AC, Savoldi M, Goldman GH (2013) Identification of glucose transporters in *Aspergillus nidulans*. PLOS One. 8: e81412.

Dos Reis TF, de Lima PB, Parachin NS, Mingossi FB, de Castro Oliveira JV, Ries LN, Goldman GH (2016) Identification and characterization of putative xylose and cellobiose transporters in *Aspergillus nidulans*. Biotechnol Biofules. 9:1–19.

Dos Reis TF, Nitsche BM, de Lima PBA, de Assis LJ, Mellado L, Harris SD, Meyer V, dos Santos RAC, Riaño-Pachón DM, Ries LNA, Goldman GH (2017) The low affinity glucose transporter HxtB is also involved in glucose signalling and metabolism in *Aspergillus nidulans*. Sci Rep. 7: 45073.

Ene IV, Adya AK, Wehmeier S, Brand AC, MacCallum DM, Gow NA, Brown AJ (2012) Host carbon sources modulate cell wall architecture, drug resistance and virulence in a fungal pathogen. Cell. Microbiol. 14: 1319–1335.

Fan J, Chaturvedi V, Shen SH (2002) Identification and phylogenetic analysis of a glucose transporter gene family from the human pathogenic yeast *Candida albicans*. J. Mol. Evol. 55: 336–346.

Fang W, Wu J, Cheng M, Zhu X, Du M, Chen C, Liao W, Zhi K, Pan W (2023) Diagnosis of invasive fungal infections: Challenges and recent developments. J. Biomed. Sci 30(1): 42.

Fernandez J, Wright JD, Hartline D, Quispe CF, Madayiputhiya N, Wilson RA (2012) Principles of carbon catabolite repression in the rice blast fungus: Tps1, Nmr1-3, and a MATE–Family Pump regulate glucose metabolism during Infection. PLOS Genet. 8: e1002673.

Freitas GJC, Gouveia-Eufrasio L, Emidio ECP, Carneiro HCS, de Matos BL, Costa MC, et al. (2022) The dynamics of *Cryptococcus neoformans* cell and transcriptional remodeling during infection. Cells. 11(23): 3896.

Gámez-Arjona FM, Vitale S, Voxeur A, Dora S, Müller S, Sancho-Andrés G, et al. (2022) Impairment of the cellulose degradation machinery enhances *Fusarium oxysporum* virulence but limits its reproductive fitness. Sci. Adv. 8(16): eabl9734.

Garbe E, Gerwien F, Driesch D, Müller T, Böttcher B, Gräler M, et al. (2022) Systematic metabolic profiling identifies de novo sphingolipid synthesis as hypha associated and essential for *Candida albicans* filamentation. mSystems. 7(6): e0053922.

García-Salcedo R, Lubitz T, Beltran G, Elbing K, Tian Y, Frey S, Wolkenhauer O, Krantz M, Klipp E, Hohmann S (2014) Glucose de-repression by yeast AMP-activated protein kinase SNF1 is controlled via at least two independent steps. FEBS J. 281: 1901–1917.

Ghanaat F, Tayek JA (2017) Weight loss and diabetes are new risk factors for the development of invasive aspergillosis infection in non-immunocompromised humans. Clin. Practice (London, England). 14: 296.

Gu Q, Zhang C, Liu X, Ma Z (2015) A transcription factor FgSte12 is required for pathogenicity in *Fusarium graminearum*. Mol. Plant Pathol 16(1): 1–13.

Hardie DG (2007) AMP-activated/SNF1 protein kinases: Conserved guardians of cellular energy. Nat. Rev. Mol. Cell Biol. 8: 774–785.

Harpf V, Kenno S, Rambach G, Fleischer V, Parth N, Weichenberger CX, et al. (2022) Influence of glucose on *Candida albicans* and the relevance of the complement FH-binding molecule Hgt1 in a murine model of candidiasis. Antibiotics (Basel). 11(2): 257.

Hu Z, Parekh U, Maruta N, Trusov Y, Botella JR (2015) Down-regulation of *Fusarium oxysporum* endogenous genes by host-delivered RNA interference enhances disease resistance. Front. Chem. 3: 1.

Ichinose S, Tanaka M, Shintani T, Gomi K (2014) Improved α-amylase production by *Aspergillus oryzae* after a double deletion of genes involved in carbon catabolite repression. Appl. Microbiol. Biotechnol. 98: 335–343.

Jonkers W, Rep M (2009) Mutation of CRE1 in *Fusarium oxysporum* reverts the pathogenicity defects of the FRP1 deletion mutant. Mol. Microbiol. 74: 1100–1113.

Kadry AA, El-Ganiny AM, Mosbah RA, Kaminskyj SG (2018) Deletion of *Aspergillus nidulans* GDP-mannose transporters affects hyphal morphometry, cell wall architecture, spore surface character, cell adhesion, and biofilm formation. Med. Mycol. 56: 621–630.

Kayikci O, Nielsen J (2015) Glucose repression in *Saccharomyces cerevisiae*. FEMS Yeast Res. 15: fov068.

Khosravi C, Benocci T, Battaglia E, Benoit I, de Vries RP (2015) Chapter one: Sugar catabolism in *Aspergillus* and other fungi related to the utilization of plant biomass. Adv. Appl. Microbiol. 90: 1–28.

Kim JH, Johnston M (2006) Two glucose-sensing pathways converge on Rgt1 to regulate expression of glucose transporter genes in *Saccharomyces cerevisiae*. J. Biol. Chem. 281: 26144–26149.

Kim JH, Rodriguez R (2021) Glucose regulation of the paralogous glucose sensing receptors Rgt2 and Snf3 of the yeast *Saccharomyces cerevisiae*. Biochim. Biophys. Acta-Gen. Subj. 1865(6): 129881.

Kim JH, Roy A, Jouandot D, Cho KH (2013) The glucose signaling network in yeast. Biochim. Biophys. Acta-Gen. Subj. 1830: 5204–5210.

Kubicek CP, Mikus M, Schuster A, Schmoll M, Seiboth B (2009) Metabolic engineering strategies for the improvement of cellulase production by *Hypocrea jecorina*. Biotechnol. Biofuels. 2: 19.

Lagree K, Woolford CA, Huang MY, May G, McManus CJ, Solis NV et al., (2020) Roles of *Candida albicans* Mig1 and Mig2 in glucose repression, pathogenicity traits, and SNF1 essentiality. PLOS Genet. 16: e1008582

Larcombe DE, Bohovych IM, Pradhan A, Ma Q, Hickey E, Leaves I, et al. (2023) Glucose-enhanced oxidative stress resistance-a protective anticipatory response that enhances the fitness of *Candida albicans* during systemic infection. PLOS Pathog. 19(7): e1011505.

Lengyel S, Rascle C, Poussereau N, Bruel C, Sella L, Choquer M, Favaron F (2022) Snf1 kinase differentially regulates botrytis cinerea pathogenicity according to the plant host. Microorganisms 10(2): 444.

Liu TB, Wang Y, Baker GM, Fahmy H, Jiang L, Xue C (2013) The glucose sensor like protein Hxs1 is a high-affinity glucose transporter and required for virulence in *Cryptococcus neoformans*. PLOS One. 8: e64239.

Lockington RA, Kelly JM (2001) Carbon catabolite repression in *Aspergillus nidulans* involves deubiquitination. Mol. Microbiol. 40: 1311–1321.

Lockington RA, Kelly JM (2002) The WD40-repeat protein CreC interacts with and stabilizes the deubiquitinating enzyme CreB in vivo in *Aspergillus nidulans*. Mol. Microbiol. 43: 1173–1182.

Luberto C, Toffaletti DL, Wills EA, Tucker SC, Casadevall A, Perfect JR, et al. (2001). Roles for inositol-phosphoryl ceramide synthase 1 (IPC1) in pathogenesis of *C. neoformans*. Genes Dev. 15: 201–212

Mandal SM, Mahata D, Migliolo L, Parekh A, Addy PS, Mandal M, Basak A (2014) Glucose directly promotes antifungal resistance in the fungal pathogen, *Candida spp*. J. Biol. Chem. 289: 25468–25473.

Matar KAO, Chen X, Chen D, Anjago WM, Norvienyeku J, Lin Y, Chen M, Wang Z, Ebbole DJ, Lu G (2017) WD40-repeat protein MoCreC is essential for carbon repression and is involved in conidiation, growth and pathogenicity of *Magnaporthe oryzae*. Curr. Genet. 63: 685–696.

Meyer V, Andersen MR, Brakhage AA, Braus GH, Caddick MX, Cairns TC, de Vries RP, Haarmann T, Hansen K, Hertz-Fowler C, Krappmann S (2016) Current challenges of research on filamentous fungi in relation to human welfare and a sustainable bio-economy: a white paper. Fungal Biol Biotechnol. 3: 1–17.

Mohamed RA, Ren K, Mou YN, Ying SH, Feng MG (2021) Genome-wide insight into profound effect of carbon catabolite repressor (Cre1) on the insect-pathogenic lifecycle of *Beauveria bassiana*. J. Fungi. (Basel) 7(11): 895.

Mondal S, Baksi S, Koris A, Vatai G (2016) Journey of enzymes in entomopathogenic fungi. Pac. Sci. Rev. A Nat. Sci. Eng. 18: 85–99.

New AM, Cerulus B, Govers SK, Perez-Samper G, Zhu B, Boogmans S, Xavier JB, Verstrepen KJ (2014) Different levels of catabolite repression optimize growth in stable and variable environments. PLOS Biol. 12: e1001764.

Ng TS, Desa MNM, Sandai D, Chong PP, Than LTL (2016) Growth, biofilm formation, antifungal susceptibility and oxidative stress resistance of *Candida glabrata* are affected by different glucose concentrations. Infect. Genet. Evol. 40: 331–338.

Palomino A, Herrero P, Moreno F (2006) Tpk3 and Snf1 protein kinases regulate Rgt1 association with *Saccharomyces cerevisiae* HXK2 promoter. Nucleic Acids Res. 34: 1427–1438.

Peeters K, Thevelein JM (2014) Glucose sensing and signal transduction in *Saccharomyces cerevisiae*. In Molecular Mechanisms in Yeast Carbon Metab; Springer: Berlin/Heidelberg, Germany, pp. 21–56.

Peeters T, Versele M, Thevelein JM (2007) Directly from Gα to protein kinase A: The kelch repeat protein bypass of adenylate cyclase. Trends Biochem. Sci. 32: 547–554.

Pentland DR, Piper-Brown E, Mühlschlegel FA, Gourlay CW (2018) Ras signalling in pathogenic yeasts. Microb Cell 18; 5(2): 63–73.

Price MS, Betancourt-Quiroz M, Price JL, Toffalett DL, Vora H, Hu G, et al. (2011) *Cryptococcus neoformans* requires a functional glycolytic pathway for disease but not persistence in the host. mBio, 2(3): e00103–e111.

Qadri H, Qureshi MF, Mir MA, Shah AH (2021) Glucose-the X factor for the survival of human fungal pathogens and disease progression in the host. Microbiol. Res. 247: 126725.

Ray A, Aayilliath KA, Banerjee S, Chakrabarti A, Denning DW (2022) Burden of serious fungal infections in India. Open Forum Infect. Dis. 9(12): ofac603.

Ries LN, Beattie SR, Espeso EA, Cramer RA, Goldman GH (2016) Diverse regulation of the CreA carbon catabolite repressor in *Aspergillus nidulans*. Genetics. 203: 335–352.

Rodaki A, Bohovych IM, Enjalbert B, Young T, Odds FC, Gow NA, Brown AJ (2009) Glucose promotes stress resistance in the fungal pathogen *Candida albicans*. Mol. Biol. Cell. 20: 4845–4855.

Rolland F, Winderickx J, Thevelein JM (2002) Glucose-sensing and-signalling mechanisms in yeast. FEMS Yeast Res. 2: 183–201.

Roy A, Kim JH (2016) Endocytosis and vacuolar degradation of the yeast cell surface glucose sensors Rgt2 and Snf3. J. Biol. Chem. 291: 14913.

Ruijter GJ, Visser J (1997) Carbon repression in *Aspergilli*. FEMS Microbiol. Lett. 151: 103–114.

Sabina J, Brown V (2009) Glucose sensing network in *Candida albicans*: A sweet spot for fungal morphogenesis. Eukaryotic Cell. 8: 1314–1320.

Screen S, Bailey A, Charnley K, Cooper R, Clarkson J (1997) Carbon regulation of the cuticle-degrading enzyme PR1 from Metarhizium anisopliae may involve a trans-acting DNA-binding protein CRR1, a functional equivalent of the *Aspergillus nidulans* CREA protein. Curr. Genet. 31: 511–518.

Sexton JA, Brown V, Johnston M (2007) Regulation of sugar transport and metabolism by the *Candida albicans* Rgt1 transcriptional repressor. Yeast. 24: 847–860.

Shahnazi A, Mansouri N, Malek A, Sepehri Z, Mansouri SD (2010) Is Pulmonary Aspergillosis Common in Diabetes Mellitus Patients? TANAFFOS. 9(3): 69–74.

Shashkova S, Welkenhuysen N, Hohmann S (2015) Molecular communication: Crosstalk between the Snf1 and other signaling pathways. FEMS Yeast Res 15(4): fov026.

Soulard A, Cremonesi A, Moes S, Schütz F, Jenö P, Hall MN (2010) The rapamycin-sensitive phosphoproteome reveals that TOR controls protein kinase A toward some but not all substrates. Mol. Biol. Cell. 21: 3475–3486.

Suzuki M, Hikita M (1989) Anomalous field-induced resistive transition in La2-xSrxCuO4. Jpn. J. Appl. Phys. 28: L1368.

Tannous J, Kumar D, Sela N, Sionov E, Prusky D, Keller NP (2018) Fungal attack and host defence pathways unveiled in near-avirulent interactions of *Penicillium expansum* creA mutants on apples. Mol. Plant. Pathol. 19(12): 2635–2650.

Tarinejad A (2015) Glucose signalling in *Saccharomyces cerevisiae*. Agric. Commun. 3: 7–15.

Tucey TM, Verma J, Harrison PF, Snelgrove SL, Lo TL, Scherer AK, et al. (2018) Glucose homeostasis is important for immune cell viability during *Candida* challenge and host survival of systemic fungal infection. Cell Metab. 27: 988–1006. e1007

Van Ende M, Wijnants S, Van Dijck P (2019) Sugar sensing and signaling in *Candida albicans* and *Candida glabrata*. Front. Microbiol. 10: 99.

Varma A, Singh BB, Karnani N, Lichtenberg-Fraté H, Hofer M, Magee B, Prasad R (2000) Molecular cloning and functional characterisation of a glucose transporter, CaHGT1, of *Candida albicans*. FEMS Microbiol. Lett. 182: 15–21.

Wang Y, Pawar S, Dutta O, Wang K, Rivera A, Xue C (2022) Macrophage mediated immunomodulation during *Cryptococcus* pulmonary infection. Front. Cell Infect. Microbiol. 12: 859049.

Weinhandl K, Winkler M, Glieder A, Camattari A (2014) Carbon source dependent promoters in yeasts. Microb. Cell Factor. 13: 5.

Welkenhuysen N, Borgqvist J, Backman M, Bendrioua L, Goksör M, Adiels CB, Cvijovic M, Hohmann S (2017) Single-cell study links metabolism with nutrient signaling and reveals sources of variability. BMC Syst. Biol. 11: 59.

Wen H, Meng S, Xie S, Shi H, Qiu J, Jiang N, Kou Y (2023) Sucrose non-fermenting protein kinase gene UvSnf1 is required for virulence in *Ustilaginoidea virens*. Virulence 14(1): 2235460.

Williams V, Del Poeta M (2011) Role of glucose in the expression of *Cryptococcus neoformans* antiphagocytic protein 1. Appl. Eukaryotic Cell. 10: 293–301.

Wilson RA, Gibson RP, Quispe CF, Littlechild JA, Talbot NJ (2010) An NADPH-dependent genetic switch regulates plant infection by the rice blast fungus. Proc. Natl. Acad. Sci. USA. 107: 21902–21907.

Xu XW, Zhao R, Xu XZ, Tang L, Shi W, Chen D, et al. (2022) MoSnf5 regulates fungal virulence, growth, and conidiation in *Magnaporthe oryzae*. J. Fungi (Basel, Switzerland), 9(1): 18.

Yao Y, Tsuchiyama S, Yang C, Bulteau AL, He C, Robison B, Tsuchiya M, Miller D, Briones V, Tar K (2015) Proteasomes, Sir2, and Hxk2 form an interconnected aging network that impinges on the AMPK/Snf1-regulated transcriptional repressor Mig1. PLOS Genet. 11: e1004968.

Index

9781032633022